THE FRONTIERS COLLECTION

# THE FRONTIERS COLLECTION



The books in this collection are devoted to challenging and open problems at the forefront of modern science, including related philosophical debates. In contrast to typical research monographs, however, they strive to present their topics in a manner accessible also to scientifically literate non-specialists wishing to gain insight into the deeper implications and fascinating questions involved. Taken as a whole, the series reflects the need for a fundamental and interdisciplinary approach to modern science. Furthermore, it is intended to encourage active scientists in all areas to ponder over important and perhaps controversial issues beyond their own speciality. Extending from quantum physics and relativity to entropy, consciousness and complex systems – the Frontiers Collection will inspire readers to push back the frontiers of their own knowledge.

## *Other Recent Titles*

**Weak Links**
The Universal Key to the Stability of Networks and Complex Systems
By P. Csermely

**Entanglement, Information, and the Interpretation of Quantum Mechanics**
By G. Jaeger

**Particle Metaphysics**
A Critical Account of Subatomic Reality
By B. Falkenburg

**The Physical Basis of the Direction of Time**
By H.D. Zeh

**Mindful Universe**
Quantum Mechanics and the Participating Observer
By H. Stapp

**Decoherence and the Quantum-To-Classical Transition**
By M.A. Schlosshauer

**The Nonlinear Universe**
Chaos, Emergence, Life
By A. Scott

**Symmetry Rules**
How Science and Nature Are Founded on Symmetry
By J. Rosen

**Quantum Superposition**
Counterintuitive Consequences of Coherence, Entanglement, and Interference
By M.P. Silverman

For all volumes see back matter of the book

**...but thou hast ordered all things in measure and number and weight**

*The Wisdom of Solomon 11:20*

# Preface

This book is concerned with the physical properties of matter and the diverse processes that occur in nature at ultrahigh energy densities, which correspond to extreme pressures and temperatures. It is based on the lecture course I hold at the Moscow Institute of Physics and Technology, as well as reviews and invited papers written for scientific conferences and symposia.

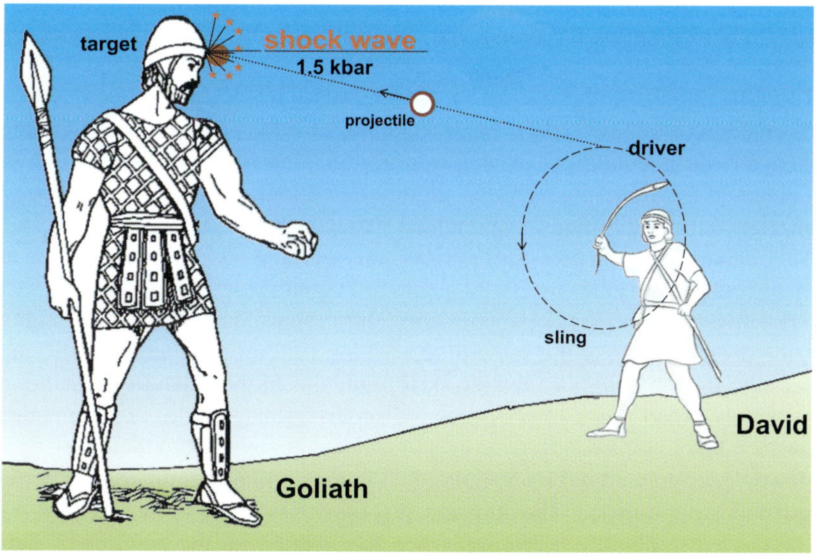


**Fig. 0.1** The battle between David and Goliath [1].

Interest in the physics of extreme states has always been strong, owing to the natural desire of humans to achieve more and to operate with record parameters, as well as owing to the wealth of applications to astrophysics, energy production, and defense. It was indeed military application that fostered the first successful experiment involving extreme states, which was conducted more than 3000 years ago – during the battle between David and Goliath (Fig. 0.1). According to the Old Testament [1], the high-velocity impact of a stone, which was shot from David's sling, on Goliath's head killed him. It gave rise to a shock wave in Goliath's head with an amplitude pressure of $\approx 1.5$ kbar, which was more than twice the strength of Goliath's frontal bone and determined the outcome of the duel, to the great joy of the army and people of Israel. Discovered to be successful at that time, this scheme of action serves as the basis for all subsequent experiments in the area of dynamic high-energy-density physics.

The application of more powerful and highly sophisticated energy cumulation systems – chemical and nuclear explosives, powder, light-gas, and electrodynamic "guns", charged-particle fluxes, laser and X-ray radiation – has enabled the velocity of "thrown" projectiles to be raised by three or four orders of magnitude, since the time of David, and the pressure in the shock wave by six to eight orders of magnitude, thereby reaching the megabar–gigabar pressure range and "nuclear" energy densities in substances.

In the 20th century, mainstream high-energy-density physics was closely related to the arrival of the atomic and space era in our civilization. In nuclear charges, the high energy densities generated by intense shock waves serve to initiate chain nuclear reactions in compressed nuclear fuel. In thermonuclear charges and micro-targets for controlled fusion, high-energy states are the main instrument for compressing and heating the thermonuclear fuel and initiating thermonuclear reactions in it.

Recently, interest in the science of extreme states has been rekindled as a result of the emergence of new experimental techniques for the generation of high-energy states in terrestrial conditions as well as intriguing new observational astrophysical data obtained by modern-generation ground-based and space telescopes operating at different wavelengths and by unmanned space stations.

Although the limiting pressures in laboratory plasmas so far differ from the maximum astrophysical values by 20–30 orders of magnitude, this gap is rapidly shrinking. The physical processes in a laboratory and in space quite frequently exhibit an amazing diversity and, at the same time, striking similarities, thereby testifying, at the least, to the uniformity of the physical principles of the behavior of matter over an extremely broad range of pressures (42 orders of magnitude) and temperatures (up to $10^{13}$ K).

However, as pointed out by Voltaire, "... in nature this phenomenon is perfectly natural and commonplace. The domains of some rulers in Germany and Italy, which can be circled in about a half hour, when compared with the empires of Turkey, Moscow, or China, give only a faint idea of the remarkable contrasts that are hidden in all of nature" (Voltaire, *Le Micromégas*, Paris 1752) [3].

It is significant that the range of technical problems related to the physics of extreme states is progressively broadening along with the range of the fundamental ones. These states of matter underlie the operation of pulsed thermonuclear reactors with inertial confinement of the hot plasma, high-power explosive-driven magnetic generators, power installations and rocket propulsors with gas-phase nuclear reactors, plasmachemical and microwave reactors, plasmatrons, and high-power sources of optical radiation and X-rays. In the energy installations of the future, strongly compressed and heated plasmas will be employed as the working medium, like water vapor in modern thermal power stations.

Extreme states emerge when a substance is subjected to intense shock, detonation, or electroexplosion waves, concentrated laser radiation, electron and ion beams, in powerful chemical and nuclear explosions, in the pulsed vaporization of liners in pinches and magnetic cumulation generators, in the hypersonic motion of bodies in dense planetary atmospheres, in high-velocity impacts, and in many other situations characterized by ultrahigh pressures and temperatures. The physics of near-electrode, contact, and electroexplosion processes in vacuum breakdown is intimately related to the high-energy plasma that defines the operation of high-power pulsed accelerators, microwave radiation generators, and plasma switches.

High energy densities determine the behavior of matter in a vast domain of the phase diagram that occupies the range from a solid and a fluid to a neutral gas, covers the phase boundaries of melting and boiling, and also the metal–dielectric transition domain. The metal–dielectric transition problem has been much explored in experiments involving the multiple (quasi-isentropic) shock-wave compression of dielectrics, their metallization, and the recently discovered plasma phase transitions in the megabar pressure range, as well as dielectrization of strongly compressed simple metals.

Scientific interest in high-energy plasmas has also been rekindled along with pragmatic interest, because the great bulk of matter in the universe is in precisely this exotic state. For about 95% of the mass (neglecting dark matter), according to modern estimates, is the plasma of ordinary and neutron stars, pulsars, black holes, and giant planets of the solar system, as well as the recently discovered hundreds of planets beyond the solar system.

Prior to becoming a star, the matter of the universe experiences the most diverse physical transformations: from quarks and elementary particles to complex molecules and again to atoms and particles; from relativistic energies to absolute zero and again to the state of high-energy-density plasma; from enormous densities to ultrahigh vacuum and again to the densities of atomic nuclei and quarks. And so our fundamental knowledge of the structure, evolution, and history of the universe is directly dependent on our understanding of the behavior of matter in all of its transformations up to ultrahigh energy densities, which forms not only specific physical models, but also the world-outlook of modern natural science because, by steadily increasing the high energy densities attainable by investigations on Earth and in space, we delve into the past, using a "time machine", looking for the singular conditions of the Big Bang – the instant of the universe's inception $\approx 15$ billion years ago.

Today we can clearly see that the study of matter in extreme states is one of the "hottest" and most rapidly developing basic scientific disciplines, situated at the interface between plasma physics, nonlinear optics, condensed-matter physics, nuclear, atomic and molecular physics, relativistic and magnetic hydrodynamics, involving a wealth of compression- and heating-stimulated physical effects and a constantly widening variety of objects and states in which the plasma nonideality is critically important. Despite the extraordinary diversity of the objects and experimental and astrophysical situations, they all share the common property that high energy densities play a decisive role in their physical behavior.

These circumstances are a permanent steadfast incentive for intensive theoretical and experimental investigations, which have recently resulted in a great body of new and, most important, reliable information about the thermodynamic, structural, gasdynamic, optical, electrophysical, and transport properties of matter under extreme conditions. These specific data are contained in a massive flow of original publications, some of which are not easily accessible to Russian scientists, especially young ones.

This book attempts to systematize, generalize, and set forth from a unified viewpoint the theoretical and experimental material related to this new realm of science, and to show, following Titus Lucretius Carus, "Thus from the mixture of the elements there emerge infinite multitudes of creatures, which are Strange and highly diversified in appearance" (Titus Lucretius Carus) [2]. In it I endeavor to discuss the maximally broad range of problems related to high-energy-density physics. That is why many interesting astrophysical, laser, and nuclear-physical problems as well as technical applications are briefly outlined and the reader is referred to original papers and monographs. In doing this, I did not aim to include everything known about extreme states to date. Emphasis was placed on the issues which appeared to be most interesting to me and which I happened to work on directly. Realizing that the material touched upon in this book will steadily broaden, become defined more precisely, and inevitably become obsolete owing to the extremely rapid development of high-energy-density physics, I would be thankful to readers who send me critical remarks and suggestions.

It is hoped that this book will be beneficial to a wide readership of scientists, postgraduates, and students in the natural sciences by offering them access to original papers and being a helpful guide through the exciting problems of modern Extreme State Physics.

In the preparation of the manuscript I benefited from the assistance, stimulating discussions, and advice of Yu.Yu. Balega, N.E. Andreev, A.Ya. Faenov, S.I. Blinnikov, M.B. Kozintsova, A.N. Starostin, V.S. Imshennik, I.L. Iosilevskii, N.S. Kardashev, V.G. Sultanov, A.N. Sissiakian, B.Yu. Sharkov, and V.A. Yakovleva, to whom I express my sincere gratitude.

Abramtsevo, May 2009. *Vladimir Fortov*

## References

[1] Bible. Old Testament, 1 Samuel, 17: 34, 40, 43, 48–51
[2] Lucretius: The Nature of Things. Penguin, London (2007)
[3] Voltaire: Micromégas and Other Short Fiction. Penguin, London (2002)

# Contents

# Chapter 1
# Introduction

"We already know the laws that govern
the behavior of matter under all but
the most extreme conditions"

*S. Hawking. A Brief History of Time (p. 168).*

The states of matter at extremely high temperatures and densities, and hence with extraordinarily high energy densities, have always attracted researchers, owing to the possibility of reaching record parameters, advancing to new domains of the phase diagram, and producing in the laboratory the exotic states that gave birth to our universe through the big bang and which now account for the great bulk (90–95%) of the mass of baryon (visible) matter – in stellar and interstellar objects, planets, and exoplanets [25, 14, 11, 16, 33, 19, 18, 40, 10, 39]. That is why the study of these states of matter – so exotic for us in terrestrial conditions and yet so typical for the rest of the universe – is of great cognitive importance, forming our modern notions of the surrounding world. Furthermore, a constant pragmatic incentive for such investigations is the application of high-energy-density states in nuclear, thermonuclear, and impulse power engineering, high-voltage and high-power electrophysics, for the synthesis of superhard materials, for strengthening and welding materials, for the impact protection of space vehicles, and, of course, in defense. For the nuclear devices with controllable (inertial confinement fusion, ICF) and quasi-controllable (atomic and hydrogen bombs) energy release rely on the initiation of nuclear reactions in a strongly compressed and heated nuclear fuel.

The revolutionary discoveries in astronomy of recent decades [25, 33, 19, 18, 40, 10, 39] (neutron stars, pulsars, black holes, wormholes, $\gamma$–ray bursts, exoplanets) furnish new examples of extreme states, which call for investigation in order to solve the most fundamental problems of modern astrophysics.

Beginning in the mid-1950s in the framework of nuclear defense projects, high-energy-density research has made considerable progress owing to the advent of new devices for the generation of high energy densities, such as lasers, charged-particle beams, high-current Z-pinches, explosion and electric explosion generators of intense shock waves, multistage light-gas guns and diamond anvils. These complicated and expensive technical devices have made it possible to substantially advance

V.E. Fortov, *Extreme States of Matter*, The Frontiers Collection,
DOI 10.1007/978-3-642-16464-4_1, © Springer-Verlag Berlin Heidelberg 2011

along the scale of energy density attainable in a physical experiment and to obtain in laboratory or quasi-laboratory conditions the states that range into the megabar–gigabar pressure domain, which is unattainable with the traditional techniques of experimental physics. The use of new generators for the cumulation of high energy densities leads to a variety of fascinating physical effects, such as a radical restructuring of the energy spectrum and composition of a compressed and heated material [25, 14, 11, 16, 33, 17], new cooperative effects and diverse instabilities in the interaction of directed energy fluxes with a dense plasma, its transient motion under the conditions of substantial radiative energy transfer, and relativistic, gravitational, and nuclear phenomena as well as a number of other exotic effects, which may be predicted, if at all, only in the most general form [33, 19, 18].

By now high-energy-density physics has turned into an extensive and rapidly developing branch of modern science that makes use of the most sophisticated means of generation, diagnostic techniques, and numerical simulations with high-power supercomputers.

It is no accident that half of the 30 problems of "the physics minimum at the beginning of the XXIst century" proposed by Academician V.L. Ginzburg [19] are to a greater or lesser degree dedicated to high-energy-density physics.

The term "high" is conventionally [25, 14, 11, 16, 33, 10] used to refer to in-substance energy densities exceeding $\approx 10^4$–$10^5$ J/cm$^3$, which corresponds to the binding energy of condensed media (for instance, high explosives (HEs), $H_2$, or metals) and the pressure level of millions of atmospheres. For comparison: the pressure at the center of the Earth is equal to $\approx$3.6 Mbar, of Jupiter to $\approx$40 Mbar, and at the center of the Sun to $\approx$200 Gbar. We mention that the term "substance" is considered to mean a lengthy continuous medium of macroscopic size. The "ultra-extreme" states inside the atomic nucleus and the conditions occurring in microscopic volumes in the individual collisions of the nuclei speeded up in accelerators to subluminal velocities, which are the subject of relativistic nuclear physics [35], are therefore beyond the scope of detailed consideration.

As a rule, a substance under high-energy-density conditions is in the plasma state — an ionized state arising from thermal- and/or pressure-induced ionization. In astrophysical objects, such compression and heating is effected by gravitational forces and nuclear reactions, and in laboratory conditions by intense shock waves, which are excited by a wide variety of "drivers", ranging from two-stage gas guns [34] to lasers [22, 2] and high-current Z-pinches [27, 21] with powers of hundreds of terawatts[1] However, while the lifetime of extreme states in astrophysical objects ranges from milliseconds to billions of years, making it possible to conduct detailed observations and measurements of them with the help of space probes and orbital and ground-based telescopes of different wavelengths, in terrestrial conditions we have to do with the microsecond–femtosecond duration range [14, 11, 16, 2], which calls for the application of ultrafast specific diagnostic techniques.

At present, every large-scale physical facility that generates extremely high pressures and temperatures is engaged in work programs (not infrequently international)

[1] The total power of terrestrial electric power plants amounts to about 3.5 TW.

on the basic physics of high energy density, in addition to having practical, applied tasks in impulse energetics or defense. It is significant that the experimental capabilities of these devices are extending rapidly. In particular, modern short-pulse laser systems (NIKE; OMEGA; TRIDENT, LULI-200, the first module of the National Ignition Facility (NIF), USA; Laser Mégajoule (LMJ), France; GEKKO-XII, Japan; VULKAN, Great Britain; Iskra-6, Russia; etc. (see Tables 2.1 and 3.1)) are capable of releasing 1–40 kJ in a volume of the order of 1 $mm^3$ in several nanoseconds to produce pressures of tens to hundreds of megabars. The projected launch of the NIF (Livermore) [22, 31] and LMJ (France) [7, 20] laser systems would raise the at-focus energy liberation to the megajoule level.

Furthermore, the Z-pinch technology is now exhibiting considerable progress: at the Sandia facility (USA), $\approx 1.8$ MJ soft X-ray radiation was obtained in the collapse of plasma liners during 5–15 ns in a region measuring about 1 $cm^3$ [38, 9, 37]. Supplemented by experiments with diamond anvils, explosion and electric explosion devices, and light-gas guns in the megabar pressure range, these data at record parameters are now the source of new and sometimes unexpected information about the behavior of extreme-parameter plasmas [11, 16].

Interestingly, when conducting experiments on extreme-state laboratory plasma, even today it is possible to partly reproduce on a small scale many parameters, effects, and processes occurring in astrophysical objects, information on which has become accessible by use of ground- and space-based means of observation. These are the data on hydrodynamic mixing and instabilities, shock-wave phenomena, strongly radiating and relativistic streams and jets, solitons, relativistic phenomena, equations of state, and the composition and spectra of nonideal plasmas, as well as the characteristics of interstellar cosmic plasma, dust, and a number of other effects.

The physics of high-energy densities is intimately related to several branches of science, for example plasma physics and condensed-matter physics, relativistic physics, the physics of lasers and charged-particle beams, nuclear, atomic, and molecular physics, radiative, gas and magnetic hydrodynamics, and astrophysics. In this case, the distinguishing features of high-energy-density physics are an extreme complexity and strong nonlinearity of the plasma processes occurring in it, the significance of collective interparticle interaction, and relativity. This makes the study of phenomena occurring in this area a fascinating and absorbing task, which attracts a constantly increasing number of researchers.

For all these reasons, the National Research Council of the USA National Academies of Science formulated a large-scale national program of research [33] in the area of high-energy-density physics and gave it high priority. In our subsequent discussions we shall orient ourselves with respect to this document in several aspects [33]. Similar work programs are being vigorously pursued in many developed countries capable of making the requisite experimental devices and having qualified personnel in sufficient number.

The aim of this book is to discuss the current state-of-the-art in high-energy-density physics and its trends, the advantages and limitations of different experimental techniques for generating and diagnosing dense plasmas, and to briefly analyze the results achieved. Because of the vastness and dissimilarity of the material, in several cases the presentation will be abstract-like in form, referring the reader to the corresponding reviews and monographs [25, 14, 11, 16, 33, 19, 18, 40, 10, 39, 3, 26, 12, 23, 13, 28, 4, 29, 36, 32, 24, 5, 6, 8, 1, 15, 30].

After the Introduction (Chap. 1), the second chapter presents a classification of the states of matter at high energy densities and discusses the general form of the phase diagram, dimensionless parameters, and the physical conditions corresponding to terrestrial and astrophysical objects.

The means of generating extreme states available to experimenters are presented in Chapter 3. The properties of the interaction of high-power fluxes of directed energy with plasmas are the concern of Chapters 4 and 5. Numerous technical applications of high-energy-density physics are briefly outlined in Chapter 6. The monograph concludes (Chapter 7) with a discussion of the most typical astrophysical objects and phenomena related to the realization of extreme states of matter.

The author endeavored to introduce the reader to the rather broad scope of diverse problems of high-energy-density physics, and therefore many topics are discussed by no means in sufficient detail, for the aim of this book is to help readers orient himselves in the vast sea of information accumulated to date, see the trend, and to help young scientists choose the directions of future investigations rather than provide answers to all steadily arising questions.

## References

[1] Al'tshuler, L.V., Krupnikov, K.K., Fortov, V.E., Funtikov, A.I.: Origins of megabar pressure physics. Herald Russ. Acad. Sci. **74**(6), 613 (2004)

[2] Anisimov, S.I., Prokhorov, A.M., Fortov, V.E.: Application of high-power lasers to study matter at ultrahigh pressures. Sov. Phys. – Usp. **27**(3), 181–205 (1984). DOI 10.1070/PU1984v027n03ABEH004036. URL http://stacks.iop.org/0038-5670/27/181

[3] Ashkroft, N.A.: Condensed matter at higher densities. In: G.L. Chiarotti, R.J. Hemley, M. Bernasconi, L. Ulivi (eds.) High Pressure Phenomena, Proceedings of the International School of Physics "Enrico Fermi" Course CXLVII, p. 151. IOS Press, Amsterdam (2002)

[4] Atzeni, S., Meyer-ter-Vehn, J.: The Physics of Inertial Fusion. Oxford University Press, Oxford (2004)

[5] Avrorin, E.N., Vodolaga, B.K., Simonenko, V.A., Fortov, V.E.: Intense shock waves and extreme states of matter. Phys. Usp. **36**(5), 337–364 (1993). DOI 10.1070/PU1993v036n05ABEH002158. URL http://stacks.iop.org/1063-7869/36/337

[6] Calderola, P., Knopfel, H. (eds.): Physics of High Energy Density. Academic, New York (1971)

[7] Cavailler, C.: Inertial fusion with the LMJ. Plasma Phys. Control. Fusion **47**(12B), B389–B403 (2005). DOI 10.1088/0741-3335/47/12B/S28

[8] Chen, F.F.: Introduction to Plasma Physics and Controlled Fusion, Vol. 1, 2nd edn. Springer, New York (1984)
[9] Cuneo, M.E., Vesey, R.A., Bennett, G.R., et al.: Progress in symmetric ICF capsule implosions and wire-array Z-pinch source physics for double-pinch-driven hohlraums. Plasma Phys. Control. Fusion **48**(2), R1–R35 (2006). DOI 10.1088/0741-3335/48/2/R01
[10] Drake, R.P.: High-Energy-Density Physics. Springer, Berlin, Heidelberg (2006)
[11] Fortov, V., Iakubov, I., Khrapak, A.: Physics of Strongly Coupled Plasma. Oxford University Press, Oxford (2006)
[12] Fortov, V.E. (ed.): Entsiklopediya nizkotemperaturnoi plazmy (Encyclopedia of Low-Temperature Plasma). Nauka, Moscow (2000)
[13] Fortov, V.E.: Intense Shock Waves and Extreme States of Matter. Bukos, Moscow (2005)
[14] Fortov, V.E.: Intense shock waves and extreme states of matter. Phys. Usp. **50**(4), 333 (2007). DOI 10.1070/PU2007v050n04ABEH006234. URL `http://ufn.ru/en/articles/2007/4/c/`
[15] Fortov, V.E., Al'tshuler, L.V., Trunin, R.F., Funtikov, A.I.: High-Pressure Shock Compression of Solids VII: Shock Waves and Extreme States of Matter. Springer, New York (2004)
[16] Fortov, V.E., Khrapak, A.G., Yakubov, I.T.: Fizika neideal'noi plazmy (Physics of Nonideal Plasma). Fizmatlit, Moscow (2004)
[17] Fortov, V.E., Ternovoi, V.Y., Zhernokletov, M.V., et al.: Pressure-produced ionization of nonideal plasma in a megabar range of dynamic pressures. JETP **97**(2), 259–278 (2003). DOI 10.1134/1.1608993
[18] Ginzburg, V.L.: The Physics of a Lifetime: Reflections on the Problems and Personalities of 20th Century Physics. Springer, Berlin, Heidelberg (2001)
[19] Ginzburg, V.L.: On superconductivity and superfluidity (what I have and have not managed to do), as well as on the "physical minimum" at the beginning of the XXI century (December 8, 2003). Phys. Usp. **47**(11), 1155 (2004). DOI 10.1070/PU2004v047n11ABEH001825. URL `http://ufn.ru/en/articles/2004/11/d/`
[20] Giorla, J., Bastian, J., Bayer, C., et al.: Target design for ignition experiments on the laser Mégajoule facility. Plasma Phys. Control. Fusion **48**(12B), B75–B82 (2006). DOI 10.1088/0741-3335/48/12B/S0
[21] Grabovskii, E.V., Vorob'ev, O.Y., Dyabilin, K.S., et al.: Excitation of intense shock waves by soft x radiation from a Z-pinch plasma. JETP Lett. **60**(1), 1 (1994)
[22] Hammel, B.A., National Ignition Campaign Team: The NIF ignition program: progress and planning. Plasma Phys. Control. Fusion **48**(12B), B497–B506 (2006). DOI 10.1088/0741-3335/48/12B/S47
[23] Hemley, R.J., Mao, H.K.: Overview of static high pressure science. In: R.J. Hemley, G.L. Chiarotti, M. Bernasconi, L. Ulivi (eds.) High Pressure Phenomena, Proceedings of the International School of Physics "Enrico Fermi" Course CXLVII, p. 3. IOS Press, Amsterdam (2002)

[24] Hogan, W.J. (ed.): Energy from Inertial Fusion. IAEA, Vienna, Austria (1995)
[25] Kirzhnits, D.A.: Extremal states of matter (ultrahigh pressures and temperatures). Sov. Phys. – Usp. **14**(4), 512–523 (1972). DOI 10.1070/PU1972v014n04ABEH004734. URL http://stacks.iop.org/0038-5670/14/512
[26] Kirzhnits, D.A., Lozovik, Y.E., Shpatakovskaya, G.V.: Statistical model of matter. Sov. Phys. – Usp. **18**(9), 649–672 (1975). DOI 10.1070/PU1975v018n09ABEH005199. URL http://stacks.iop.org/0038-5670/18/649
[27] Knudson, M.D., Hanson, D.L., Bailey, J.E., et al.: Equation of state measurements in liquid deuterium to 70 GPa. Phys. Rev. Lett. **87**(22), 225501 (2001). DOI 10.1103/PhysRevLett.87.225501. URL http://link.aps.org/abstract/PRL/v87/e225501
[28] Kruer, W.L.: The Physics of Laser Plasma Interactions. Addison-Wesley, Reading, MA (1988)
[29] Lindl, J.D.: Inertial Confinement Fusion. Springer, New York (1998)
[30] Maksimov, E.G., Magnitskaya, M.V., Fortov, V.E.: Non-simple behavior of simple metals at high pressure. Phys. Usp. **48**(8), 761 (2005). DOI 10.1070/PU2005v048n08ABEH002315. URL http://ufn.ru/en/articles/2005/8/a/
[31] Moses, E.I., Bonanno, R.E., Haynam, C.A., et al.: The National Ignition Facility: path to ignition in the laboratory. Eur. Phys. J. D **44**(2), 215–218 (2006). DOI 10.1140/epjd/e2006-00106-3
[32] Mourou, G.A., Tajima, T., Bulanov, S.V.: Optics in the relativistic regime. Rev. Mod. Phys. **78**(2), 1804–1816 (2006). DOI 10.1103/RevModPhys.78.309. URL http://link.aps.org/abstract/RMP/v78/p309
[33] National Research Council: Frontiers in High Energy Density Physics. National Academies Press, Washington, DC (2003)
[34] Nellis, W.J.: Dynamic compression of materials: metallization of fluid hydrogen at high pressures. Rep. Prog. Phys. **69**(5), 1479–1580 (2006). DOI 10.1088/0034-4885/69/5/R05
[35] Okun', L.B.: Leptony i kvarki, 2nd edn. Nauka, Moscow (1990). [English Transl.: Leptons and Quarks. North-Holland, Amsterdam (1982)]
[36] Pukhov, A.: Strong field interaction of laser radiation. Rep. Prog. Phys. **66**(1), 47–101 (2003). DOI 10.1088/0034-4885/66/1/202
[37] Quintenz, J., Sandia's Pulsed Power Team: Pulsed power team. In: Proc. 13th Int. Conf. on High Power Particle Beams. Nagaoka, Japan (2000)
[38] Spielman, R.B., Deeney, C., Chandler, G.A., et al.: Tungsten wire-array Z-pinch experiments at 200 TW and 2 MJ. Phys. Plasmas **5**(5), 2105–2111 (1998). DOI 10.1063/1.872881
[39] Vacca, J.R. (ed.): The World's 20 Greatest Unsolved Problems. Prentice Hall PTR, Englewood Cliffs, NJ (2004)
[40] Zasov, A.V., Postnov, K.A.: Obshchaya astrofizika (General Astrophysics). Vek 2, Fryazino (2006)

# Chapter 2
# Matter under Extreme Conditions: Classification of States

Figures 2.1 and 2.2 show the diagrams for the extreme conditions that are realized in a number of natural physical objects and in laboratory experiments. Of course, the parameters given are no better than an estimate and provide only the most general impression of the order of magnitude of the quantities under discussion.

The emergence of extreme states in nature is caused by the forces of gravity, which are inherently long range and unscreened, unlike Coulomb forces (in plasmas). These forces compress and heat the substance either directly or by stimulating exothermic nuclear reactions in massive astrophysical objects and during the early stages of the evolution of the universe [35, 29, 28].

The scale of extreme states realized in nature defies the most vivid imagination, see Fig. 2.3. At the bottom of the Mariana Trough in the Pacific Ocean (11 km) the water pressure amounts to 1.2 kbar (1 kbar = 100 MPa), at the Earth's center the pressure is 3.6 Mbar, the temperature[1] $T \approx 0.5$ eV, and the density $\rho \approx 10$–$20$ g/cm$^3$; at the center of Jupiter $P \approx 40$–$60$ Mbar, $\rho \approx 30$ g/cm$^3$, $T \approx 2 \times 10^4$ K; at the center of the Sun $P \approx 240$ Gbar, $T \approx 1.6 \times 10^3$ eV, $\rho \approx 150$ g/cm$^3$; in the cooling-down stars – white dwarfs – $P \approx 10^{10}$–$10^{16}$ Mbar, $\rho \approx 10^6$–$10^9$ g/cm$^3$, and $T \approx 10^3$ eV. In the targets for controlled fusion with inertial plasma confinement $P \approx 200$ Gbar, $\rho \approx 150$–$200$ g/cm$^3$, and $T \approx 10^8$ eV. Neutron stars, which are elements of pulsars, black holes, $\gamma$-ray bursts and magnetars, have what are thought to be the record parameters $P \approx 10^{19}$ Mbar, $\rho \approx 10^{11}$ g/cm$^3$, $T \approx 10^4$ eV for the mantle and for the core $P \approx 10^{22}$ Mbar, $\rho \approx 10^{14}$ g/cm$^3$, $T \approx 10^4$ eV for a great magnetic field of $10^{11}$–$10^{16}$ Gs (1 Gs = $10^{-4}$ T).

As noted above, by the lower bound of "high energy densities" we mean an energy density comparable to the binding energy of condensed matter, $10^4$–$10^5$ J/cm$^3$, which corresponds to the valence electron binding energy (of several electron volts) and to pressures of ca. 100 kbar–1 Mbar. These pressures far exceed the ultimate mechanical strength of materials and call for the inclusion of their compressibility and hence hydrodynamic motion under pulsed energy liberation.

[1] In certain fields, such as plasma physics, the electronvolt is used as a unit of temperature. The conversion to kelvins is defined using $k$, the Boltzmann constant: 1 eV/$k \approx 11.6$ kK (kilokelvin).

V.E. Fortov, *Extreme States of Matter*, The Frontiers Collection,
DOI 10.1007/978-3-642-16464-4_2, 

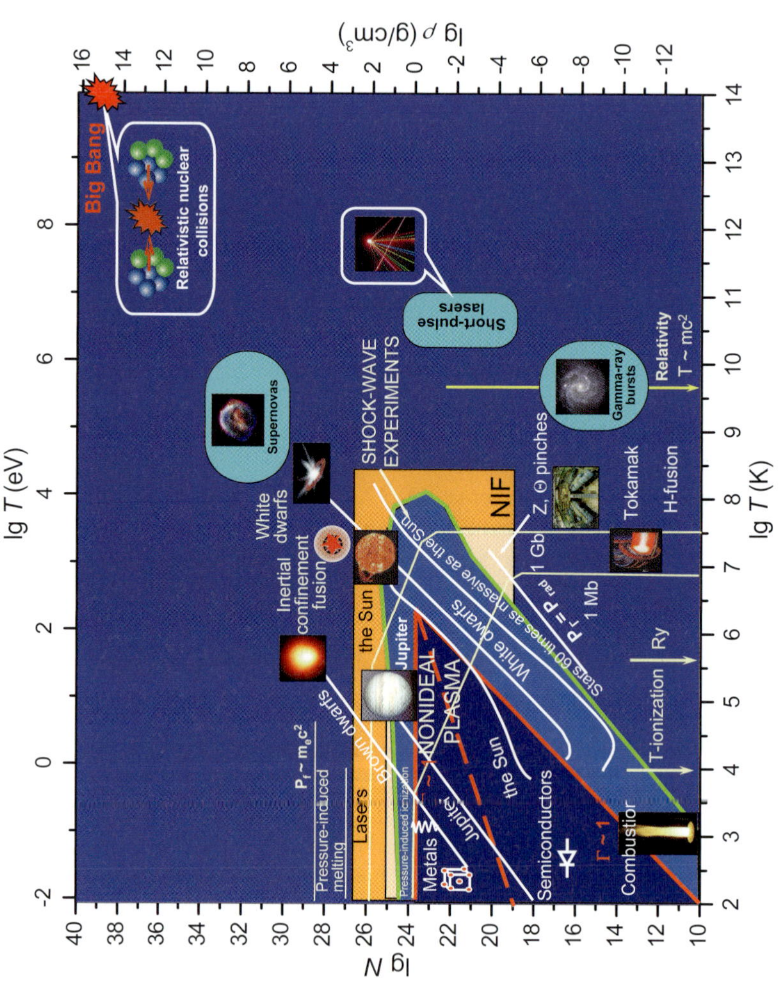


Fig. 2.1 Matter phase diagram [21, 49].

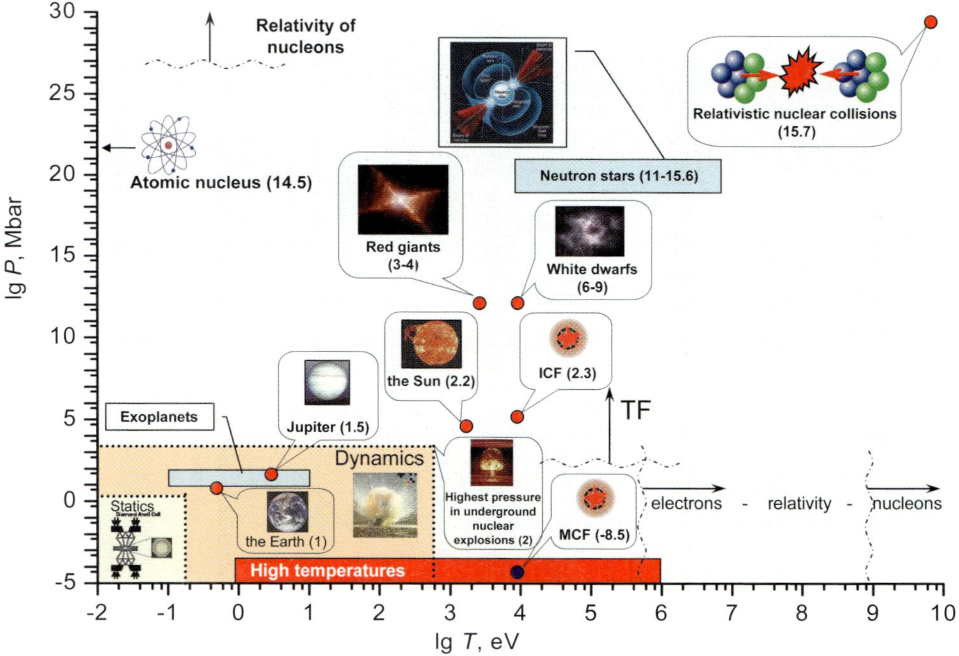


**Fig. 2.2** Extreme states [35] in nature and in the laboratory. The figures in parentheses indicate the logarithm of density (in g/cm$^3$). The "statics" domain corresponds to static techniques for high-pressure production, the "dynamics" domain to the dynamic ones, and "high temperatures" to high-temperature experiments.

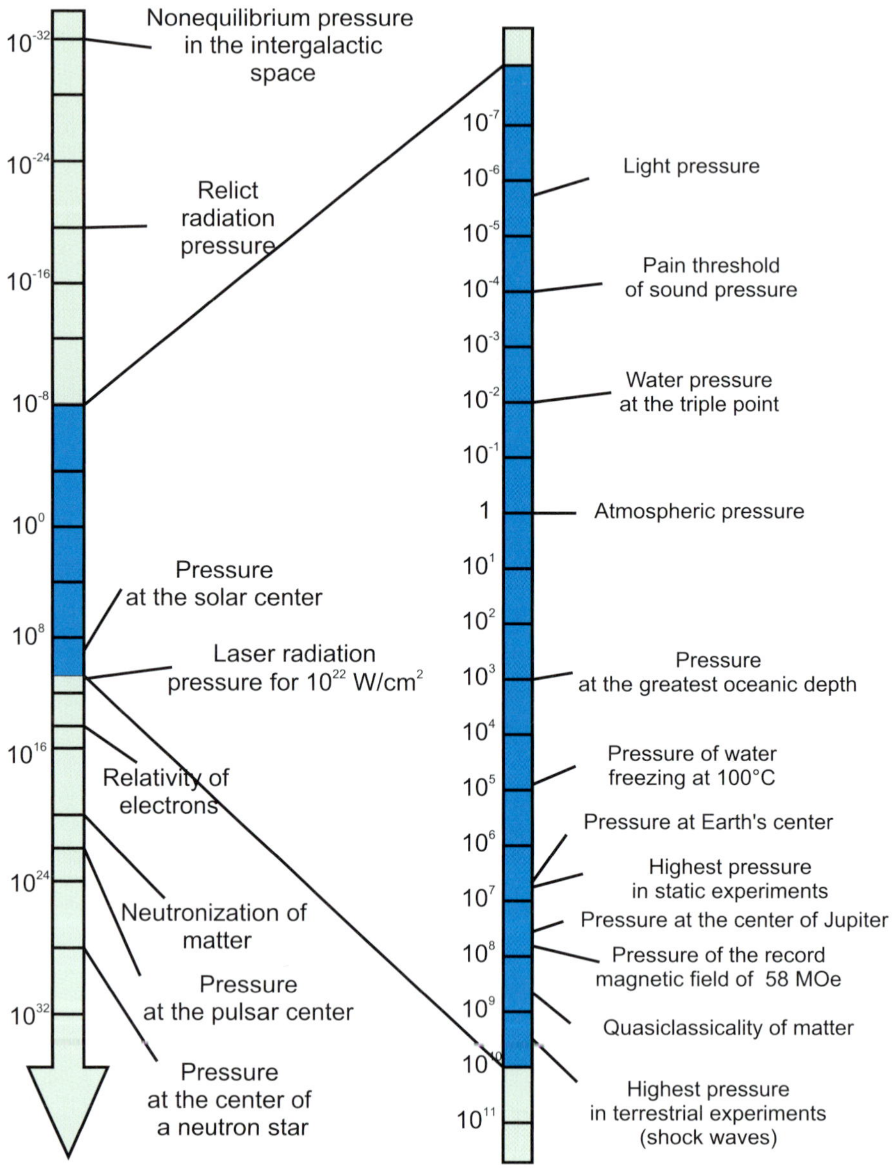


**Fig. 2.3** The pressure scale (1 atm $\approx 10^5$ Pa) in nature encompasses $\approx$ 64 orders of magnitude. The right-hand scale relates to experimentally attainable conditions [4].

In the domain of low pressures and temperatures, matter exhibits the exceptional diversity of properties and structures that we encounter under normal conditions day after day [35].

The physical, chemical, structural, and biological properties of substances are sharp nonmonotonic functions of composition. The classification of these "low-energy" states is complicated and cumbersome. They are defined by the position, properties, and occupation features of the electron levels of atoms, ions, and molecules and ultimately underlie the amazing wealth of the forms and manifestations of organic and inorganic nature on Earth.

Laser and evaporative cooling techniques (Fig. 2.4) enable ultralow ($10^{-9}$ K) ion temperatures to be reached and interesting quantum phenomena such as Bose–Einstein condensation, Rydberg matter, and Coulomb condensation to be studied.

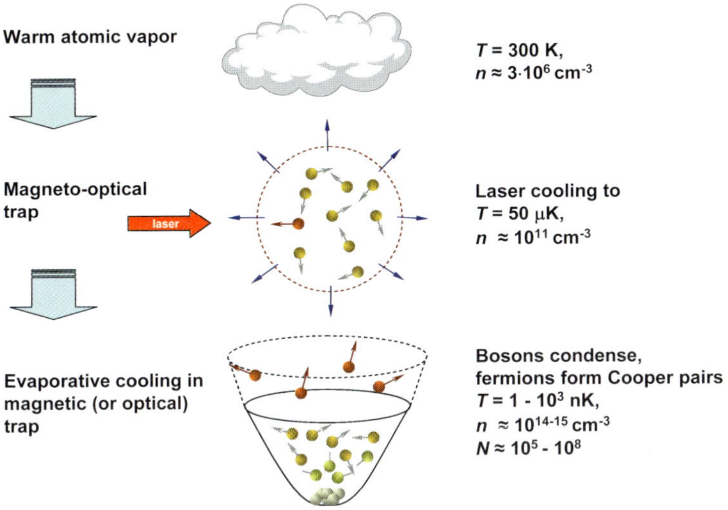


**Fig. 2.4** Methods of obtaining extremely low temperatures.

With increasing energy density ($P$ and $T$), substances acquire an increasingly universal structure [35, 36]. The distinctions between the neighboring elements of the periodic system smooth out and the properties of a substance become progressively smoother functions of its composition. Owing to an increase in energy density there occurs an obvious "universalization" or simplification of the substance properties. The growth of pressure and temperature ruptures molecular complexes to form atomic states, which next lose outer-shell electrons, which are responsible for the chemical individuality of the substance, due to thermal and/or pressure-induced ionization. Atomic and ion electron shells restructure to acquire an increasingly regular level occupation, and a crystal lattice after a number of polymorphic transformations (this ordinarily takes place for $P < 0.5$ Mbar) transforms to a close-packed body-centered cubic arrangement common to all substances.

These substance "simplification" processes take place at energy densities comparable to the characteristic energies of the aforementioned "universalization" pro-

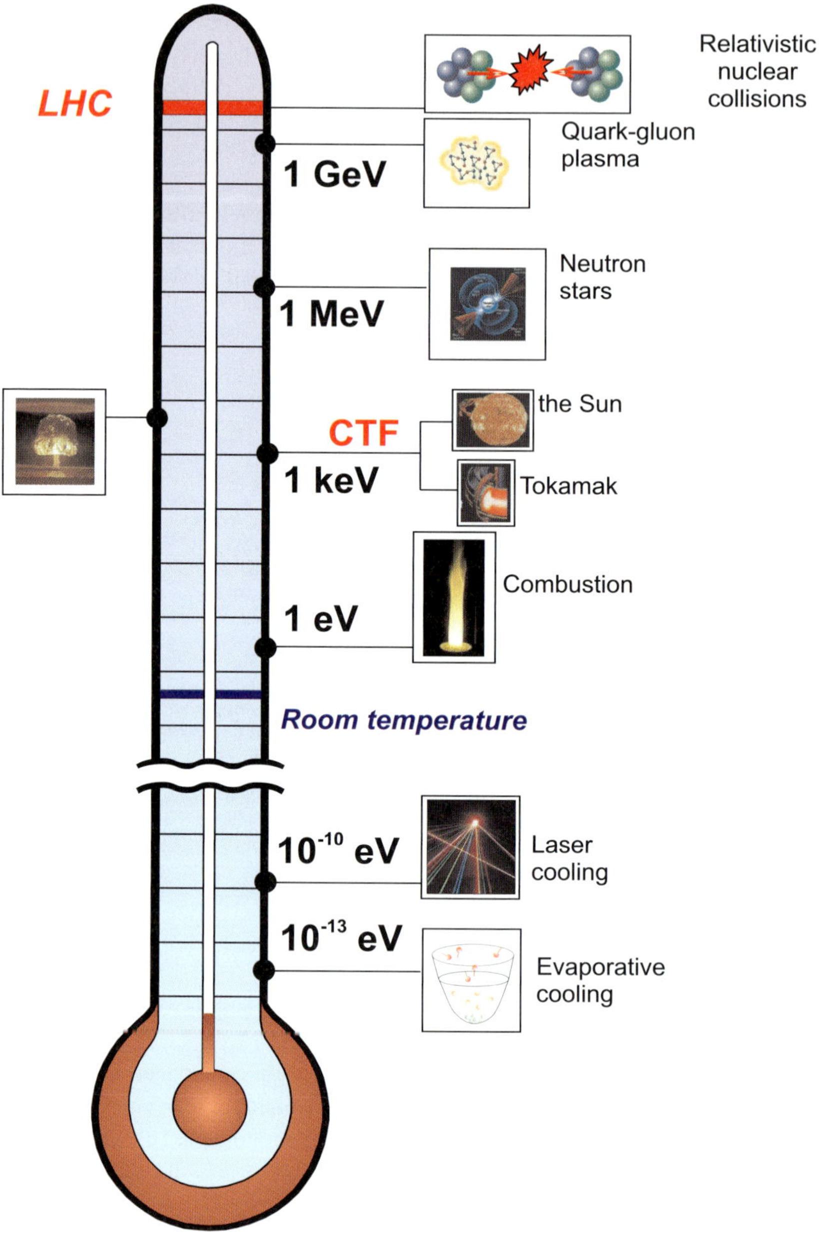


**Fig. 2.5** Experimentally attainable temperatures.

cesses. When the characteristic energy density is of the order of the valence shell energies, $e^2/a_0^4 \approx 3\times10^{14}$ erg/cm$^3$ ($a_0 = \hbar/(mc^2) = 5.2\times10^{-9}$ cm is the Bohr radius, 1 erg $= 1^{-7}$ J), the order of magnitude of the lower substance "universalizing" bound $T \approx 10$ eV, $P \approx 300$ Mbar, is reached. The exact quantitative determination of these bounds is an important task of the experimental physics of high energy densities (see Chap. 6), especially as theory [36] also predicts a highly varied behavior of substances in the ultramegabar pressure range (shell effects [36], electron and plasma phase transitions [2, 1, 19, 17, 16, 7, 24, 42, 33, 64, 25, 14], etc.).

The upper bound of the domain of extreme states is defined by the contemporary level of knowledge in high-energy physics [28, 63, 51] and observational astrophysical data, and is expected to be limited only by our imagination. The ultraextreme matter parameters accessible using modern physical concepts are defined by the so-called Planck quantities [28, 51], which are combinations of the fundamental constants Planck's constant $\hbar$, the velocity of light $c$, the gravitational constant $G$, and the Boltzmann constant $k$:

— the length $l_\mathrm{p} = \sqrt{\dfrac{\hbar G}{c^3}} = \dfrac{\hbar}{m_\mathrm{p} c} \approx 1.62\times10^{-33}$ cm;

— the mass (the so-called "maximon" mass) $m_\mathrm{p} = \sqrt{\dfrac{\hbar c}{G}} = 2.18\times10^{-5}$ g;

— the time $t_\mathrm{p} = \dfrac{l_\mathrm{p}}{c} = \dfrac{\hbar}{m_\mathrm{p} c^2} = \sqrt{\dfrac{\hbar G}{c^5}} = 5.39\times10^{-44}$ s;

— the temperature $T_\mathrm{p} = \dfrac{m_\mathrm{p} c^2}{k} = \sqrt{\dfrac{\hbar c^5}{G k^2}} = 1.42\times10^{32}$ K;

— the energy $W_\mathrm{p} = m_\mathrm{p} c^2 = \dfrac{\hbar}{t_\mathrm{p}} = \sqrt{\dfrac{\hbar c^5}{G}} = 1.96\times10^{9}$ J;

— the density $\rho_\mathrm{p} = \dfrac{m_\mathrm{p}}{l_\mathrm{p}^3} = \dfrac{\hbar t_\mathrm{p}}{l_\mathrm{p}^5} = \dfrac{c^5}{\hbar G^2} = 5.16\times10^{93}$ g/cm$^3$;

— the force $F_\mathrm{p} = \dfrac{W_\mathrm{p}}{l_\mathrm{p}} = \dfrac{\hbar}{l_\mathrm{p} t_\mathrm{p}} = \dfrac{c^4}{G} = 1.21\times10^{44}$ N;

— the pressure $p_\mathrm{p} = \dfrac{F_\mathrm{p}}{l_\mathrm{p}^2} = \dfrac{\hbar}{l_\mathrm{p}^3 t_\mathrm{p}} = \dfrac{c^7}{\hbar G^2} = 4.63\times10^{113}$ Pa;

— the charge $q_\mathrm{p} = \sqrt{\hbar c 4\pi\varepsilon_0} = 1.78\times10^{-18}$ C;

— the power $P_\mathrm{p} = \dfrac{W_\mathrm{p}}{t_\mathrm{p}} = \dfrac{\hbar}{t_\mathrm{p}^2} = \dfrac{c^5}{G} = 3.63\times10^{52}$ W;

— the circular frequency $\omega_\mathrm{p} = \sqrt{\frac{c^5}{\hbar G}} \approx 1.85 \times 10^{43}\ \mathrm{s}^{-1}$;

— the electric current $I_\mathrm{p} = \frac{q_\mathrm{p}}{t_\mathrm{p}} = \sqrt{\frac{c^6 4\pi\varepsilon_0}{G}} = 3.48 \times 10^{25}$ A;

— the voltage $V_\mathrm{p} = \frac{W_\mathrm{p}}{q_\mathrm{p}} = \frac{\hbar}{t_\mathrm{p}} = \sqrt{\frac{c^4}{G 4\pi\varepsilon_0}} = 1.05 \times 10^{27}$ V;

— the impedance $Z_\mathrm{p} = \frac{V_\mathrm{p}}{I_\mathrm{p}} = \frac{\hbar}{q_\mathrm{p}^2} = \frac{1}{4\pi\varepsilon_0 c} = \frac{Z_0}{4\pi} = 29.98\ \Omega$;

— the electric field intensity $E_\mathrm{p} = \frac{V_\mathrm{p}}{l_\mathrm{p}} = \frac{1}{G}\sqrt{\frac{c^7}{4\pi\varepsilon_0 \hbar}} = 6.4 \times 10^{59}$ V/cm;

— the magnetic field intensity

$$H_\mathrm{p} = \frac{1}{G}\sqrt{\frac{c^9 4\pi\varepsilon_0}{\hbar}} = 2.19 \times 10^{60}\ \mathrm{A/m} = 2.75 \times 10^{58}\ \mathrm{Oe}.$$

These superextreme matter parameters, at which the known physical laws are thought to be no longer valid, might have been realized early in the Big Bang or at the singularity in the collapse of black holes. In this field physical models are discussed that assume that our space has more than three dimensions and that ordinary matter is in a three-dimensional manifold – the "3-brane world" [55] – embedded in this many-dimensional space. For the present, the capabilities of modern experiments in the area of high-energy-density physics are far from these "Planck" values and allow the properties of elementary particles to be elucidated up to energies on the order of 0.1–10 TeV and down to distances $\approx 10^{-16}$ cm. The launch of the proton–proton collider at CERN with a proton collision energy of $7 \times 7$ TeV is an important milestone, opening up the teraelectronvolt energy range.

Considering (following Kirzhnits [35]) the energy range $mc^2 \approx 1$ GeV, which is amenable to a more substantial analysis and is nonrelativistic for nucleons, we obtain a boundary temperature of $10^9$ eV, an energy density of $10^{37}$ erg/cm$^2$, and a pressure of $\approx 10^{25}$ Mbar, although it is not unlikely that even more extreme states of matter are realized in the cores of massive pulsars and could be found early in the evolution of the universe.

Although our experimental capabilities are progressing rapidly, of course they are only partly able to enter the province of ultraextreme astrophysical states. Material strengths abruptly limit the use of static techniques for investigating high energy densities, because the overwhelming majority of constructional materials are unable to withstand the pressures of interest to us. An exception is diamond — the record holder in hardness ($\sigma_n \approx 500$ kbar); its use in diamond anvils enables a pressure of 3–5 Mbar to be reached in static experiments [42, 33, 32].

**Table 2.1** Physical conditions corresponding to high energy densities of $10^4$–$10^5$ J/cm$^3$ [49].

| Physical conditions | Values of physical parameters |
|---|---|
| Energy density<br>Pressure | $\approx 10^4$–$10^5$ J/cm$^3$<br>$\approx 0.1$–1 Mbar |
| Condensed explosives<br>pressure<br>temperature<br>density<br>velocity of detonation | $W \approx 10^4$ J/cm$^3$<br>$\approx 400$ kbar,<br>$\approx 4000$ K,<br>$\approx 2.7$ g/cm$^3$,<br>$\approx 9 \times 10^5$ cm/s |
| Impact of aluminum plate on aluminum, velocity<br>Impact of molybdenum plate on molybdenum, velocity | 5–13.2 $\times 10^5$ cm/s<br>3–7.5 $\times 10^5$ cm/s |
| Electromagnetic radiation<br>laser, intensity $q$ ($W \sim q$)<br>blackbody temperature $T$ ($p \sim T^4$) | <br>$2.6\times10^{15}$–$3 \times 10^{15}$ W/cm$^2$<br>$2 \times 10^2$–$4 \times 10^2$ eV |
| Electric intensity $E$ ($W \sim E^2$)<br>Magnetic induction $B$ ($W \sim B^2$) | $0.5 \times 10^9$–$1.5 \times 10^9$ V/cm<br>$1.6 \times 10^2$–$5 \times 10^2$ T |
| Plasma density for a temperature $T = 1$ keV<br>($p = nkT$) | <br>$6\times10^{19}$–$6 \times 10^{20}$ cm$^{-3}$ |
| Laser radiation intensity $q$, for $\lambda = 1$ $\mu$m, $W \sim q^{2/3}$<br>blackbody temperature $T$ ($p \sim T^{3.5}$) | $0.86 \times 10^{12}$–$4\times10^{12}$ W/cm$^2$<br>66–75 eV |

Currently superior are dynamic techniques [21, 18, 26, 27, 50, 2, 1, 64], which rely on the pulsed cumulation of high energy densities in substances. The lifetime of such high-energy states is defined by the time of inertial plasma expansion, typically in the range $10^{-10}$–$10^{-6}$ s, which calls for the application of sophisticated fast diagnostic techniques. Table 2.1 gives the physical conditions corresponding to the lower bound of the states of interest [49, 18, 26, 9]. One can see that the production of high energy densities in plasmas imposes high demands on the means of generation, necessitating efficient spatial and temporal power compression.

The matter phase diagram corresponding to high energy densities is depicted in Fig. 2.1 [21, 27, 18, 26, 49], which indicates the conditions existing in astrophysical objects as well as in technical and laboratory experimental devices. Being the most widespread state of matter in nature (95% of the mass of the universe, neglecting dark matter), plasma occupies virtually all the domains of the phase diagram as seen in Fig. 2.1. In this case, special difficulties in the physical description of such a medium are presented by the nonideal plasma domain, where the Coulomb interparticle interaction energy $e^2n^{1/3}$ is comparable to or higher than the kinetic energy $E_k$ of particle motion. In this domain, $\Gamma = e^2n^{1/3}/E_k > 1$, plasma nonideality effects cannot be described by perturbation theory [35, 18, 26], while the application of computer parameter-free Monte-Carlo or molecular dynamics methods [19, 42] is fraught with the difficulties of selection of adequate pseudopotentials and correct inclusion of quantum effects.

The effects of electron relativity in the equation of state and transport plasma properties, when $m_ec^2 \approx kT$, correspond to $T \approx 0.5$ MeV $\approx 6 \times 10^6$ K. Above this temperature, the matter becomes unstable with respect to spontaneous electron–positron pair production.

Quantum effects are defined by the degeneracy parameter $n\lambda^3$ ($\sqrt{\hbar^2/2mkT}$ is the thermal de Broglie wavelength). For a degenerate plasma, $n\lambda^3 \gg 1$, the kinetic energy scale is the Fermi energy $E_F \sim \hbar^2 n^{2/3}/2m$, which increases with plasma density, making the plasma more and more ideal as it contracts, $n \to \infty$; $\Gamma = me^2/(\hbar^2 n^{1/3}) \to 0$. The relativity condition, which corresponds to the condition $m_e c^2 \sim E_F \approx 0.5$ MeV, yields a density of $\approx 10^6$ g/cm$^3$.

Similar asymptotes are also found in the other limiting case $T \to 0$ of a classical ($n\lambda^3 \ll 1$) plasma, where $E_k \approx kT$ and the plasma becomes increasingly ideal, $\Gamma \approx e^2 n^{1/3}/(kT)$, under heating. One can see that the periphery of the matter phase diagram is occupied by ideal ($\Gamma \ll 1$), Boltzmann ($n\lambda^3 \ll 1$), or degenerate ($n\lambda^3 \gg 1$) plasmas, which are adequately described by currently available physical models [35, 21, 18, 26, 19, 17, 16, 7, 24].

The electron plasma in metals and semiconductors corresponds to the degenerate case with an interaction energy $E_{\text{int}} \sim e^2/r_0$, $r_0 \sim \hbar/k_f$, $E_k \sim k_f^2/m$; $\Gamma \sim e^2/\hbar v_f \approx$ 1–5, where $v_f \approx 10^{-2}$–$10^{-3}$ s and the subscript f indicates parameters at the Fermi limit.

For a quark–gluon plasma $E_{\text{int}} \sim g^2/r_0$, $r_0 \sim 1/T$, $E_k \sim T$; $\Gamma \sim g^2 \approx 1$–50. The most significant challenge to the theory is presented by the vast domain of nonideal plasmas $\Gamma \geqslant 1$ occupied by numerous technical applications (semiconductor and metal plasma, impulse energetics, explosions, arcs, electric discharges, etc.), where theory predicts qualitatively new physical effects (metallization, "cool" ionization, dielectrization, plasma phase transitions, etc. [21, 18, 26, 20]). Their study requires substantial experimental and theoretical efforts.

Of special interest are plasma phase transitions in strongly nonideal Coulomb systems: the crystallization of dust plasmas (Fig. 2.6) [24, 25] and ions in electrostatic traps [14] and cyclotrons [58, 59], electrolytes [52, 56] and colloidal systems [48], two-dimensional electron systems on the surface of liquid helium [61], exciton condensation in semiconductors [16], etc. Special mention should be made of the recently discovered phase transition in thermal deuterium plasma [23, 17, 16, 7] quasi-adiabatically compressed to megabar pressures by a series of reverberating shock waves.

The quest for qualitatively new effects of this kind in the nonideal parameter domain is a powerful and permanent incentive to investigate substances at high energy densities [21, 18, 26, 49].

Another characteristic property of a high-energy-density plasma is the collective nature of its behavior and the strong nonlinearity of its response to external energy actions such as shock and electromagnetic waves, solitons, laser radiation, and fast particle fluxes. In particular, the propagation of electromagnetic waves in plasma excites several parametric instabilities (Raman, Thomson, and Brillouin radiation scattering) and is accompanied by the self-focusing and filamentation of radiation, the development of inherently relativistic instabilities, the production of fast particles and jets, and – at higher intensities – the "boiling" of the vacuum with electron–positron pair production [49, 3, 39, 5, 41, 38, 20, 53, 11, 45, 43, 57, 47, 8].

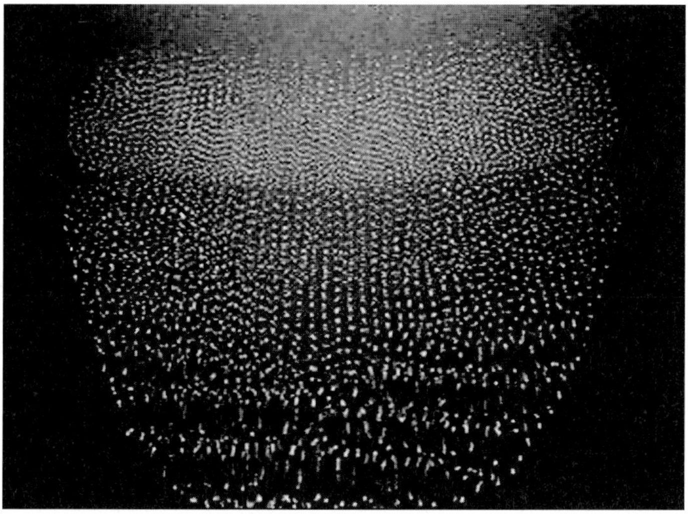

**Fig. 2.6** Dust plasma crystal and plasma liquid.

Of special interest in the exposure to extreme energy actions are transient hydrodynamic phenomena such as the instabilities of shock waves and laminar flows [38], transition to the turbulent mode [15], turbulent mixing, and jet and soliton dynamics [15, 13].

Figure 2.7, which is taken from [49], shows the domains of the dimensionless parameters Reynolds number $Re \sim Ul/\nu$ and Mach number $M = v/c$ (where $c$ is the velocity of sound, $l$ is the characteristic dimension, and $\nu$ is the kinematic viscosity) in which the different hydrodynamic phenomena related to high-energy-density physics occur. To astrophysical applications there correspond the flow modes, where $Re > 10^4$ and $M > 0.5$.

All these fascinating and inherently nonlinear phenomena manifest themselves in both astrophysical and laboratory plasmas and, despite the enormous difference in spatial scale, have much in common and make up the subject of "laboratory astrophysics" [13, 57].

Laboratory astrophysics [13, 57] permits the states of matter and processes with high energy densities characteristic for astrophysical objects to be reproduced in microscopic volumes. These are the effects of instability and hydrodynamic mixing; ordinary and magnetohydrodynamic turbulence; the dynamics of intense radiative, soliton, and shock waves; expansion waves; magnetically compressed and fast relativistic jets; and strongly radiating flows.

Of considerable interest is information about the equation of state, composition, optical and transport properties, emission and absorption spectra, cross sections of elementary processes, radiative thermal conductivity coefficients, and properties of relativistic plasma. This makes it possible to study and model the physical conditions, stationary and pulsed processes in astrophysical objects and phenomena

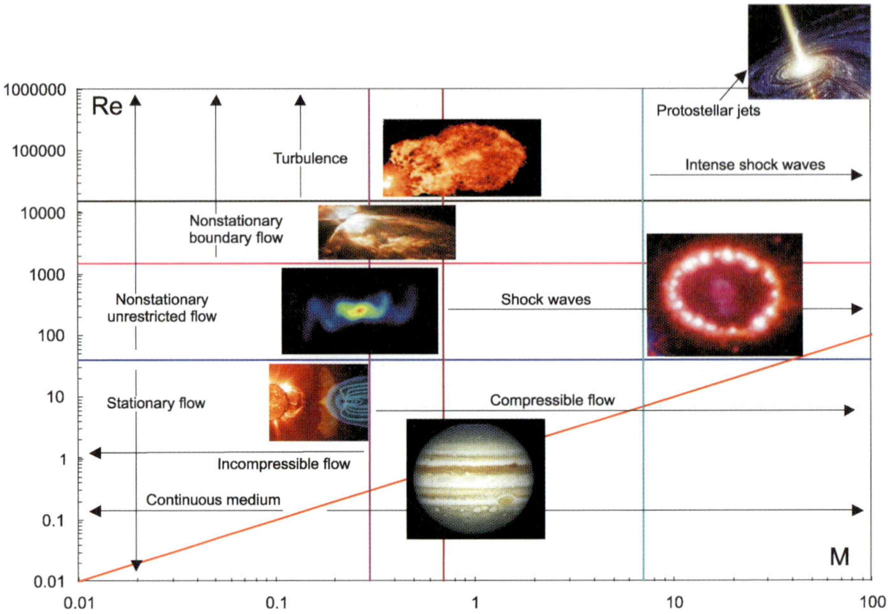


**Fig. 2.7** Hydrodynamic modes related to high-energy-density physics [49]. In the explosion of a type Ia supernova, the Mach number ranges from 0.01 in the region of thermonuclear combustion to 100 in the shock wave of its surface explosion. In the majority of astrophysical phenomena $Re > 10^6$ [49].

such as giant planets and exoplanets, stellar evolution and supernova explosions, gamma-ray burst structure, substance accretion dynamics in black holes, processes in binary and neutron stars as well as in the radiative motion of molecular interplanetary clouds, collisionless shock wave dynamics, and charged-particle acceleration to ultrahigh energies.

Nonlinear optical effects in the interaction of high-power short laser pulses with plasmas enable a major advance in the scales of energy density, specific power, pressure, and temperature by generating extreme-parameter relativistic plasmas in laboratory conditions and attaining record electric and magnetic intensities in terrestrial conditions, which lead, in particular, to the production of high-energy electron and ion flows [53, 47]. These investigations have already provided a wealth of new information for plasma physics, controlled fusion, and astrophysics, as well as technological and defense applications. The progress of this research may result in radically new strong-gradient electron accelerators (see Sect. 6.3), new sources of high-frequency radiation and light (Sect. 6.4), diagnostic techniques, and novel thermonuclear fusion schemes (Sects. 6.1 and 6.2).

Short-pulse electron beams with a density exceeding the plasma density may generate intense plasma waves, which, in combination with ponderomotive forces [53, 47], expel electrons from a plasma channel and produce plasma lenses for the energy transport to the center of compressed fusion targets. Furthermore, relativistic electron beams of tremendous peak power emerge inside the plasma under the action of intense intraplasma fields. In particular, with the help of lasers with a petawatt peak power it has been possible to obtain incredibly high electron and ion currents, which exceed the Alfvén limit by several orders of magnitude, and charged-particle flows of the megavolt energy range [53, 11, 45, 43, 47].

The interaction of intense electron beams with high-power short laser pulses in turn furnishes the possibility of conducting laboratory experiments in the area of quantum electrodynamics, leading to electron–positron pair production and giving rise to Compton X-ray radiation [53, 47, 8, 6]. This opens up interesting practical opportunities for the development of new-generation compact X-ray sources with record radiation brightness for materials science and medicine (see Sect. 6.3).

Considerable progress has been recently achieved in the technology of high-current Z-pinches [37, 31, 40] produced by magnetodynamic compression of multiple-wire ($\varnothing \approx 6\,\mu$m) and gas liners by passing current pulses (100 ns, $\approx 20$ MA) through them. These experiments enabled the radiative magnetic hydrodynamics of a dense plasma to be studied and produce, in the axial plasma compression, a 5–15 ns soft X-ray pulse with an energy of 1.8 MJ.

At the same facility, a current pulse with a megaampere amplitude was employed for the time-programmable electrodynamic acceleration of a metallic liner to a velocity of 12 km/s for the purpose of generating intense shock and adiabatic compression waves. These waves were used to study the equations of substance states in the magabar pressure range [37].

In experiments at the high-current ANGARA facility, soft (with $T \approx 100$ eV) X-ray radiation was used for the highly symmetrical generation of intense shock and radiative waves, as well as for Propelling metallic liners with velocities up to 10 km/s [31]. At the MAGPIE electrophysical facility [44] supersonic plasma jets with characteristics close to the astrophysical ones were obtained.

Experiments on high-energy plasmas yield valuable information about the spectral properties and radiative energy transfer in compressed plasmas. Of special interest are the findings in the strongly nonideal parameter domain [21, 18, 26], where the interparticle interaction effects are responsible for a radical distortion of atomic and ion spectra, partial "transparentization" of compressed plasma, and substantial shifts and broadening of spectral lines (more details are given by Ryutov et al. [57]).

Pulsed energy cumulation in plasma gives rise to diverse hydrodynamic effects, which are comprehensively studied in the corresponding laboratory experiments [49, 13]. These are the Rayleigh–Taylor, Richtmyer–Meshkov, and other instabilities in nonlinear regimes; high-velocity ($M \approx 15$–20) shock, magnetohydrodynamic, and radiative (thermal) waves; and turbulence and turbulent mixing in compressible emitting media [13].

The application of shock waves generated by sources of locally high energy densities makes it possible to greatly extend the range of experimentally attainable pressures and temperatures and penetrate the parameter domain intermediate between the parameters of chemical HEs and the unique conditions in underground nuclear explosions. Even in the first experiments involving laser-, beam-, and electrodynamically driven shock waves it was possible to obtain interesting experimental data about the equations of state of hydrogen, deuterium, copper, iron, carbon, and water, and to employ them in the structure analysis of the giant planets of the Solar system and exoplanets [21, 18, 26, 27, 50, 3, 37, 31, 20].

Finally, research in the area of high-energy-density physics provides the necessary scientific basis for inertial confinement fusion research [39, 5, 41, 34, 60]. The objective of this work is to ignite the thermonuclear fusion reaction in deuterium–tritium microtargets under high-intensity laser irradiation (the NIF, USA [18, 26, 46], and LMJ, France [10, 30], facilities), under exposure to the soft X-ray radiation of high-current Z-pinches (Zeta, USA) [62, 12, 54] or relativistic heavy-ion beams [60, 22].

## References

[1] Al'tshuler, L.V.: Use of shock waves in high-pressure physics. Sov. Phys. – Usp. **8**(1), 52–91 (1965). DOI 10.1070/PU1965v008n01ABEH003062. URL http://stacks.iop.org/0038-5670/8/52

[2] Al'tshuler, L.V., Trunin, R.F., Urlin, V.D., et al.: Development of dynamic high-pressure techniques in Russia. Phys. Usp. **42**(3), 261 (1999). DOI 10.1070/PU1999v042n03ABEH000545. URL http://ufn.ru/en/articles/1999/3/c/

[3] Anisimov, S.I., Prokhorov, A.M., Fortov, V.E.: Application of high-power lasers to study matter at ultrahigh pressures. Sov. Phys. – Usp. **27**(3), 181–205 (1984). DOI 10.1070/PU1984v027n03ABEH004036. URL http://stacks.iop.org/0038-5670/27/181

[4] Ashkroft, N.A.: Condensed matter at higher densities. In: G.L. Chiarotti, R.J. Hemley, M. Bernasconi, L. Ulivi (eds.) High Pressure Phenomena, Proceedings of the International School of Physics "Enrico Fermi" Course CXLVII, p. 151. IOS Press, Amsterdam (2002)

[5] Atzeni, S., Meyer-ter-Vehn, J.: The Physics of Inertial Fusion. Oxford University Press, Oxford (2004)

[6] Bamber, C., Boege, S.J., Koffas, T., et al.: Studies of nonlinear QED in collisions of 46.6 GeV electrons with intense laser pulses. Phys. Rev. D **60**(9), 092004 (1999). DOI 10.1103/PhysRevD.60.092004. URL http://link.aps.org/abstract/PRD/v60/e092004

[7] Bezkrovniy, V., Filinov, V.S., Kremp, D., et al.: Monte Carlo results for the hydrogen Hugoniot. Phys. Rev. E **70**(5), 057401 (2004). DOI 10.1103/PhysRevE.70.057401. URL http://link.aps.org/abstract/PRE/v70/e057401

[8] Burke, D.L., Field, R.C., Horton-Smith, G., et al.: Positron production in multiphoton light-by-light scattering. Phys. Rev. Lett. **79**(9), 1626–1629 (1997). DOI 10.1103/PhysRevLett.79.1626. URL http://link.aps.org/abstract/PRL/v79/p1626

[9] Calderola, P., Knopfel, H. (eds.): Physics of High Energy Density. Academic, New York (1971)

[10] Cavailler, C.: Inertial fusion with the LMJ. Plasma Phys. Control. Fusion **47**(12B), B389–B403 (2005). DOI 10.1088/0741-3335/47/12B/S28

[11] Clark, E.L., Krushelnick, K., Davies, J.R., et al.: Measurements of energetic proton transport through magnetized plasma from intense laser interactions with solids. Phys. Rev. Lett. **84**(4), 670–673 (2000). DOI 10.1103/PhysRevLett.84.670. URL `http://link.aps.org/abstract/PRL/v84/p670`

[12] Cuneo, M.E., Vesey, R.A., Bennett, G.R., et al.: Progress in symmetric ICF capsule implosions and wire-array Z-pinch source physics for double-pinch-driven hohlraums. Plasma Phys. Control. Fusion **48**(2), R1–R35 (2006). DOI 10.1088/0741-3335/48/2/R01

[13] Drake, R.P.: High-Energy-Density Physics. Springer, Berlin, Heidelberg (2006)

[14] Dubin, D.H.E., O'Nail, T.M.: Trapped nonneutral plasmas, liquids and crystals (the thermal equilibrium states). Rev. Mod. Phys. **71**, 87 (1999)

[15] Faber, T.E.: Fluid Dynamics for Physicists. Cambridge University Press, Cambridge (1977)

[16] Filinov, V.S., Bonitz, M., Levashov, P., et al.: Plasma phase transition in dense hydrogen and electron–hole plasmas. J. Phys. A **36**(22), 6069–6076 (2003). DOI 10.1088/0305-4470/36/22/332

[17] Filinov, V.S., Levashov, P.R., Bonitz, M., Fortov, V.E.: Calculation of the shock Hugoniot of deuterium atpressures above 1 Mbar by the path-integral Monte Carlo method. Plasma Phys. Rep. **31**(8), 700–704 (2005). DOI 10.1134/1.2031631

[18] Fortov, V., Iakubov, I., Khrapak, A.: Physics of Strongly Coupled Plasma. Oxford University Press, Oxford (2006)

[19] Fortov, V.E. (ed.): Entsiklopediya nizkotemperaturnoi plazmy (Encyclopedia of Low-Temperature Plasma). Nauka, Moscow (2000)

[20] Fortov, V.E.: Intense Shock Waves and Extreme States of Matter. Bukos, Moscow (2005)

[21] Fortov, V.E.: Intense shock waves and extreme states of matter. Phys. Usp. **50**(4), 333 (2007). DOI 10.1070/PU2007v050n04ABEH006234. URL `http://ufn.ru/en/articles/2007/4/c/`

[22] Fortov, V.E., Hoffmann, D.H.H., Sharkov, B.Y.: Intense ion beams for generating extreme states of matter. Phys. Usp. **51**(2), 109 (2008). DOI 10.1070/PU2008v051n02ABEH006420. URL `http://ufn.ru/en/articles/2008/2/a/`

[23] Fortov, V.E., Ilkaev, R.I., Arinin, V.A., et al.: Phase transition in a strongly nonideal deuterium plasma generated by quasi-isentropical compression at megabar pressures. Phys. Rev. Lett. **99**(18), 185001 (2007). DOI 10.1103/PhysRevLett.99.185001. URL `http://link.aps.org/abstract/PRL/v99/e185001`

[24] Fortov, V.E., Ivlev, A.V., Khrapak, S.A., et al.: Complex (dusty) plasma: current status, open issues, perspectives. Phys. Rep. **421**(1), 1–103 (2005). DOI 10.1016/j.physrep.2005.08.007
[25] Fortov, V.E., Khrapak, A.G., Khrapak, S.A., et al.: Dusty plasmas. Phys. Usp. **47**(5), 447 (2004). DOI 10.1070/PU2004v047n05ABEH001689. URL `http://ufn.ru/en/articles/2004/5/b/`
[26] Fortov, V.E., Khrapak, A.G., Yakubov, I.T.: Fizika neideal'noi plazmy (Physics of Nonideal Plasma). Fizmatlit, Moscow (2004)
[27] Fortov, V.E., Ternovoi, V.Y., Zhernokletov, M.V., et al.: Pressure-produced ionization of nonideal plasma in a megabar range of dynamic pressures. JETP **97**(2), 259–278 (2003). DOI 10.1134/1.1608993
[28] Ginzburg, V.L.: The Physics of a Lifetime: Reflections on the Problems and Personalities of 20th Century Physics. Springer, Berlin, Heidelberg (2001)
[29] Ginzburg, V.L.: On superconductivity and superfluidity (what I have and have not managed to do), as well as on the "physical minimum" at the beginning of the XXI century (December 8, 2003). Phys. Usp. **47**(11), 1155 (2004). DOI 10.1070/PU2004v047n11ABEH001825. URL `http://ufn.ru/en/articles/2004/11/d/`
[30] Giorla, J., Bastian, J., Bayer, C., et al.: Target design for ignition experiments on the laser Mégajoule facility. Plasma Phys. Control. Fusion **48**(12B), B75–B82 (2006). DOI 10.1088/0741-3335/48/12B/S0
[31] Grabovskii, E.V., Vorob'ev, O.Y., Dyabilin, K.S., et al.: Excitation of intense shock waves by soft x radiation from a Z-pinch plasma. JETP Lett. **60**(1), 1 (1994)
[32] Hemley, R.J., Ashcroft, N.W.: The revealing role of pressure in the condensed matter sciences. Phys. Today **51**(8), 26–32 (1998). DOI 10.1063/1.882374
[33] Hemley, R.J., Mao, H.K.: Overview of static high pressure science. In: R.J. Hemley, G.L. Chiarotti, M. Bernasconi, L. Ulivi (eds.) High Pressure Phenomena, Proceedings of the International School of Physics "Enrico Fermi" Course CXLVII, p. 3. IOS Press, Amsterdam (2002)
[34] Hogan, W.J. (ed.): Energy from Inertial Fusion. IAEA, Vienna, Austria (1995)
[35] Kirzhnits, D.A.: Extremal states of matter (ultrahigh pressures and temperatures). Sov. Phys. – Usp. **14**(4), 512–523 (1972). DOI 10.1070/PU1972v014n04ABEH004734. URL `http://stacks.iop.org/0038-5670/14/512`
[36] Kirzhnits, D.A., Lozovik, Y.E., Shpatakovskaya, G.V.: Statistical model of matter. Sov. Phys. – Usp. **18**(9), 649–672 (1975). DOI 10.1070/PU1975v018n09ABEH005199. URL `http://stacks.iop.org/0038-5670/18/649`
[37] Knudson, M.D., Hanson, D.L., Bailey, J.E., et al.: Equation of state measurements in liquid deuterium to 70 GPa. Phys. Rev. Lett. **87**(22), 225501 (2001). DOI 10.1103/PhysRevLett.87.225501. URL `http://link.aps.org/abstract/PRL/v87/e225501`

[38] Konyukhov, A.V., Likhachev, A.P., Oparin, A.M., et al.: Numerical modeling of shock-wave instability in thermodynamically nonideal media. JETP **98**(4), 811–819 (2004). DOI 10.1134/1.1757680

[39] Kruer, W.L.: The Physics of Laser Plasma Interactions. Addison-Wesley, Reading, MA (1988)

[40] Lebedev, S.V., Ciardi, A., Ampleford, D.J., et al.: Magnetic tower outflows from a radial wire array Z-pinch. Month. Notices R. Astronom. Soc. **361**(1), 97–108 (2005). DOI 10.1111/j.1365-2966.2005.09147.x. URL `http://www3.interscience.wiley.com/journal/118689245/abstract`

[41] Lindl, J.D.: Inertial Confinement Fusion. Springer, New York (1998)

[42] Loubeyre, P., Occelli, F., Le Toulec, R.: Optical studies of solid hydrogen to 320 GPa and evidence for black hydrogen. Nature **416**(6881), 613–617 (2002). DOI 10.1038/416613a

[43] Mackinnon, A.J., Borghesi, M., Hatchett, S., et al.: Effect of plasma scale length on multi-MeV proton production by intense laser pulses. Phys. Rev. Lett. **86**(9), 1769–1772 (2001). DOI 10.1103/PhysRevLett.86.1769. URL `http://link.aps.org/abstract/PRL/v86/p1769`

[44] MAGPIE Project: MAGPIE Project Home Page. URL `http://dorland.pp.ph.ic.ac.uk/magpie/`

[45] Maksimchuk, A., Gu, S., Flippo, K., et al.: Forward ion acceleration in thin films driven by a high-intensity laser. Phys. Rev. Lett. **84**(18), 4108–4111 (2000). DOI 10.1103/PhysRevLett.84.4108. URL `http://link.aps.org/abstract/PRL/v84/p4108`

[46] Moses, E.I., Bonanno, R.E., Haynam, C.A., et al.: The National Ignition Facility: path to ignition in the laboratory. Eur. Phys. J. D **44**(2), 215–218 (2006). DOI 10.1140/epjd/e2006-00106-3

[47] Mourou, G.A., Tajima, T., Bulanov, S.V.: Optics in the relativistic regime. Rev. Mod. Phys. **78**(2), 1804–1816 (2006). DOI 10.1103/RevModPhys.78.309. URL `http://link.aps.org/abstract/RMP/v78/p309`

[48] Murray, C.A., Wenk, R.A.: Observation of order–disorder transitions and particle trajectories in a model one-component plasma: time resolved microscopy of colloidal spheres. In: H.M. Van Horn, S. Ichimaru (eds.) Strongly Coupled Plasma Physics, p. 367. University of Rochester Press, Rochester, NY (1993)

[49] National Research Council: Frontiers in High Energy Density Physics. National Academies Press, Washington, DC (2003)

[50] Nellis, W.J.: Dynamic compression of materials: metallization of fluid hydrogen at high pressures. Rep. Prog. Phys. **69**(5), 1479–1580 (2006). DOI 10.1088/0034-4885/69/5/R05

[51] Okun', L.B.: Leptony i kvarki, 2nd edn. Nauka, Moscow (1990). [English Transl.: Leptons and Quarks. North-Holland, Amsterdam (1982)]

[52] Pieranski, P.: Colloidal crystals. Contemp. Phys. **24**(1), 25–70 (1983). DOI 10.1080/00107518308227471

[53] Pukhov, A.: Strong field interaction of laser radiation. Rep. Prog. Phys. **66**(1), 47–101 (2003). DOI 10.1088/0034-4885/66/1/202

[54] Quintenz, J., Sandia's Pulsed Power Team: Pulsed power team. In: Proc. 13th Int. Conf. on High Power Particle Beams. Nagaoka, Japan (2000)

[55] Rubakov, V.A.: Large and infinite extra dimensions. Phys. Usp. **44**(9), 871 (2001). DOI 10.1070/PU2001v044n09ABEH001000. URL http://ufn.ru/en/articles/2001/9/a/

[56] Russel, W.B., Saville, D.A., Schowalter, W.R.: Colloidal Dispersions. Cambridge University Press, Cambridge (1989)

[57] Ryutov, D.D., Remington, B.A., Robey, H.F., Drake, R.P.: Magnetodynamic scaling: from astrophysics to the laboratory. Phys. Plasmas **8**(5), 1804–1816 (2001). DOI 10.1063/1.1344562

[58] Schatz, T., Schramm, U., Habs, D.: Crystalline ion beams. Nature **412**(6848), 717–720 (2001). DOI 10.1038/35089045

[59] Schramm, U., Schatz, T., Bussmann, M., Habs, D.: Cooling and heating of crystalline ion beams. J. Phys. B **36**(3), 561–571 (2003). DOI 10.1088/0953-4075/36/3/314

[60] Sharkov, B.Y. (ed.): Yadernyi sintez s inertsionnym uderzhaniem (Inertial Confinement Nuclear Fusion). Fizmatlit, Moscow (2005)

[61] Shashkin, A.A.: Metal—insulator transitions and the effects of electron—electron interactions in two-dimensional electron systems. Phys. Usp. **48**(2), 129 (2005). DOI 10.1070/PU2005v048n02ABEH001944. URL http://ufn.ru/en/articles/2005/2/b/

[62] Spielman, R.B., Deeney, C., Chandler, G.A., et al.: Tungsten wire-array Z-pinch experiments at 200 TW and 2 MJ. Phys. Plasmas **5**(5), 2105–2111 (1998). DOI 10.1063/1.872881

[63] Vacca, J.R. (ed.): The World's 20 Greatest Unsolved Problems. Prentice Hall PTR, Englewood Cliffs, NJ (2004)

[64] Zel'dovich, Y.B., Raizer, Y.P.: Fizika udarnykh voln i vysokotemperaturnykh gidrodinamicheskikh yavlenii, 2nd edn. Nauka, Moscow (1966). [English Transl.: Physics of Shock Waves and High-Temperature Hydrodynamic Phenomena. Dover, Mineola, NY (2002)]

# Chapter 3
# High Energy Densities in Laboratories

## 3.1 Main Lines of Research

The ultimate objective of experiments in high-energy-density physics consists in the generation of extreme material parameters, whose values are at the boundaries of modern experimental capabilities (Table 3.1). Already, plasma states with peak pressures of hundreds or thousands of megabars, temperatures up to 10 billion degrees, and energy densities of $10^9$ J/cm$^3$, which is comparable to the energy density of nuclear matter, have become the subject of laboratory investigations [29, 38, 31, 10, 99, 9]. According to the ideas developed to date [64, 8, 68, 58], to implement a controlled thermonuclear reaction with inertial plasma confinement requires an energy of several megajoules to be delivered to a spherical target in $10^{-9}$ s to generate at its center a deuterium–tritium plasma with extremely high parameters $T \approx (1\text{–}2) \times 10^8$ K, $\rho \approx 200$ g/cm$^3$, $P \approx 150\text{–}200$ Gbar, which is close to the conditions at the center of the Sun. The corresponding output laser power should exceed the total power of all terrestrial electric power plants by several orders of magnitude.

These conditions, which are required for the ignition of a controlled thermonuclear reaction, are quite extraordinary by terrestrial standards, but are quite typical for the great bulk of the universe's matter compressed by gravitational forces in the interior of stars and other astrophysical objects. In this case, in the physics of high energy densities there arise several absorbing problems, the progress in this and allied fields of knowledge depending on their solution [78].

The ignition of thermonuclear reactions with inertial plasma confinement under controlled conditions is the principal pragmatic objective of research in high-energy-density physics. Here, lasers are in the lead [54, 21, 64, 8, 68, 14, 19, 75, 21, 47], although electrodynamic techniques (Z-pinches [95, 25, 84] and heavy-ion schemes [91, 36]) are making rapid strides. The operation of such targets is basically close to supernovae explosions, allowing the vast wealth of experimental results and sophisticated computer codes for the calculation of fusion microtargets to be employed in astrophysics.

V.E. Fortov, *Extreme States of Matter*, The Frontiers Collection,
DOI 10.1007/978-3-642-16464-4_3, © Springer-Verlag Berlin Heidelberg 2011

Table 3.1: Energy sources and experimental devices employed in high-energy-density physics [29, 38].

| Primary energy source | Final form of the energy source | Energy density [MJ/cm$^3$] | Temperature [eV] | Pressure [$10^5$ Pa] | Total energy [MJ] | Duration [s] | Power [W] |
|---|---|---|---|---|---|---|---|
| Chemical HEs | Chemical HEs | $10^{-2}$ | 0.5 | $5\times10^5$ | $10^2$ | $10^{-7}$ | $10^{10}$ |
| | Metallic plates | 0.3 | 60 | $10^7$ | 3 | $10^{-6}$ | $10^{10}$ |
| | 1 MOe magnetic field | $4\times10^{-3}$ | 0.3 | $5\times10^4$ | 5 | $10^{-6}$ | $5\times10^{12}$ |
| | 25 MOe magnetic field | 2.5 | 200 | $2.5\times10^7$ | 1 | $10^{-7}$ | $10^{13}$ |
| | Explosion plasma generators | $10^{-2}$ | 60 | $10^5$ | 30 | $10^{-6}$ | $10^{12}$ |
| Nuclear explosives | Nuclear explosives | $10^4$ | $10^7$ | $10^{10}$ | $10^{11}$ | $10^{-6}$ | $10^{22}$ |
| | Neutron-induced heating | 10 | 50 | $2\times10^7$ | $10^3$ | $10^{-6}$ | $10^{15}$ |
| | Shock waves in solids | 5 | 50 | $5\times10^7$ | $10^4$ | $3\times10^{-6}$ | $10^{15}$ |
| | Shock waves in gases | 0.3 | 40 | $2\times10^5$ | $10^7$ | $10^{-5}$ | $10^{18}$ |
| Compressed gas | Adiabatic compression | $2\times10^{-5}$ | 0.3 | 150 | $10^3$ | $6\times10^{-3}$ | $10^5$ |
| | Pneumatic shock tubes | $10^{-4}$ | 1 | 250 | $10^{-2}$ | $10^{-4}$ | $3\times10^8$ |
| | Combustion-driven shock tube | $10^{-6}$ | 2 | 10 | $2\times10^{-2}$ | $3\times10^{-4}$ | $10^8$ |
| Compressed gas | Shock tubes, electric discharges | $10^{-7}$ | 2 | 1 | $10^{-2}$ | $10^4$ | $10^8$ |

*Table 3.1 continuation*

| Primary energy source | Final form of the energy source | Energy density [MJ/cm$^3$] | Temperature [eV] | Pressure [$10^5$ Pa] | Total energy [MJ] | Duration [s] | Power [W] |
|---|---|---|---|---|---|---|---|
| Capacitor | – | $10^{-8}$ | – | – | 40 | $10^{-5}$ | $10^{12}$ |
| Rotor energy-storage device | – | $10^{-3}$ | – | – | 100 | $10^{-4}$ | $10^{12}$ |
| Inductive storage device | – | $10^{-4}$ | – | – | 100 | $10^{-4}$ | $10^{12}$ |
| Storage battery | – | $5\times10^{-4}$ | – | – | 1000 | $10^{-3}$ | $10^{12}$ |
| | Fast wire explosion | $5\times10^{-2}$ | 4 | $10^5$ | $10^{-3}$ | $10^{-6}$ | $10^9$ |
| | Slow wire explosion | $2\times10^{-2}$ | 0, 5 | $5\times10^2$ | $10^{-3}$ | $10^{-4}$ | $10^7$ |
| | Pulsed discharges | $10^{-3}$ | 10 | $10^4$ | $10^{-4}$ | $10^{-3}$ | $10^9$ |
| | Plasma focus | $10^{-2}$ | $10^3$ | 10 | $10^{-4}$ | $10^{-5}$ | $10^{10}$ |
| | High-pressure arcs | $10^{-5}$ | 2 | $10^4$ | $10^{-4}$ | $\infty$ | $10^4$ |
| | Experiments involving furnaces | $10^{-3}$ | 0.3 | $5\times10^3$ | $10^{-3}$ | $\infty$ | $10^3$ |
| Laser | – | $10^{-6}$ | – | - | $0.5\times10^{-3}$ | | $10^{13}$ |
| | Target | $10^4$ | $10^6$ | $10^8$ | 0, 5 | $10^{-10}$ | $10^{13}$ |
| Electron beam | – | $10^{-6}$ | – | – | 1 | $10^{-8}$ | $10^{13}$ |
| | Target | $5\times10$ | $5\times10^3$ | $10^7$ | 0.1 | $10^{-8}$ | $10^{13}$ |

Of fundamental significance is the study of the equation of state of matter and plasma composition in a broad domain of the phase diagram, including the conditions inherent in giant planets, exoplanets, dwarfs, giants, and neutron stars, and the determination of quasi-classicality bounds (the Thomas–Fermi model [61, 62]) in thermodynamics. Of considerable interest are the properties of degenerate compressed plasmas, their thermodynamics, equilibrium, kinetic, and transport properties in the region of strong nonideality and in the presence of intense magnetic fields, as well as the properties of a quark–gluon plasma, and the existence and properties of its hypothetical phase transition.

The construction and verification of the structural and evolutional models of planets and exoplanets calls for the derivation of reliable experimental data in laboratory conditions. For Jupiter and other planets it is vital to ascertain or disprove the existence of a hard core and determine the dimensions of the domain occupied by metallic hydrogen and the metallization bound for $H_2$ and $H_2$+He. Of fundamental importance is the analysis of Jupiter's energetics with the inclusion of phase layering of the mixtures He–H, C–O, etc., as well as the study of the origin and dynamics of its magnetic field. Similar problems are also encountered in studies of giant planets and exoplanets. In this case, a large part is played by shock-wave experiments, which enable the metallization bounds to be determined and the occurrence of a plasma phase transition to be ascertained.

The intriguing question of a phase transition of the first kind in a strongly nonideal plasma has remained open for almost 80 years (see the first data on the observation of this transition in deuterium [37]).

The more distant future will see studies of the conditions for pycnonuclear reactions ("cold" fusion) and the conditions of strong Coulomb screening, which speeds up thermonuclear reactions in the interior of massive stars, as well as the production of relativistic degenerate matter with the Fermi energy $E_{\mathrm{F}} > m_e c^2$ and the investigation of its equation of state.

The methods of laboratory energy cumulation now being developed will make it possible to obtain [78, 28] relativistic jets and intense collisional, collisionless, and magnetohydrodynamic shock waves, much like those observed in astrophysical objects.

Under discussion are projects involving the production of radiation-dominated hot plasmas, like the conditions in black holes and accretion disks of neutron stars, as well as the stability of these regimes (see the experiments involving nuclear explosions [10, 99, 9], lasers [28, 83, 76], and Z-pinches [50, 95, 25, 84]). Interesting suggestions have been put forward concerning the generation of radiatively collapsing magnetohydrodynamic and collisionless shocks, fast particles, relativistic jets, and their focusing [78, 28, 83, 76].

The high-energy-density techniques under development open up the possibility of generating ultrahigh ($B > 1$ GGs) magnetic fields and investigating their effect on the physical properties of matter. Already the fields obtained in laser-produced plasmas range into hundreds of megagauss.

Ultrahigh laser power levels may bring closer the prerequisites for the observation of relativistic gravitational effects.

The list may be easily lengthened and is perhaps limited only by our imagination. At present it is hard to tell, even to within decades, which of these fascinating problems of high-energy-density physics will be solved, if at all, at the existing and projected laboratory facilities. However, according to the favorite advice of academician P.L. Kapitsa, "When going fishing there is good reason to take the rod with the largest fish hook in the hope of the largest fish".

## 3.2 Generators of High Energy Densities

The spectrum of experimental devices for generating high energy densities is highly diversified. It includes diamond anvils for static material compression, gunpowder and light-gas launching devices ("guns"), explosion generators of intense shock waves, electroexplosion devices, magnetic cumulation generators, lasers, high-current generators of high-power electric current pulses, charged-particle accelerators, and possible combinations of these devices. The characteristics of these ways of generation are collected in Table 3.1, which shows typical (not necessarily the highest) parameters of the plasma states generated.

Table 3.2 [78] (at the end of this section) compares the parameters of the highest-power facilities either in service today or under construction: lasers, pulsed electrical devices, Z-pinches, and charged-particle accelerators [85, 81]. Developed for carrying out plasma research in the interests of defense and high-energy physics, they are presently employed to advantage for research in basic plasma physics of high energy densities. Table 3.3 (also at the end of this section) and Figs. 3.1 and 3.2 give the parameters of laser devices being constructed or designed, whose commissioning would greatly enhance the capabilities of experiments in laser plasma physics. The foremost among those currently working is the OMEGA (Fig. 3.3) American laser facility (Rochester University) [14, 19], which consists of 60 laser beams with a total energy of 30 kJ at a wavelength of 0.35 $\mu$m. A broad spectrum of investigations on laser fusion is conducted at the facility with direct and soft–X-ray conversion ignition schemes, on the generation of intense shock waves, the equation of state and optical properties of dense plasmas, hydrodynamic instabilities, the generation of intense X-ray radiation, etc. The National Ignition Facility (NIF, USA) [54, 75] (Fig. 3.4) and the Laser Mégajoule (LMJ) system (France) [21, 47], which are the highest-power facilities under construction, will deliver an energy of 1.8–2 MJ in 192 beams in the fundamental harmonic and afford the conditions for the thermonuclear ignition of microtargets. These facilities will make it possible to carry out experiments with shock waves in the gigabar pressure range, thereby advancing to the domain of the quasi-classical description of matter [62], and to study plasma flows under the conditions of developed radiation effects.

Table 3.2: Facilities for generating high energy densities [78].

| | Laser facilities | | | Z-pinch | | | | | |
|---|---|---|---|---|---|---|---|---|---|
| | NIF | LMJ | Petawatt laser | Sandia | | C-300 | | Angara V | |
| | | | | current | X-rays | current | X-rays | current | X-rays |
| Energy per particle | 3.6 eV | 3.6 eV | 1.5 eV | 20 MA | 50–250 eV | 1.5–4 MA | 70 eV | 25 MA | 100 eV |
| Pulse duration | 1–20 ns | ≈ 10 ns | 0.5 ps | 100 ns(rise time) | 5–15 ns | 80 ns | 12 ns | 90 ns | 6 ns |
| Spot size | 0.3 mm | 0.3 mm | 5 $\mu$m | — | 1 mm(cylinder) | — | 2 mm | — | 2 mm |
| Pulse energy | 1.8 MJ | 2 MJ | 0.5–5 kJ | 16 MJ | 1.8 MJ | 400 kJ | 50 kJ | 600 kJ | 120 kJ |
| Intensity [W/cm$^2$] | $2\times10^{15}$ | $\approx 10^{15}$ | $10^{22}$ | — | $10^{14}$ | — | $(2–3)\times10^{12}$ | — | $(6–10)\times10^{12}$ |

| | Accelerators | | | | | |
|---|---|---|---|---|---|---|
| | Electron accelerator (SLAC) | LHC accelerator | SIS 18 | | SIS 100 | TWAC |
| | | | today | design | | |
| Energy per particle | 50 GeV | 7 TeV | 1 GeV | 1 GeV | 4 GeV | 700 MeV |
| Pulse duration | 5 ns | 0.25 ns | 200 ns | 50 ns | 20 ns | 100 ns |
| Spot size | 3 mm | 16 mm | 1 mm | 1 mm | 1 mm | 1 mm |
| Pulse energy | 150 J | 334 MJ | ≈ 1 kJ | 30 kJ | 300 kJ | 100 kJ |
| Intensity [W/cm$^2$] | $10^{20}$ | $10^{19}$ | $5\times10^{11}$ | $6\times10^{13}$ | $10^{15}$ | $10^{14}$ |

Table 3.3: Parameters of laser facilities for high-energy-density physics.

| Laser facility | Parameters | Year of commissioning | Comment |
|---|---|---|---|
| **Great Britain** | | | |
| VULCAN<br>CLF, Rutherford | 2.6 kJ, several nanoseconds, 100 TW, 0.5–1 ps, $10^{21}$ W/cm$^2$ | 1999 | The largest European Center, which runs high-power ns and fs lasers simultaneously. |
| VULCAN-PW<br>CLF, Rutherford | 1 PW, 0.5–1 ps | 2002 | Two-beam laser with a 100–1000 times higher repetition rate, 10 times higher intensity at target |
| ASTRA TA1, TA 2<br>CLF, Rutherford | Site 1:<br>10 fs, 10–1000 Hz, 1 TW<br>Site 2:<br>40 fs, 10 Hz, 12 TW, $10^{19}$ W/cm$^2$ | 2005 | |
| ASTRA-GEMINI<br>CLF, Rutherford | 15 J, 30 fs, 1 shot per minute, 0.5 PW in either of the 2 beams | 2007 | Ultrahigh-intensity ultrashort-pulse laser with a wavelength tunable from the infrared to vacuum UV, combined with a synchrotron as a diagnostic complex. Radiation shielding of the interaction chamber provided |
| VULCAN-OPCPA<br>CLF, Rutherford | 10 PW 0.5–1 ps, $10^{23}$ W/cm$^2$. | 2011 | Ultrahigh-intensity ultrashort-pulse laser. Using optical parametric approaches will make it possible to obtain a very high intensity contrast ratio at the target |
| Orion<br>Atomic Weapons Establishment | 500 J, 1 ns, 3 $\omega_0$, 10 beams<br>+<br>2 beams 500 J, 500 fs, 1 PW, $10^{21}$ W/cm$^2$ | | |

*Table 3.3 continuation*

| Laser facility | Parameters | Year of commissioning | Comment |
|---|---|---|---|
| **European Project** | | | |
| HiPER<br>Future European Facility | 40 beams with a total energy of 200 kJ<br>+<br>24 beams with a total energy of 70 kJ and a ≈10 ps pulse duration for target ignition | early in 2010 | The feasibility of future operation in a pulse-periodic mode is under consideration |
| **France** | | | |
| LMJ<br>CESTA, Bordeaux | 2 MJ, 0.35 μm, ≈10 ns, 550 TW, 240 beams | 2010–2012 | The nuclear fusion energy yield is expected to amount to 25 MJ with 1.4 MJ absorbed laser energy |
| LIL<br>CESTA, Bordeaux | ≈ 40 kJ , 0.35 μm, 9 ns, 8 beams | 2006 | The first beam launched in 2003 |
| PETAL<br>Region Aquitaine, Bordeaux | 3.5 kJ 0.5–5 ps | 2007–2008 | Ultrahigh-intensity laser, will operate in combination with 8 beams of the LIL (CESTA) laser, which is a prototype of LMJ |
| LOA<br>Laboratoire d'Optique Appliquée | 30 J, 30 fs, 100 TW, 1–10 Hz | 2004 | Electrons were accelerated to 200 MeV and monochromatic collimated high-current electron beams obtained |
| LULI<br>Laboratoire pour l'Utilisation des Lasers Intenses | Several ns, 2 beams, ≈ 1 kJ each<br>+<br>30 J, 300 fs, 100 TW with upgrade into the PW range | 2000<br>2007 | |

*Table 3.3 continuation*

| Laser facility | Parameters | Year of commissioning | Comment |
|---|---|---|---|
| **USA** | | | |
| NIF<br>LLNL, Livermore | 4.2 MJ, 192 beams, 5–25 ns, 1 $\omega_0$ | 2010 | Attainment of radiative temperature of 300 eV is expected in 2010 for a 1 MJ laser energy. The laser-to-target energy conversion ratio is expected to amount 14–16% |
| OMEGA<br>LLE Rochester | 30 kJ, several ns, 60 TW, 60 beams, 3 $\omega_0$ | 1995 | A record number ($10^{14}$) of neutrons was obtained |
| OMEGA EP<br>LLE Rochester | 6.5 kJ, 3 $\omega_0$, 4 beams, there will be 2 beams (High-Energy Petawatt laser, HEPW) with 2.6 kJ, 10 ps in 1 $\omega_0$ | 2007 | Capability of conducting cryogenic experiments |
| **Japan** | | | |
| FIREX I project<br>Laser LFEX<br>Institute of Laser Engineering, Osaka | 3.6 kJ/beam, 10 ps, 4 beams, 14.4 kJ total energy | 2007–2008 | Objective: to demonstrate that a target can be heated up to 5–10 keV |
| FIREX II project Institute of Laser Engineering, Osaka | 100 kJ, 1 Hz laser for heating and 100 kJ, 1 Hz laser for ignition | $> 2010$ | The thermonuclear reaction power is expected to amount to 10 MW and the energy gain to 50. 40% of the energy will be converted to electric energy. 2 MW of this electric energy with 10% efficiency will be used for laser operation |

*Table 3.3 continuation*

| Laser facility | Parameters | Year of commissioning | Comment |
|---|---|---|---|
| **China** | | | |
| SG-II | 6 kJ 1 $\omega_0$, 1 ns, 3 rJ in 3 $\omega_0$,<br>8 beams<br>8 TW power in $\omega_0$ for 100 ps | 2005 | |
| SG-IIU | 3 ns, 18 kJ in 3 $\omega_0$,<br>+<br>PW laser with a ps pulse duration and 1–2 kJ energy<br>+<br>3 ns, 4.5 kJ laser for X-ray backlighting | 2007 | |
| SG-III | 3 ns, 150 kJ in 3 $\omega_0$, 64 beams | 2010 | |
| **Russia** | | | |
| VNIIEF+IAP RAS | 1 PW (100 J, 100 fs) | 2006 | $10^{22}$ W/cm$^2$ on-target laser intensity achieved |
| Iskra-5<br>VNIIEF+IAP RAS | 120 TW, 0.3 ns, iodine laser,<br>12 beams, 1$\omega_0 = 1.315$ $\mu$m, 2$\omega_0$ | 2002 | A temperature $T_e \approx 180$ eV obtained |
| Iskra-6<br>VNIIEF+IAP RAS | 300 kJ, 1–3 ns, 128 beams | | |
| IHED, JIHT RAS | 10 TW, 40 fs, $10^{19}$W/cm$^2$ | 2006 | |

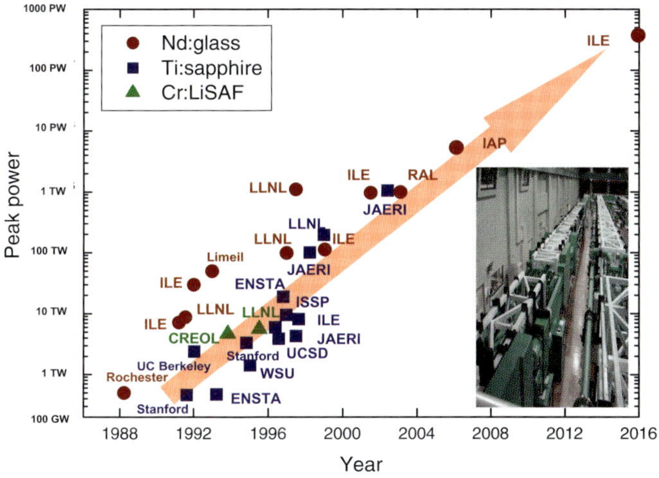


**Fig. 3.1** Peak power of laser facilities.

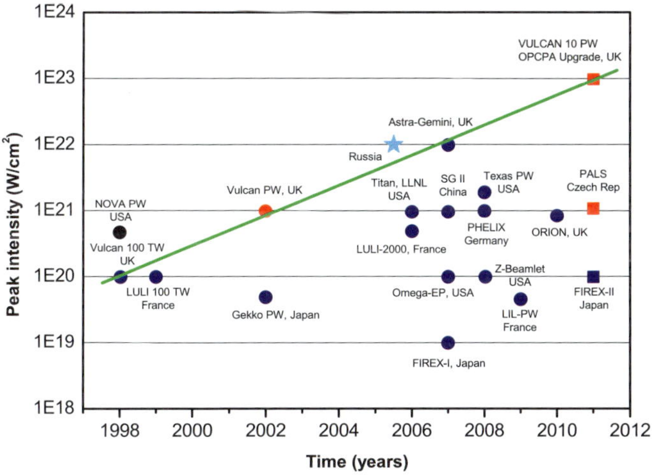


**Fig. 3.2** Development of the world's petawatt laser facilities (http://www.scitech.ac.uk/). Blue circles indicate the presently funded laser facilities and the blue squares those projected for funding. Red symbols correspond to single-beam laser facilities and the blue ones to ultrahigh-intensity laser facilities with additional capabilities. The black circle stands for the petawatt laser system that operated at the Livermore Laboratory. The green line shows the development trend of the Rutherford Laser Laboratory. The sky-blue star denotes the VNIIEF+IAP RAS laser facility commissioned late in 2006.

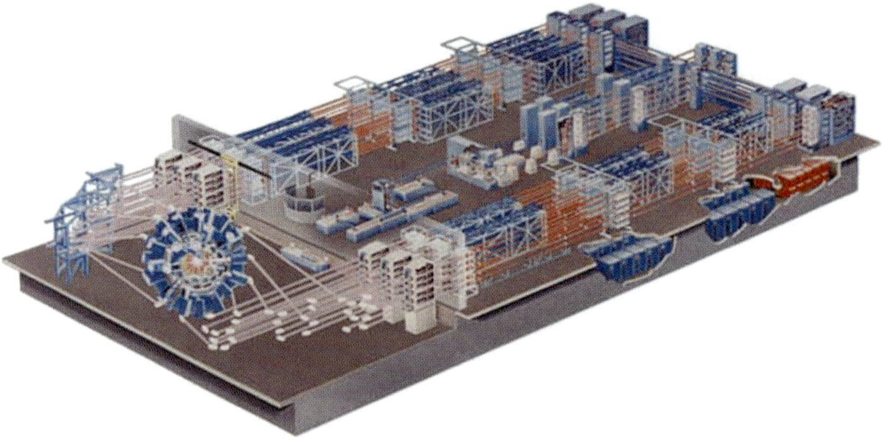

**Fig. 3.3** OMEGA laser complex.

The action of laser radiation of nanosecond duration at up to $10^{18}$ W/cm$^2$ intensities has been adequately studied and described in the literature [64, 8, 7, 83, 76, 20, 22]. The laser energy is released primarily in the vicinity of the critical density, $n_c \approx 10^{21}/(\lambda[\mu m])^2$ cm$^{-3}$ (where $n_c$ is the critical density and $\lambda$ the wavelength), located in the expanding plasma corona and is transferred to the target interior by means of electron thermal conduction. The plasma corona heated by the laser radiation moves towards the laser beam and generates an intense shock wave traveling into the target interior. The shock either compresses the target to pressures of hundreds of megabars or accelerates the outer target shell to velocities of hundreds of kilometers per second, which in turn compresses and heats the thermonuclear fuel or the substance under investigation. This scheme is the earliest of the thermonuclear schemes and is referred to as the "direct" drive scheme [64, 8, 68].

An alternative "indirect" drive scheme [64, 8, 68] relies on the conversion of laser radiation to X-ray radiation (with a brightness temperature ranging into hundreds of electronvolts), which effects the highly symmetric compression of the spherical target, as does the ablative pressure in the "direct"-drive compression scheme.

In operation in Russia (VNIIEF, Sarov) is the Iskra-5 laser facility (Fig. 3.5) [100] with an energy of $\approx 30$ kJ and a pulse duration of 0.3 ns, which is intended for inertial confinement fusion (ICF) and laser-plasma physics research. The next stage – Iskra-6 – will have 10 times higher power.

The second class of laser systems, which are less costly and therefore more easily accessible for the physics of high energy densities, corresponds to lower pulse energies and durations, but higher intensities at the target $\approx 10^{16}$–$10^{22}$ W/cm$^2$. Their action relies on the chirping effect discovered in the early 1980s, where the laser radiation of the master oscillator is stretched in time by diffraction systems, amplified in amplifier stages, and then compressed to a femtosecond duration [19].

**Fig. 3.4** NIF laser complex (USA) intended for controlled thermonuclear fusion with inertial plasma confinement.

Since such tera- to petawatt-power laser systems have become accessible to university laboratories, the work front in the plasma physics of high energy densities has broadened greatly: from relativistic plasma studies to the generation of accelerated electron, ion, and soft X-ray radiation (for lithography, medicine, and thermonuclear fusion) fluxes and electron–positron pairs, the implementation of nuclear reactions, and many other tasks.

**Fig. 3.5** Petawatt laser complex in VNIIEF.

Future progress in this direction will involve the advance of laser power to the zettawatt ($10^{21}$ W) power level (see Fig. 4.2). These fantastic power levels can be reached both by shortening the laser pulse duration to tens of attoseconds (1 as $= 10^{-18}$ s) and by raising the energies of laser beams (increasing the diameter and number of laser beams in femtosecond facilities). The feasibility of generating an isolated single-cycle 130 as long pulse has already been demonstrated [87].

Among the multitude of interesting physical experiments with such systems we mention the interaction of ultrashort laser pulses with clusters [27], in which their multiple nonequilibrium ionization knocks out electrons from a cluster and stimulates a "Coulomb explosion" with the production of kiloelectronvolt ions, laser-induced shock waves, and even thermonuclear reactions.

The shortest femto- and attosecond laser pulses are comparable to a light cycle in duration and are a tool for extremely fast action on a substance. The parameters of laser facilities in operation and under construction are collected in Table 3.3, and Fig. 3.2 shows the progress with time of the world's petawatt laser facilities.

## 3.3 Static Methods Using Diamond Anvils

Substantial progress in the area of static pressures was made in the early 1980s with the advent of the experimental technique of diamond anvils (Fig. 3.6) [70, 57]. In these facilities, two diamonds cut in a special way compress thin (10–100 $\mu$m) plane layers of the substance under investigation to the highest attainable pressures of the megabar range, their upper limit being defined by the ultimate strength of diamond of $\approx$0.5–1.0 Mbar. In a number of experiments, the compressed substance is heated by laser radiation (Fig. 3.7) [15, 16] or the material compressed in diamond anvils is the target for its subsequent compression by laser-driven shock waves (Fig. 3.8) [69].

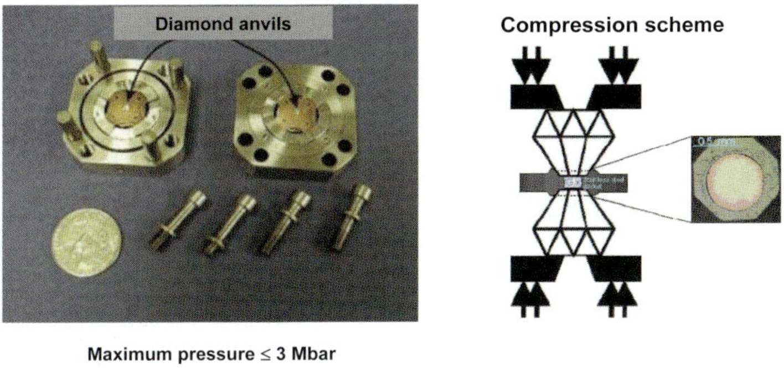


**Fig. 3.6** Facility for the static compression of substances in diamond anvils [57, 56].

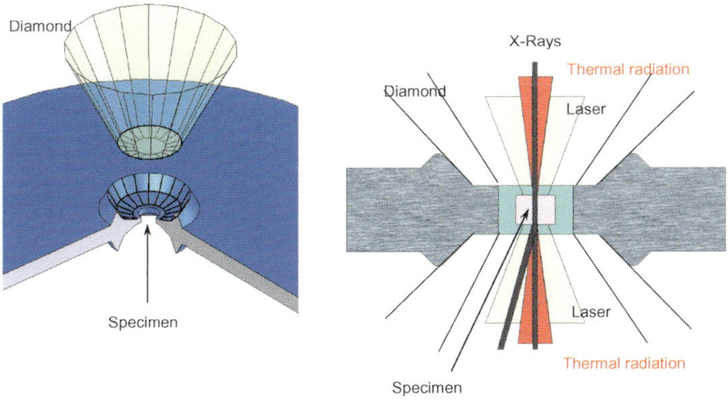


**Fig. 3.7** Schematic representation of a static experiment on the compression of a substance in a diamond anvil involving laser heating [16].

**Fig. 3.8** Laser-driven shock-wave compression of a substance precompressed in a diamond anvil at the OMEGA facility [69].

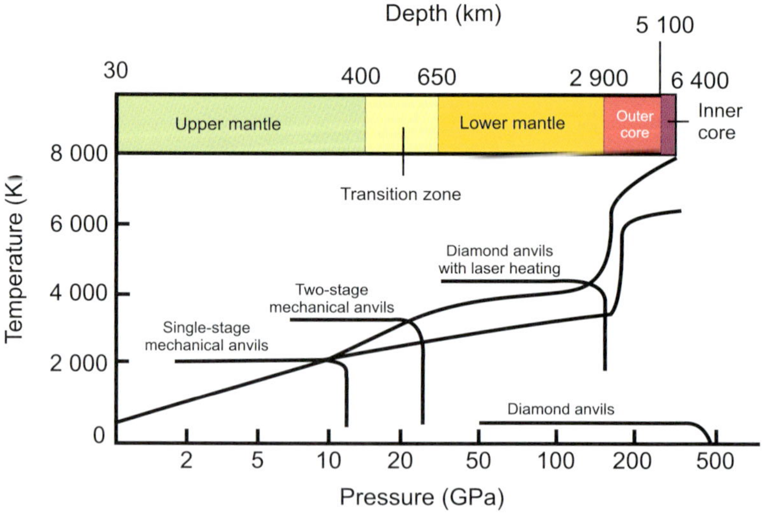


**Fig. 3.9** Comparison of the parameters [57, 56] attainable under static conditions and the physical conditions in the Earth's interior.

The unlimited time of static compression permits a wide spectrum of diagnostic tools to be employed, including different kinds of spectroscopy, and also X-ray structure analysis to be performed with the use of kiloelectronvolt X-ray and synchrotron radiation. Experiments of this kind have yielded a wealth of useful information about the mechanical properties, thermodynamics, and phase transformations in geophysical objects (Fig. 3.9) in the parameter range $P \approx 0.1$–3.6 Mbar, $T \approx 10^3$–$6 \times 10^3$ K, which is extreme for terrestrial conditions.

## 3.4 Dynamic Methods

Further advancement towards higher energy densities is related to the transition to dynamic methods of investigation ([5, 1, 104, 79, 33, 29, 38, 31, 10, 99, 9, 2, 34, 106]),which rely on pulsed energy cumulation in the substance under investigation by means of intense shock waves or by means of electromagnetic radiation or particle beams of different types. The plasma temperatures and pressures occurring in this case are far greater than the thermal and mechanical strength of the structural materials of the facilities, resulting in limitations on the characteristic plasma lifetime in dynamic experiments, which is defined by the target expansion dynamics and is equal to $\approx 10^{-10}$–$10^{-5}$ s. In the dynamic approach there are no fundamental limitations on the magnitudes of the maximum energy density and pressure: they are limited only by the power of the energy source, the "driver".

The routine tool for producing high energy densities is intense shock waves [5, 1, 104, 31], which emerge due to nonlinear hydrodynamic effects in a substance as a result of motion caused by pulsed energy liberation. In this case, a major part is played by the shock wave, a viscous compression shock, in which the kinetic energy of the oncoming flow is converted to thermal energy of the compressed and irreversibly heated plasma (Fig. 3.10).

Shock-wave techniques play a leading role in high-energy-density physics today, making it possible to produce amplitude pressures of the megabar and gigabar range for many chemical elements and compounds. This range of peak dynamic pressures is six orders of magnitude higher than the pressures occurring upon the impact of a bullet and three orders of magnitude higher than the pressure at the center of the Earth, and is close [10, 99, 9] to the pressure in the central layers of the Sun and inertial thermonuclear fusion targets. These exotic states of matter were realized during the birth of our universe, within several seconds after the big bang [78, 45, 103]. In a sense, we can say that by progressively increasing the pressure and temperature in dynamic experiments it is as if we are traveling backwards on the time axis to approach the instant of creation of the universe – the big bang (see Chap. 7, Fig. 7.105).

Not only do the shock waves compress the substance, they also heat it to high temperatures, which is of particular importance for the production of plasma, the ionized state of matter. A number of dynamic techniques are presently employed in

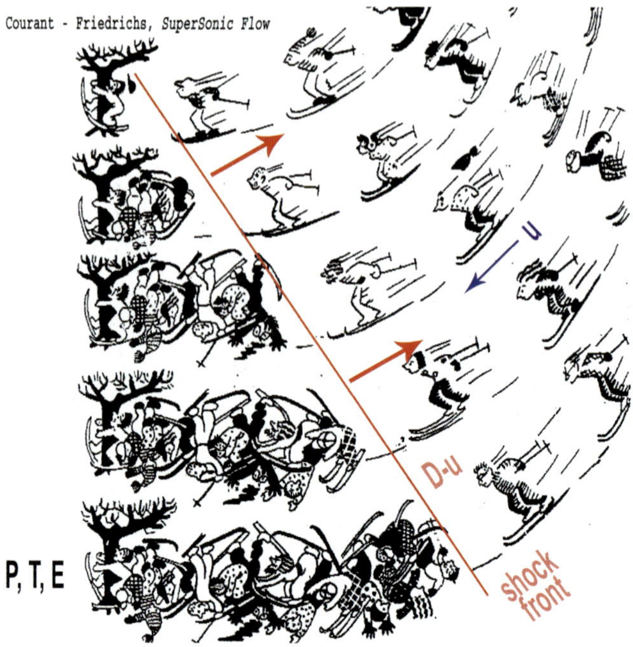


**Fig. 3.10** Shock-wave compression and heating of substance [24].

experimental studies of strongly nonideal plasmas [33, 29, 38, 5, 1, 31, 34, 106, 12], see Fig. 3.11.

The shock compression of an initially solid or liquid substance enables states of nonideal degenerate (Fermi statistics) and classical (Boltzmann statistics) plasmas to be produced behind the shock front, compressed to peak pressures of $\approx 4$ Gbar and heated to temperatures of $\approx 10^7$ K [10, 99, 9]. At these parameters, the density of the internal plasma energy is comparable to the nuclear energy density and the temperatures approach the conditions at which the energy and pressure of equilibrium radiation begin to play a significant role in the total thermodynamics of these high-energy states.

To reduce the irreversible heating effects, it is expedient to compress a material by a sequence of incident and reflected $H_k$ shock waves [79, 41, 37, 31]. As a result, this multistage compression becomes closer to the "softer" isentropic compression, making it possible to obtain substantially higher compression ratios (10–50 times) and lower temperatures ($\approx 10$ times) in comparison with a single-stage shock-wave compression. Multiple shock compression has been used validly for the experimental study of pressure-induced plasma ionization [79, 41, 31, 37] and substance dielectrization [71] at megabar pressures. Quasi-adiabatic compression has also been realized in the highly symmetric cylindrical explosive compression of hydrogen and rare gases (Fig. 3.12) [41, 37, 31]. In experiments involving "soft"

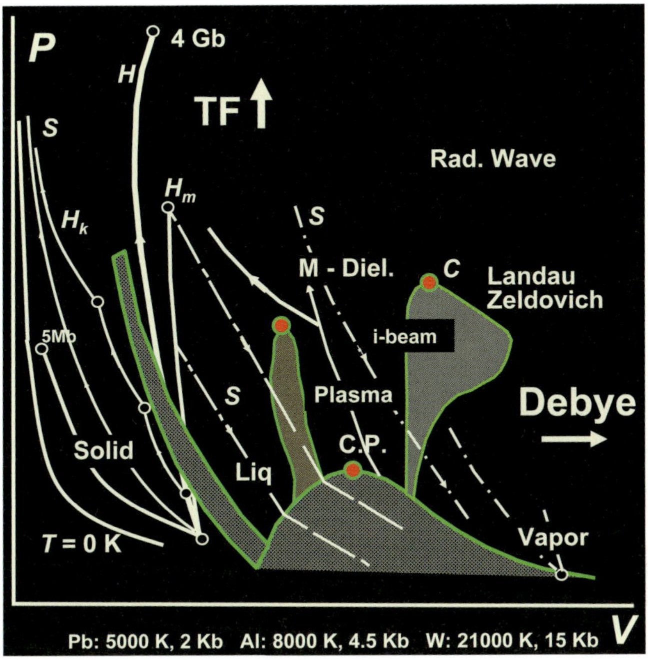


**Fig. 3.11** Thermodynamic trajectories for dynamic substance investigation techniques [29, 38]. The critical point parameters of several metals are given at the bottom.

adiabatic plasma compression, advantage was taken of the explosive compression of samples by megagauss magnetic fields [55, 82].

In another limiting case, when a high-temperature plasma is required, it is expedient to effect the shock compression of lower (in comparison with solid) density targets, for example, porous metals $H_m$ [33, 5, 1, 104, 2, 34, 52] or aerogels $H_a$ [53]. This provides a way to sharply strengthen the irreversibility effects of shock compression and thereby increase the entropy and temperature of the compressed state.

Figure 3.13 shows experimental data on the thermodynamics of high-energy states in the range of solid-state densities and high temperatures obtained by shock compression of porous nickel samples [52]. This parameter range is unconventional for plasma physics. Interestingly, these experimental data [52] correspond to the metal–dielectric transition region (Fig. 3.14), where pressure-induced and temperature ionization effects are significant for the description of plasma thermodynamics [29, 38, 52, 30].

The shock compression of rare gases and saturated alkali metal vapor by incident and reflected shock waves allows the plasma to be studied in the domain with developed thermal ionization, where the electrons obey Boltzmann statistics [33, 29, 38, 41, 31, 2] (Fig. 3.15).

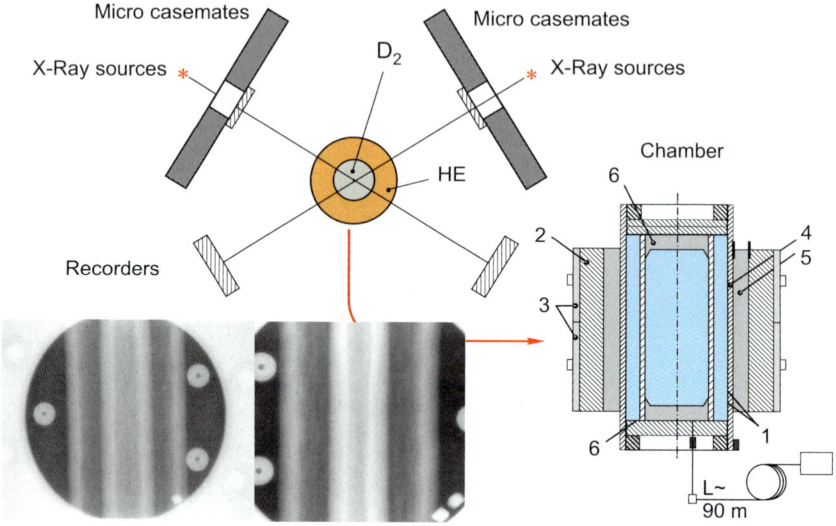


**Fig. 3.12** Cylindrical explosion devices for quasi-adiabatic plasma compression [92, 37, 35, 40]: 1 – cylindrical specimen, 2 – HE charge, 3,4 – outer and inner metal liners, 5 – X-ray radiation source, 6 – X-ray recorders.

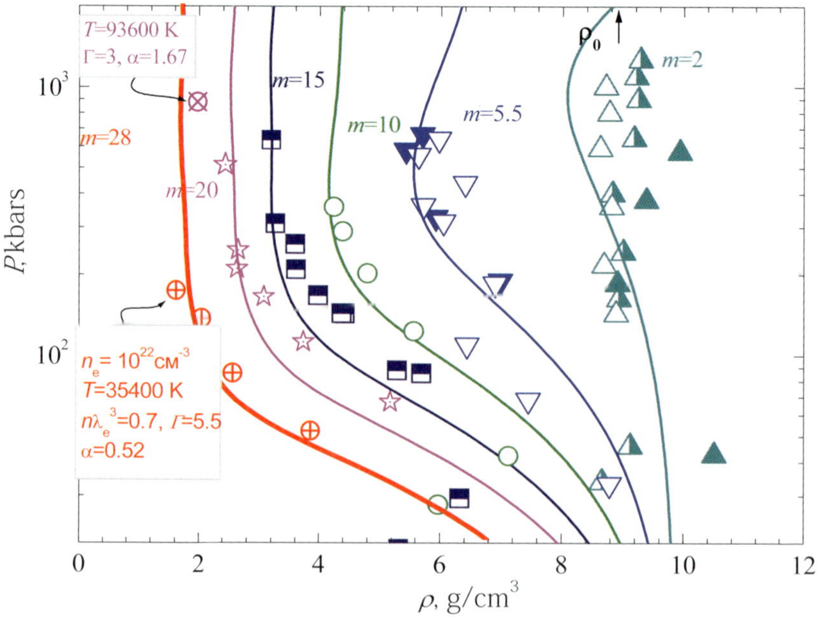


**Fig. 3.13** Thermodynamics of nonideal nickel plasma [52]. Symbols: the results of shock compression of porous ($m = \rho_0/\rho_{00}$) specimens; $\alpha$: the degree of ionization.

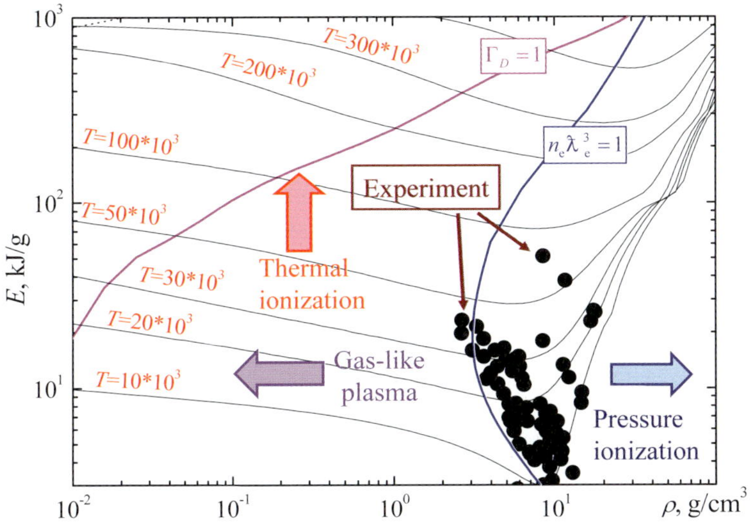


**Fig. 3.14** Energy density of shock-compressed nickel plasma [52].

A characteristic feature of the shock technique is that it permits high pressures and temperatures to be obtained in compressed media, while the low-density domain (including the boiling curve and the neighborhood of the critical point) turns out to be inaccessible to it [33, 29, 38, 31]. Investigation of the plasma states intermediate between a solid and a gas is made by the isentropic expansion technique. This tech-

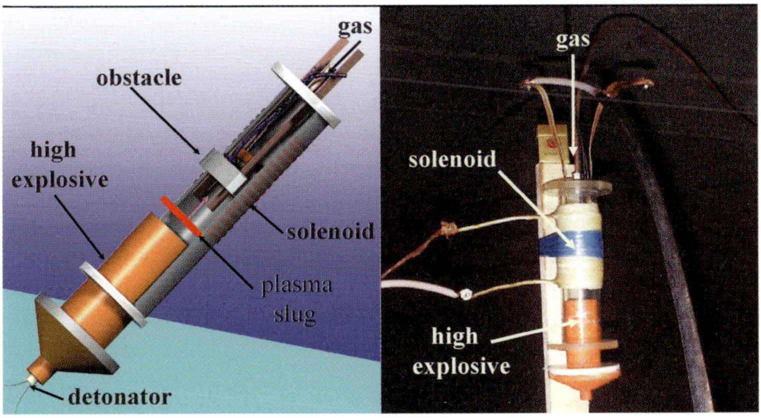


**Fig. 3.15** Explosion shock tubes for measuring the low-frequency and Hall conductivities as well as the equation of state of shock-compressed plasma [29, 38, 42].

nique involves the generation of plasma in the adiabatic expansion of a condensed substance precompressed and irreversibly preheated at the front of an intense shock wave [33, 29, 38, 31]. It was precisely this technique that was first employed for the experimental investigation of the high-temperature portions of the boiling curves, the transcritical states, and the metal–dielectric transition domains for a large number of metals (for more details, see [33, 29, 38, 105]). By way of example, Fig. 3.16 shows the domain of high-temperature vaporization of uranium [105, 39, 31], which was obtained using its adiabatic expansion data as the basis.

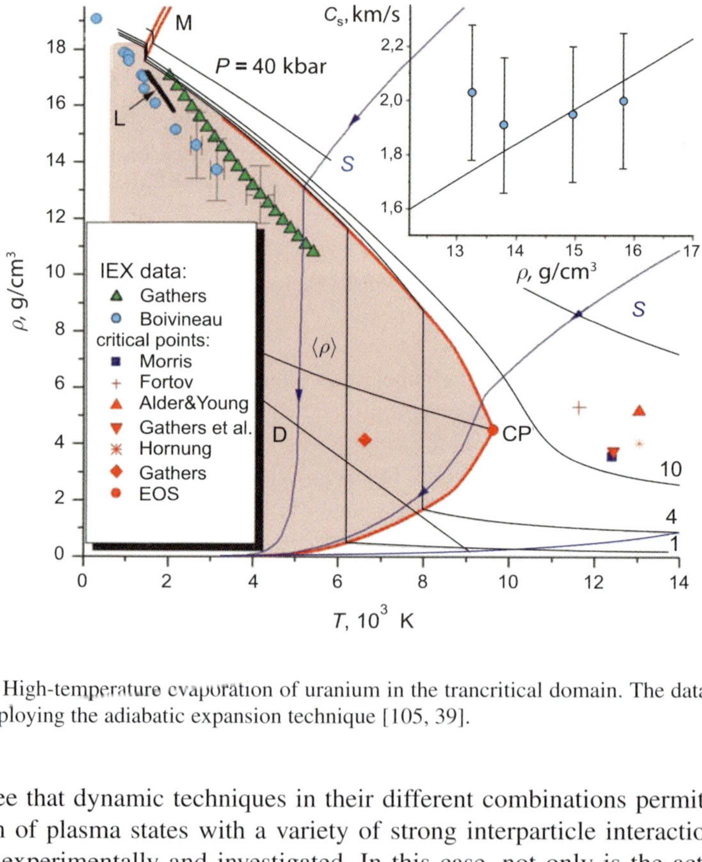


**Fig. 3.16** High-temperature evaporation of uranium in the trancritical domain. The data were obtained employing the adiabatic expansion technique [105, 39].

We see that dynamic techniques in their different combinations permit a broad spectrum of plasma states with a variety of strong interparticle interactions to be realized experimentally and investigated. In this case, not only is the actual realization of high energy density conditions possible, but so is sufficiently thorough diagnostics of these high-energy states, because shock and adiabatic waves are not merely a means of production, but also a specific tool for the diagnostics of extreme states of matter with a high energy density [33, 29, 38, 5, 1, 104]. Measurements of the mechanical parameters of the motion of shock waves and contact discontinuities make it possible to determine the thermodynamic plasma properties and, with the use of modern high-speed diagnostic techniques, many physical characteristics of extreme parameter plasmas.

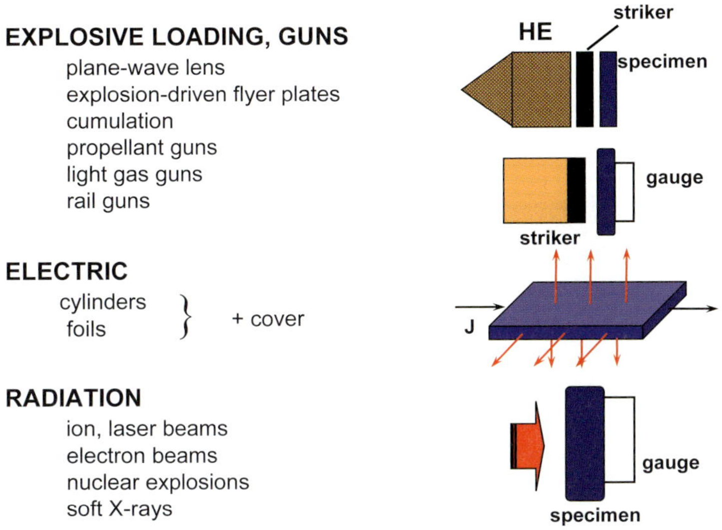


**Fig. 3.17** Schemes for intense shock generation.

At present, use is made of a substantial number of dynamic cumulation sources of high energy density in dense media (Fig. 3.17).

## 3.5 Light-Gas Guns or Chemical and Nuclear Explosions

Today the technique of intense shock waves generated by the impact of metal liners (strikers) accelerated to velocities of several kilometers per second on a target of the substance under investigation is the main source of physical information about the plasma behavior at pressures up to 10–15 Mbar. Here we shall not describe in great depth the liner acceleration technique and the means of diagnostics – they are dealt with in comprehensive reviews and monographs [33, 29, 38, 41, 79, 5, 1, 31, 10, 2, 34, 4, 96, 28]. We only note that in shock-wave experiments of this kind it is possible to carry out sufficiently ample measurements of the physical plasma properties. The equation of state is determined by electrocontact and optical recording of the time intervals in the motion of shock-wave discontinuities and contact surfaces. Pyrometric, spectroscopic, protonographic, X-ray diffraction, and adsorption measurements are performed; use is made of pulsed X-ray and synchrotron radiation sources; laser interferometric measurements are made, as are low- and high-frequency Hall conductivity measurements and the detection of piezo- and magnetoelectric effects. Unique information was gained in the study of mechanical properties of shock-compressed media: elastoplastic properties, splitting off, polymorphism, viscosity, fracture, and fragmentation [60]. Many unexpected results were also obtained in the study of chemical reactions and the kinetics of physicochemical transformations in shock and detonation waves [60].

In the USA, gunpowder and light-gas launching devices – "guns" (Fig. 3.18) [79] – were most often used, while in the USSR preference was given to explosion launching devices [4].

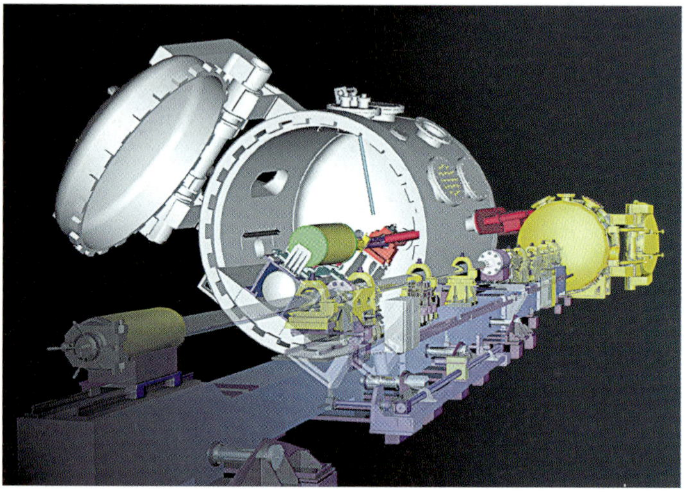

**Fig. 3.18** Schematic of a light-gas gun, Livermore, USA [79].

To increase the velocity of launching and hence the shock-compressed plasma pressure, advantage is taken of highly sophisticated gas-dynamic techniques. The technique of "gradient" cumulation (Fig. 3.19 [102, 52]) relies on a successive increase of the velocity of strikers in alternating heavy and light material planar layers. This technique is not related to the effects of geometrical energy focusing and therefore possesses a higher stability of acceleration and compression in comparison with the spherical one. The thus-obtained three-stage "layer cake" explosion [52, 4] accelerated a 100 $\mu$m striker to velocities of 5–14 km/s. The impact of such a striker excites in the target plasma a plane shock wave or a series of reverberating shock waves with a pressure amplitude ranging into the megabar region. The geometrical parameters of these experimental devices are selected in such a way as to eliminate the distorting effect of side and rear unloading waves to ensure the one-dimensionality and stationarity of gas dynamic flow in the region of recording.

Interestingly, the kinetic energy of a metal striker moving at a speed of 10 km/s is close [79] to the kinetic energy of a proton beam in the Fermilab cyclotron accelerator. The high kinetic energy in shock-wave experiments produces a strongly compressed high-temperature plasma just as a relativistic ion collision produces a quark–gluon plasma with enormous energy densities.

To increase the parameters of shock compression, in several experiments use was made of explosion generators of counter-propagating shock waves (Fig. 3.20),

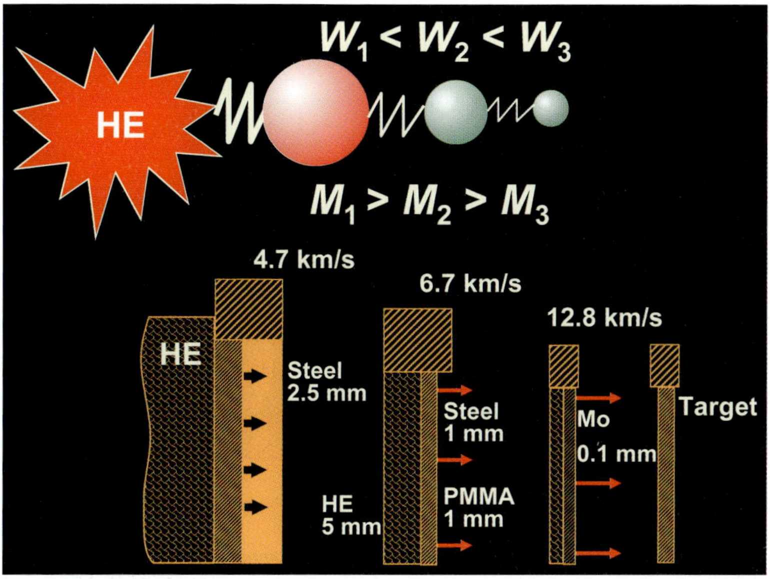


**Fig. 3.19** Principle of "gradient" cumulation [102] and three-stage "layer cake" explosion [5, 1].

where the material under investigation was loaded on both sides by the synchronous impact of steel strikers symmetrically accelerated by explosive charges [77, 43].

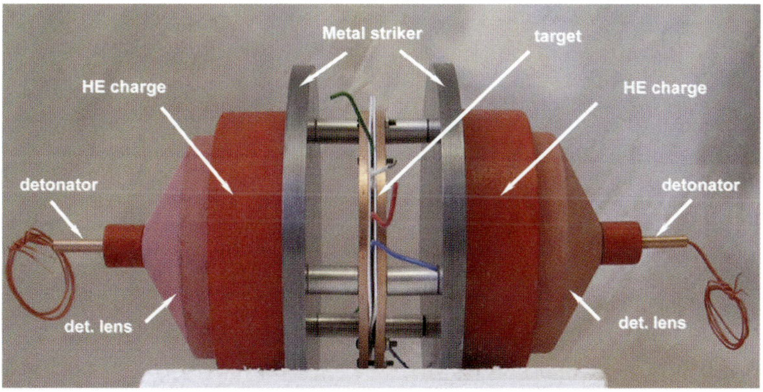


**Fig. 3.20** Explosion generator of counter-propagating shock waves [77, 43].

Very-high-precision spherical explosion generators of intense shock waves (Fig. 3.21) were made in the USSR [5, 1, 51, 4, 2] for the study of thermodynamic material properties at pressures ranging up to 10 Mbar. Using the geometrical cumulation effects in the centripetal motion (implosion) of detonation products and

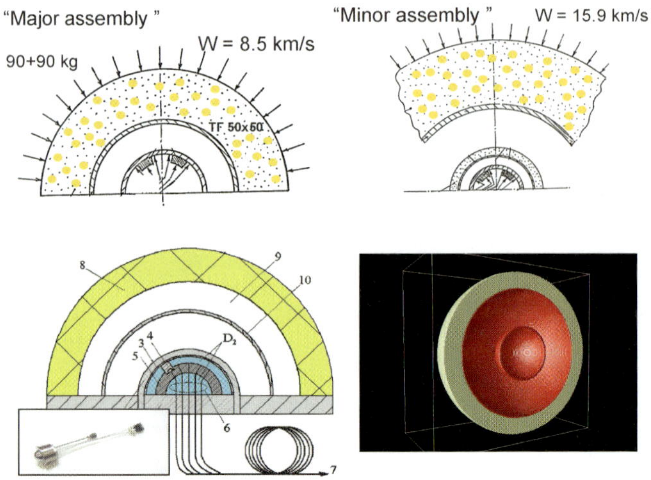


**Fig. 3.21** Explosion generators of spherically converging intense shock waves [2, 34, 4].

hemispherical shells, in devices weighing 100 kg with an energy release of 300 MJ it was possible to accelerate metal strikers to velocities of 23 km/s.

The experimental data on the thermodynamic properties of nonideal deuterium plasma obtained under cylindrical and spherical shock-wave compression are shown in Fig. 3.22, where they are compared with the data of laser (Fig. 3.4) [13] and electrodynamic (Fig. 5.6) experiments [63]. Dedicated estimates show that the shock-compressed plasma in these experiments is strongly nonideal (Debye coupling parameter $\Gamma > 1$) with developed ionization $n_e/n_\mathrm{D} \approx 1$ (where $n_e$ and $n_\mathrm{D}$ are the electron and Debye densities, respectively) and partial degeneracy $n_e \lambda_e^3 \approx 3$, where $\lambda_e$ is the thermal de Broglie wavelength. One can see (Fig. 3.22) that the models of nonideal plasmas [104, 5, 1, 34] provide a reasonable description of the data from explosion and electrodynamic experiments.

In higher-stability conical explosion generators, use was made of cumulation effects in the irregular ("Mach") convergence of cylindrical shock waves (Fig. 3.23). The combination of irregular cylindrical and "gradient" cumulation effects enabled excitation in copper of a shock wave with an amplitude of $\approx 20$ Mbar, which is comparable with pressures in the near zone of a nuclear explosion.

The capabilities of different launching techniques are compared in Fig. 3.24. Specific plasma energy densities that were records for terrestrial conditions were obtained in the near-source zone of a nuclear explosion. Several physical experiments of this kind are schematized in Figs. 3.25 and 3.26 [10, 99, 9, 34, 4, 96, 97]. The collection of experimental data on shock-compressed aluminum plasma is represented in Fig. 3.27, where the highest points correspond to record parameters in terrestrial conditions [99]. The density of the internal energy of this plasma is $E \approx 10^9$ J/cm$^3$, which is close to the energy density of nuclear matter, and the pressure $P \approx 4$ Gbar is

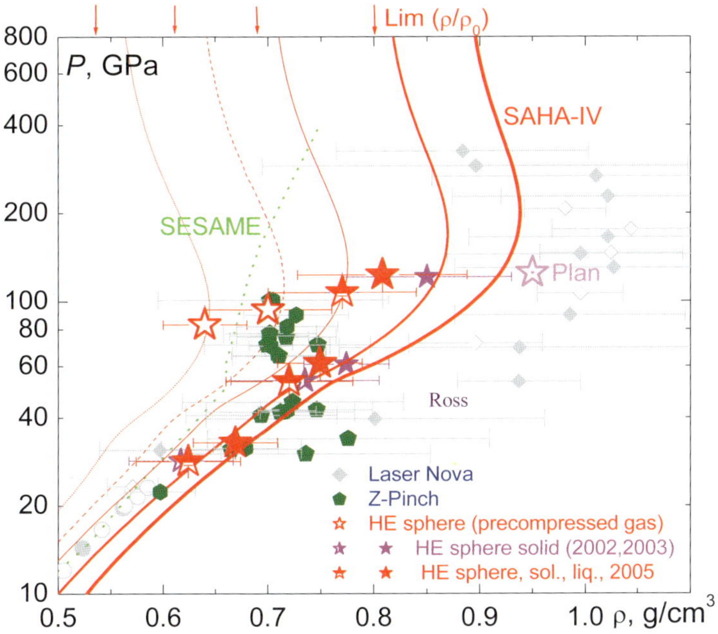


**Fig. 3.22** Shock-wave compression of deuterium plasma [63, 26, 13, 17].

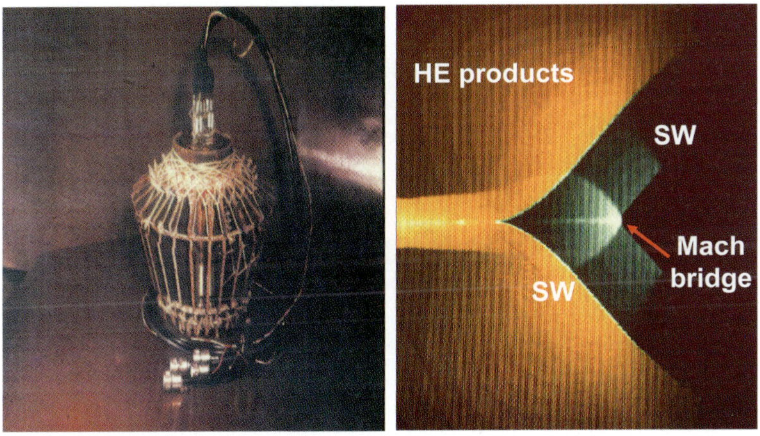


**Fig. 3.23** Explosion conic generators of "Mach" shock waves [12]. Shown at the right are the results of a two-dimensional hydrodynamic simulation.

close to the pressure in internal layers of the Sun. The plasma under these conditions ($n_e \approx 4 \times 10^{24}\text{cm}^3$, $T \approx 8 \times 10^6$ K) is nondegenerate, $n\lambda^3 \approx 0.07$, it is twelve times ionized, and the nonideality parameter is small ($\Gamma \approx 0.1$), which is an experimental illustration of the thesis of Chap. 2 about simplification of the physical plasma

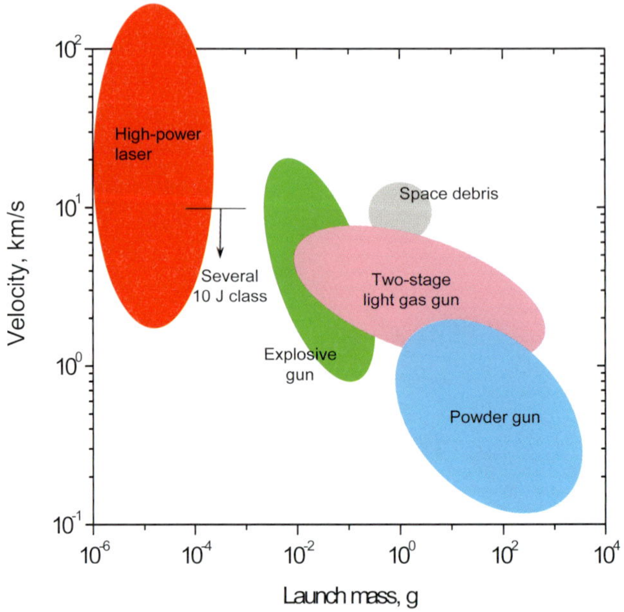


**Fig. 3.24** Capabilities of different high-velocity launching techniques.

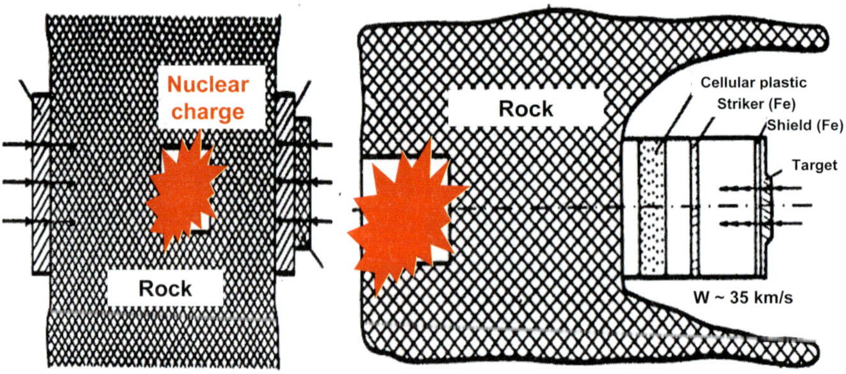


**Fig. 3.25** Scheme of experiments on the generation of intense shock waves in the immediate vicinity of a nuclear explosion [10, 9, 96, 97].

properties in the limit of ultrahigh energy density. Interestingly, the parameter range investigated is adjacent to the domain where the energy and pressure of equilibrium radiation make an appreciable contribution to the thermodynamics of the system:

$$E_{\mathrm{R}} = 4\sigma T^4/c; \qquad P_{\mathrm{R}} = E_{\mathrm{R}}/3 = \frac{4}{3}\sigma T^4/c.$$

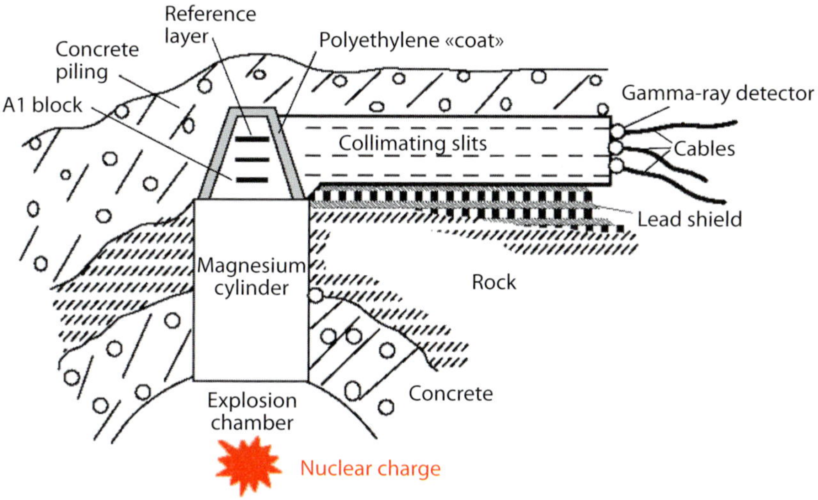


**Fig. 3.26** Schematic representation of experiments with an underground nuclear explosion using a gamma-active reference layer. Reprinted, with permission, from [9].

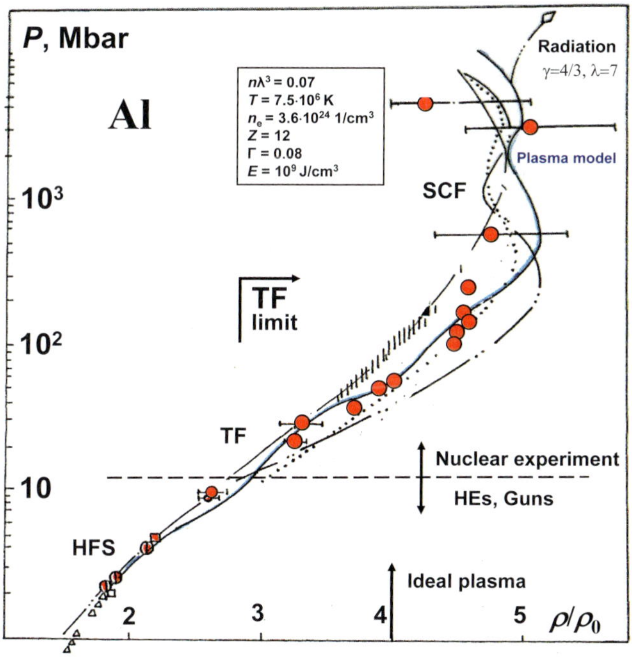


**Fig. 3.27** Shock-wave compression of aluminum to gigabar pressures [10, 99, 9].

Therefore the plasma dynamics regime is realized close to the radiative gas-dynamic one [28].

## 3.6 Quasi-classical Model of a Substance

The pressures realized by means of nuclear explosions [10, 99, 9, 4, 96, 2, 34] (Fig. 3.27) belong to the multimegabar domain and are close to the characteristic "physical" pressure, which may be found from dimensionality considerations $P \approx e^2/a_B^4 \approx 300$ Mbar ($a_B = \hbar^2/(me^2)$ is the Bohr electron radius). Beginning with these pressures, the Thomas–Fermi model [61, 62] applies, which implies a simplified quantum-statistical description of a strongly compressed substance and the "self-similarity" of its physical properties. This model [62], which relies on the quasi-classical approximation to the self-consistent field method, is a substantial simplification of the many-particle quantum-mechanical problem and therefore enjoys wide use in the solution of astrophysical and special problems.

As discussed in Chap. 2, the description of bound electron states in the high-density domain, $\rho \leq \rho_0$, is radically simplified at extremely high pressures $P \gg e^2/a_B^4$ or temperatures $T \leqslant 1$ Ry $\approx 10^5$ K, when electron shells are "crushed" and their properties are described by the quasi-classical approximation to self-consistent field theory – the Thomas–Fermi theory. In this model, the description of a system in terms of wave functions and energy eigenvalues is replaced by a simplified statistical representation in terms of the average electron density $\rho(x)$, which obeys the relationships of a quasi-uniform degenerate electron gas (in the subsequent discussion we use the atomic system of units):

$$n(x) = \frac{1}{\sqrt{2}\pi} T^{3/2} f_{1/2}\left(\frac{p_F^2(x)}{2T}\right) \underset{T\to 0}{\longrightarrow} \frac{p_F^2(x)}{3\pi^2},$$

where $f_{1/2}(\alpha)$ is the Fermi function and $\mu$ is the chemical potential,

$$\mu = \frac{p_F^2(x)}{2} + U(x).$$

To simplify calculations, in the quasi-classical model a substance is divided into electroneutral spherical Wigner–Seitz cells of radius $R$, which contain a nucleus and Z electrons surrounding it, making it possible to go over from the multicentered problem to a one-centered spherically symmetric problem. The electrons are in the self-consistent potential $U(r)$ which satisfies the Poisson equation:

$$-\nabla^2 U = 4\pi n,$$

$$\left(U(r) \underset{r\to 0}{\sim} \frac{1}{r}, \qquad U(r) \underset{r\to R}{\sim} (r-R)^2\right),$$

where the cell radius $R$ is defined by the electroneutrality condition $\int n(x)\,dx = Z$. Numerical integration of this equation enables one to determine the electron density $\rho(x)$, which underlies the determination of all thermodynamic functions of the electron gas of the atomic cell. To obtain the net thermodynamic characteristics of the substance, account must be taken, along with the electronic terms, of the nuclear motion, which is commonly described in either the ideal-gas or quasi-harmonic approximation. The equations of the Thomas–Fermi model are inherently self-similar with respect to the nuclear charge: upon introduction of the variables $ZV$, $Z^{-4/3}T$, $Z^{-10/3}P$, $Z^{-7/3}E$ they do not contain $Z$ explicitly and their solution applies to any element, which simplifies the practical employment of this model.

The Thomas–Fermi model is the quasi-classical limit with respect to the Hartree equations of the self-consistent field, and therefore modifications of this model involve taking correlation, quantum-mechanical, and relativistic effects into account in more detail [62]. Correlation corrections exist because the self-consistent Hartree field is different from the true field inside an atomic cell. These corrections arise from the antisymmetry of electron wave functions and are interpreted as exchange-correlation effects. In addition, owing to inexactness of the picture of independent particles accepted in the model, there emerge force correlation effects.

Quantum-mechanical corrections stem from the use of a quasi-classical formalism and are divided into the regular (in $\hbar^2$) part (termed quantal), which reflects the nonlocality of the relation between $n(x)$ and $U(x)$ due to the uncertainty principle, and the nonregular correction, which reflects the nonmonotonicity of physical quantities arising from discreteness of the energy spectrum [62]. It is significant that the introduction of the oscillatory correction characterizes the most up-to-date versions of the Thomas–Fermi model [62], while the inclusion of exchange, correlation, and quantum corrections is conventional practice for high-energy-density physics.

The relative magnitude of correlation and quantum effects is determined by the dimensionless parameters [62]

$$\delta_{\text{corr}} \sim \delta_0^{\nu} \quad \text{and} \quad \delta_{\text{quant}} \sim \delta_{\text{exch}} \sim n/p_{\text{F}}^4.$$

In the degeneracy domain ($n^{2/3} \gg T$, $p_{\text{F}} \sim n^{1/3}$, $\delta_0 \sim n^{-1/3}$, $\nu = 2$) these parameters are

$$\delta_{\text{corr}} \sim n^{-2/3}, \qquad \delta_{\text{exch}} \sim n^{-1/3},$$

and in the classical domain ($n^{2/3} \ll T$, $p_{\text{F}} \sim T^{1/2}$, $\delta_0 \sim n^{1/3}/T$, $\nu = 3/2$)

$$\delta_{\text{corr}} \sim n^{1/2}/T^{3/2}, \qquad \delta_{\text{exch}} \sim n/T^2.$$

A drawback of the Thomas–Fermi model is an incorrect description of the electron density at the cell periphery and near the nucleus due to violation of the quasi-classicality condition there. Kirzhnits et al. [62] came up with a method for removing this drawback by employing the method of successive approximations in the solution of the Thomas–Fermi equations with quantum corrections without expanding in a series in a small parameter. This approach forms the basis of the quantum-statistical model, in which the solution rapidly converges near the nucleus and exhibits a qualitatively correct quantum-mechanical behavior away from the nucleus.

However, the resultant formulae are no longer self-similar with respect to $Z$. Numerical calculations have to be carried out anew for each element, and the equations themselves are much more arduous for numerical calculations.

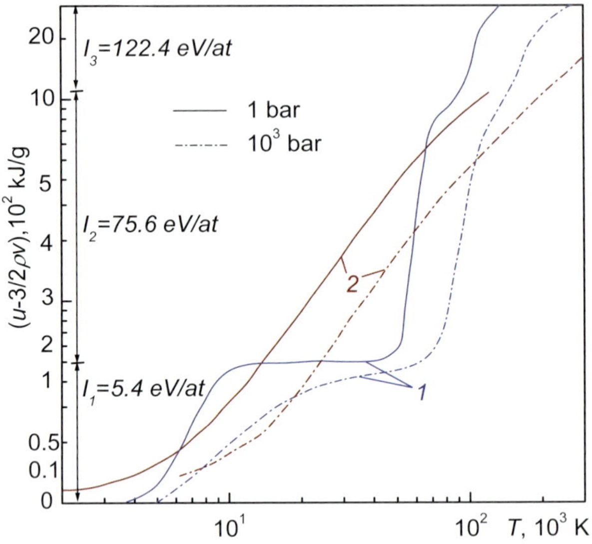


**Fig. 3.28** Equation of state of lithium plasma [59]. Calculation: *1* chemical model, *2* Thomas–Fermi model with quantum and exchange corrections. $I_1$, $I_2$, $I_3$ sequential ionization potentials.

Several other modifications [62] to the quasi-classical model in the vicinity of the nucleus were also proposed, which differ in the way the corrections are introduced. However, in the thermodynamic description the difference between these models and the Thomas–Fermi model with corrections is noticeable only outside of the domain of their formal applicability, which underlies the preference for the simpler Thomas–Fermi model to carry out specific calculations.

When use is made of the quasi-classical method of description, it is well to bear in mind the specific inaccuracies introduced by the cell model itself. All electron correlations in this model are automatically limited to the dimensions of the atomic cell and therefore cannot exceed the average internuclear distance, and nucleus–nucleus correlation effects are absent. These circumstances obviously limit the applicability of the Thomas–Fermi model for the description of plasma in conditions typical for this state of matter, when the screening sphere contains a substantial number of nuclei, whose correlations make the main contribution to the Debye correction [54]. That is why this model does not encompass the Debye limit in the plasma domain and, furthermore, its extrapolation properties become worse as the plasma becomes more tenuous (see [29, 38]), because the model does not describe the stepwise nature of thermodynamic functions in gas plasmas owing to ionization effects (Fig. 3.28).

The divergence between the calculations made in the framework of the Thomas–Fermi model and more exact quantum-mechanical calculations by the Hartree–Fock–Slater model in the parameter range characteristic for the plasma state is considered in [29, 38, 34].

Nuclear correlation effects may also be significant in the condensed phase, since the cell model ignores the deviation of the real cell volume from the average one due to nuclear motion, which is valid only for ordered systems. For nonideality parameter values $\Gamma \approx 100$ typical for the condensed state and dense plasmas, in the equation of state it is necessary to take into account the motion of nuclei, which also leads to atomic volume fluctuations. For the Thomas–Fermi model an estimate of this effect using the example of $SiO_2$ showed [74] that the inclusion of nuclear motion in the range $1 < \Gamma < 100$ increases the pressure by $\approx 15\%$.

The physical conditions for the validity of the quasi-classical model, as noted above, correspond to extremely high pressures $P \gg 300$ Mbar and temperatures$T \gg 10^5$ K, which are realized in different astrophysical objects but are hard to attain by experimental techniques in terrestrial conditions. The presently attainable highest-pressure and highest-temperature states are realized, as is evident from the foregoing, with the help of dynamic methods that make use of the techniques of intense shock waves. Although the majority of shock-wave experimental data do not correspond exactly to quantum-statistical conditions, they permit the properties of the quasi-classical models to be extrapolated beyond the limits of their formal applicability defined [62] by the smallness of the corresponding literal criteria. These constructions suggest that the introduction of quantum, exchange, and correlation corrections (oscillation corrections were not included) improves extrapolation, which becomes possible in this case, in the view of Al'tshuler et al. [3], to pressures $P \geqslant 300$ Mbar for zero temperature and to $\approx 50$ Mbar for $T \geqslant 10^4$ K. At the same time, the data of comparative measurements made in underground nuclear explosions allow ambiguous interpretations and do not provide an unequivocal answer to the question about the superiority of one or other version of the quasi-classical model, and are at variance with the data of absolute measurements [46, 9].

It is also noteworthy that experiments on the shock compression of high-porosity copper performed for pressures of 10–20 Mbar and temperatures up to $2 \times 10^6$ K (see [10, 33, 31]) and of solid materials for pressures up to 160 Mbar [3, 97, 10, 33, 88, 89, 31] are indicative of significant shell effects in the domain that was previously described by tradition by the standard Thomas–Fermi model.

On the strength of initial simplifications, the quasi-classical model with quantum and exchange corrections is inapplicable at low pressures, but it provides [5, 1] (Fig. 3.29) a reasonable averaging of elemental atomic volumes over the periodic system. We emphasize that the Thomas–Fermi model itself, due to the absence of cohesion forces in it, for a zero pressure leads to an infinite radius of an atomic cell and therefore to an incorrect – zero – substance density, and that the finite substance density is not obtained until the corrections are introduced. As noted in [5, 1], with increasing pressure the oscillations of atomic volume decrease and approach the calculations by modified Thomas–Fermi models. However, this is partly attributable to

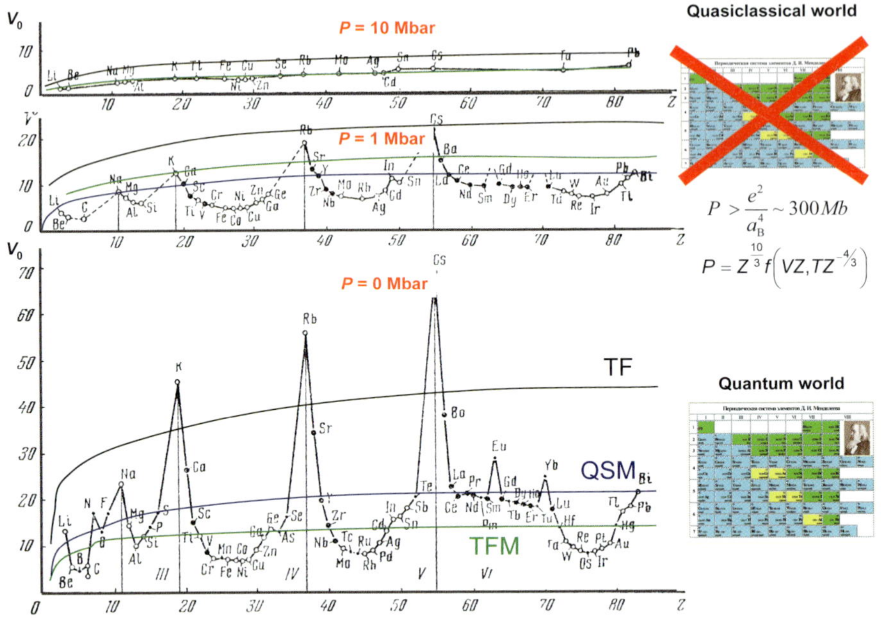


**Fig. 3.29** Variation of elemental atomic volumes with increasing pressure [5, 1].

the fact that the comparison at high pressures was made with the results of extrapolation of experimental data to quasi-classical calculations themselves.

In reality the tendency for substance properties to become simpler is violated at higher pressures due to the existence of inner atomic shells in atoms. As determined in [62], shell effects may be qualitatively described in the framework of the quasi-classical approximation by including the correction irregular in $\hbar^2$ that corresponds to the oscillatory part of the electron density, which was previously disregarded by mistake. In this case it is significant that the shell effects are described even in the lowest quasi-classical approximation for the wave function and should therefore be taken into account along with the regular corrections considered above. The shell effects reflect the irregularities of substance characteristics arising from the discrete energy spectrum and emerge in the quasi-classical model as a result of the interference of de Broglie waves [62].

Therefore, the quasi-classical model in its most up-to-date version [62] is much more substantive than considered before. It turns out that this model not only describes the averaged behavior of electrons in heavy and strongly compressed atoms, but it also qualitatively reproduces the atomic inner-shell structure and in many cases yields results that are close to the data calculated by more accurate quantum-mechanical models, being profitably different from them in simplicity and obviousness.

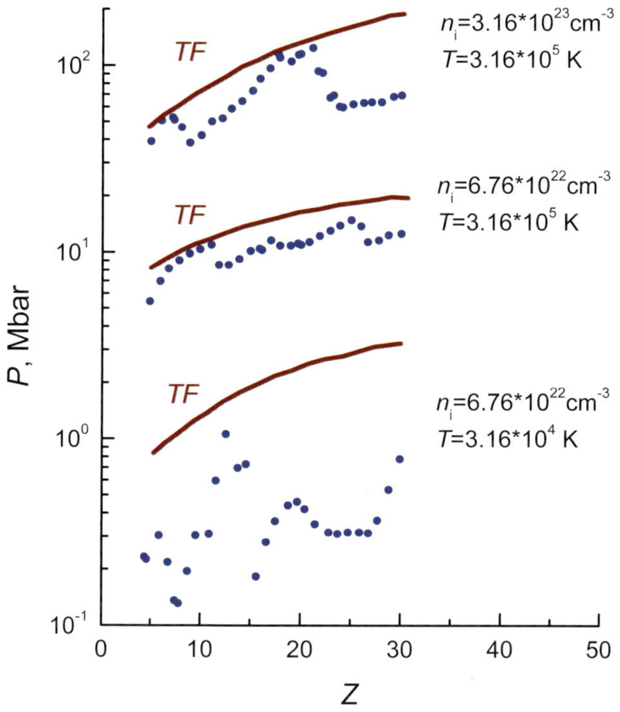


**Fig. 3.30** Pressure as a function of atomic number for fixed densities and temperatures [66]. The solid lines represent calculations by the Thomas–Fermi model and the points correspond to the inclusion of shell effects in the central field approximation.

The inclusion of shell effects appreciably changes the equation of state of substances, giving rise to discontinuities in the atomic volume curve $V(Z)$ in the high-pressure domain (where this approximation is justified) [62, 29, 38]. In this case, in the equation of state there emerge characteristic nonmonotonic features caused by electron phase transitions, when the energy levels are expelled from the discrete spectrum into the continuum. One might expect these features to smooth out with increasing temperature under the influence of resonance electrons, however, the data calculated in the central-field approximation [66] testify (Fig. 3.30) to a significant influence of the shell structure even for a strongly heated substance. The significance of the contribution from the shell effects at hyper-megabar pressures also follows from the data of quantum-mechanical calculations in the more exact Hartree–Fock–Slater approximation [93, 94] and by the augmented plane wave method [72]. These effects are predicted for a broad parameter range and should vanish for $n \gg Z^4$ in the homogeneity domain [61], when all atomic energy levels pass into the continuum. In this connection we note that the question of the asymptotic behavior of the quasi-classical model is not trivial, for it has been shown [67] that this model corresponds to the exact solution of the Schrödinger equation only for $Z \to \infty$, but not in the high-density limit.

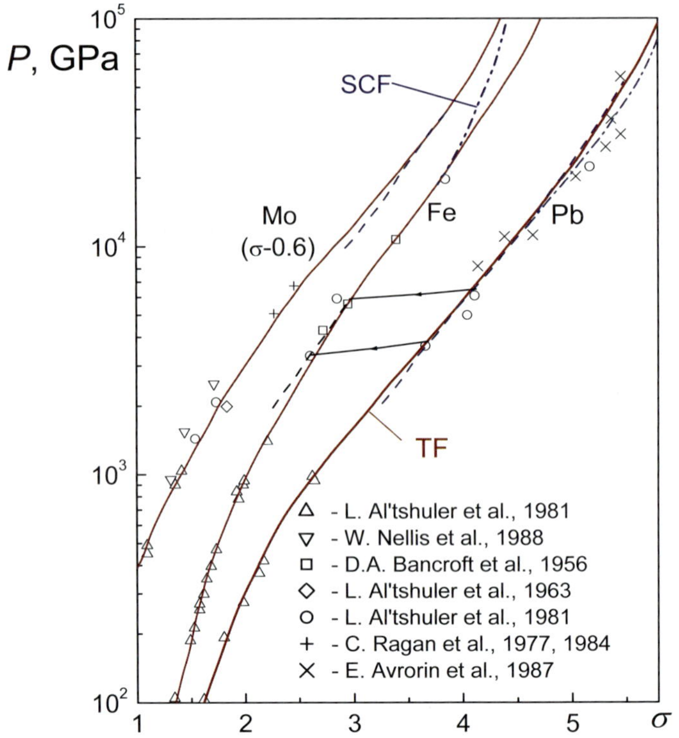


**Fig. 3.31** Extrapolation properties of the quasi-classical (Thomas–Fermi) model [5, 1]; $\sigma = \rho/\rho_0$; SCF, self-consistent field.

Therefore, the question of the limits of validity of the quasi classical model is still, to a large extent, open, and the behavior of substances in the $P > 300$ Mbar domain turns out to be more varied than assumed on the basis of earlier simplified ideas [62]. Experimental verification of the predictions of the quasi-classical shell model is presently the most interesting problem of ultrahigh-pressure physics; solving this problem is thought to call for new experimental techniques relying on intense directional energy fluxes (see the following sections).

For the time being, only the underground nuclear explosion techniques [10, 99, 9, 34, 4] furnish the opportunity to approach the multimegabar pressure domain, making no more than an estimate of the lower bounds of validity of the quasi-classical model possible (Fig. 3.31). It was determined [5, 1] that this model applies beginning from pressures of about 100 Mbar in the Hugoniot curve, whereas its validity range becomes substantially narrower with increasing temperature (the Hugoniot curves of a porous material).

The thermodynamics of superdense plasmas in the ultramegabar dynamic pressure range invites further investigation. Figure 3.32 shows the pressures thus far attained under controllable conditions with the help of shock waves and diamond anvils. One can see that going to pressures above 10 Mbar calls for the application of unconventional techniques of shock wave generation (Sect. 3.7), primarily laser-based techniques (Chap. 4).

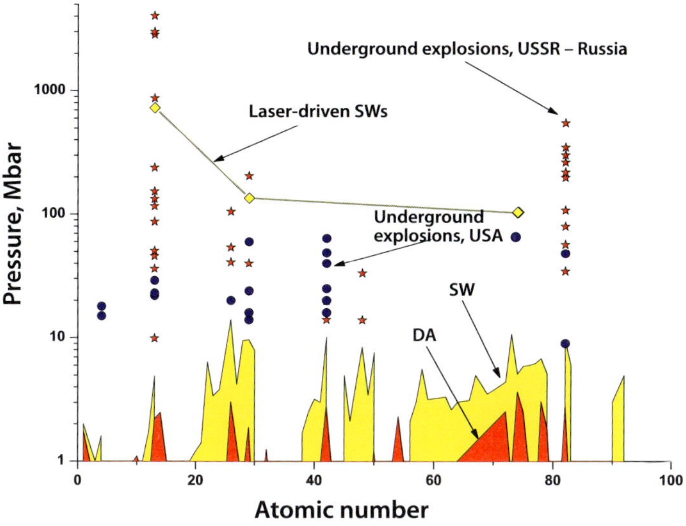


**Fig. 3.32** Characteristic pressures realized in terrestrial experiments. DA – static technique involving diamond anvils, SW – shock waves driven by light-gas guns and chemical explosives.

The reader interested in ultrahigh-pressure problems is referred to [9, 10, 99, 4, 96, 31, 33].

## 3.7 Devices of High-Current Impulse Energetics

Devices of high-current ($10^5$–$10^7$ A) impulse energetics are employed to produce high-energy-density plasmas in various kinds of experimental facilities. Electric energy may effect the direct pulsed Joule heating (electroexplosion) of conductors or the magnetohydrodynamic compression and heating of plasma objects. The stored energy may be used to produce intense bursts of soft X-ray radiation (with a radiation temperature of 200–300 eV) with the subsequent generation of intense shock or radiative thermal waves by this radiation, or for the electrodynamic generation of

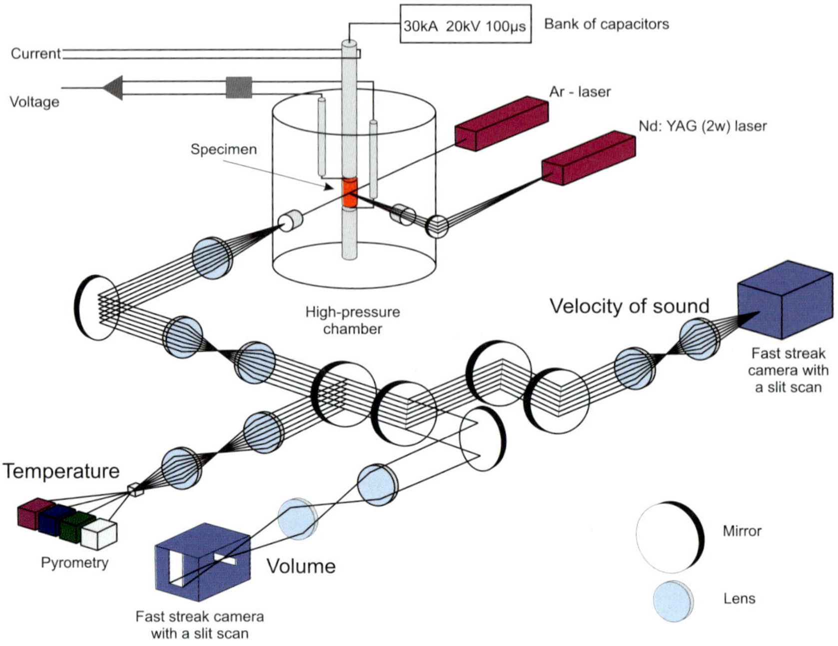


**Fig. 3.33** Schematic representation of an exploding-wire experiment [65].

shock waves, as well as for the electrodynamic acceleration of metallic liners. The energy capabilities of electrodynamic devices of this kind are, as a rule (with the exception of the NIF and MJL), several orders of magnitude higher than for lasers, making it possible to conduct experiments with thicker targets, raise the accuracy of measurements, and relax the time-resolution requirements ($10^5$–$10^7$ s) on the means of diagnostics.

The electroexplosion of conductors and metal foils by a pulsed current of $\approx$ 50–200 kA is the traditional direction of research into the thermophysical properties of refractory materials in the domain of condensed matter [65] for characteristic energy densities of $\approx$ 10 kJ/cm$^3$ (Fig. 3.33). This range was recently extended to 20–30 kJ/cm$^3$ with the attainment of strongly supercritical metal states, which permits, in particular, the "metal–dielectric" transition to be studied in the continuous supercritical expansion of metal plasmas.

The highest plasma parameters have been obtained in high-power Z-pinches of the terawatt power range, in which the electric energy of capacitors, after the corresponding peaking, effects the electrodynamic plasma acceleration followed by the focusing of its kinetic energy on the cylinder axis [98, 44]. In this way an approximately 10 ns long burst of soft X-ray 150–200 eV radiation with an energy of $\approx$ 1.8 MJ and a power of $\approx$ 230 TW was produced in the Z-pinch facility at the Sandia National Laboratories, USA [95, 25, 84]. In these experiments a cylindrical plasma shell was produced by the electric explosion of hundreds of thin (6–50 $\mu$m)

**Fig. 3.34** Pulsed Angara-5 generator, TRINITI, Russia, intended for controlled thermonuclear fusion and experiments on high-energy-density plasmas and shock and thermal waves [50].

tungsten conductors by a 20 MA current with a rise time of $\approx 100$ ns. A tungsten plasma with an ion density of $\approx 10^{20}$ cm$^{-3}$ and a degree of ionization higher than 50 was obtained in the implosion on the axis.

The second interesting application of this facility (Fig. 5.6) involves the electrodynamic generation of intense shock waves [63]. To this end, a high-power electric current pulse effected the electrodynamic acceleration of a metal liner to velocities of $\approx 20$ km/s and the generation of megabar-range shock-wave pressures at its impact on a target. In this case there is the possibility of controlling the current parameters and realizing a shock-free ("soft") compression of the target to a pressure of $\approx 3$ Mbar. Laser-assisted recording of the parameters of the quasi-isentropic compression was used, and it was possible to observe compressed states with lower temperatures and entropies than for shock-wave heating.

In experiments in the Angara facility [50] (Fig. 3.34), a pulsed current of $\approx 4$ MA accelerated a xenon liner to a velocity of $\approx 500$ km/s. The highly symmetric impact of this liner on the surface of a cylindrical highly porous target excited in it a thermal radiative wave, which emitted soft X-ray radiation with a temperature of $\approx 100$ eV. This high-intensity X-ray radiation from the cylindrical cavity was employed for highly symmetric production of plane shock waves with a pressure amplitude of $\approx 5$ Mbar, for the excitation of thermal radiative waves with a propagation velocity of 100 km/s, as well as for the acceleration of metallic liners to velocities of $\approx$ 10–12 km/s.

Interesting plasma parameters were obtained in Z-pinches with an initially gaseous shell [98, 44] of centimeter size, which is ionized by 100–200 ns long megampere current and implodes to a millimeter-sized column, increasing the plasma density by factors of 20–50 (to $10^{20}$ cm$^{-3}$) for an electron temperature of several electronvolts.

A high-energy plasma is generated in the X-pinch geometry produced by crossing two current-carrying wires [48, 23]. In the micropinch formed at the intersection point (less than 1 mm in size, with a duration of 50–80 ns), the ion density amounts to $10^{20}$–$10^{22}$ cm$^{-3}$ and the electron temperature to 2 keV. Devices of this kind hold considerable promise as X-ray radiation sources for microlithography and other applications.

In the operation of modern high-current pinches, in the course of production of high energy densities there occur developed magnetohydrodynamic flows in which a crucial role is played by radiation [86], which is of significance in its own right in radiative gas-dynamic simulations of astrophysical objects.

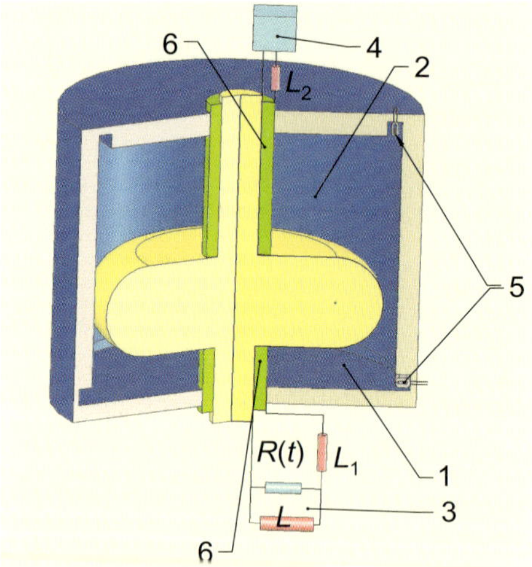


**Fig. 3.35** Schematic diagram of the MAGO plasma generator [73]. 1, 2 – plasma chamber; 3, 4 – power sources, explosion magnetic generators, 5 – sensors, 6 – insulators.

Promise may be held by the line of research in the generation of high-energy-density plasmas reliant on the quasi-adiabatic cylindrical compression of plasma structures [6, 101] that initially contain preformed closed magnetic configurations – so-called "inverse" magnetic field structures [23]. Additional compression and heating of this plasma is effected by an external metallic liner electrodynamically heated by a megampere current. Estimates show [78] that the 30 MA current of the ATLAS facility is capable of producing a several-centimeter-long plasma column 1 cm in diameter with a plasma pressure of ≈ 1 Mbar, an ion density of $10^{19}$ cm$^{-3}$, and a temperature of ≈ 10 keV.

In the MAGO project (Fig. 3.35) [73], a pulsed current of 7 MA delivered by an explosion magnetic generator [32, 18] compresses and heats the preliminarily prepared magnetized plasma to parameters close to thermonuclear conditions, $\rho \approx 20$ g/cm$^3$, $T \approx$ 3–4 keV, which furnished a deuteriem–tritium (DT) reaction neutron yield of (3–5)$\times 10^{13}$. In this case, an energy density of ≈ $10^7$ J/cm$^3$ was realized in the plasma-compressing metallic liner.

The devices of high-current impulse energetics considered above show good promise as regards improving the main parameters and developing the elements and schemes of power compression.

The Baikal facility [11, 49] projected in Russia should become a source of soft X-ray radiation with an output energy of ≈ 10 MJ intended for experiments with indirect-drive thermonuclear targets and other problems of high-energy-density physics (Fig. 3.36). In this Project, 4 GJ of electric energy is stored in inductive energy storage devices and, after the corresponding power peaking, will feed a fast liner system with a power up to 500–1000 TW.

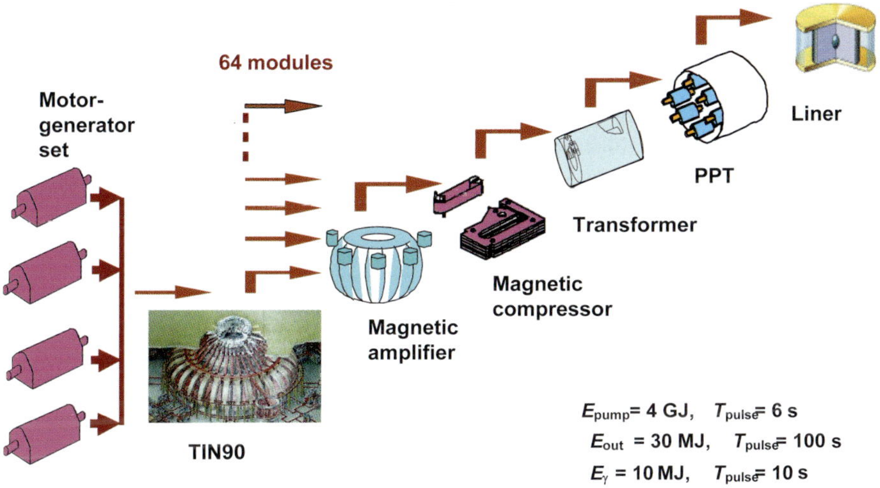


**Fig. 3.36** Schematic representation of the Baikal facility [11].

It is planned to develop the success achieved on the Z facility at Sandia, USA, employing the X-1 facility, which feeds two pinch units, each generating 7 MJ of soft X-ray radiation with a power of ≈ 1000 TW. This X-ray power will be delivered to a thermonuclear target 4 mm in diameter exposed to X-ray radiation with a temperature of more than 225 eV for 10 ns. The thermonuclear energy yield of the X-1 facility is anticipated to be 200–1000 MJ.

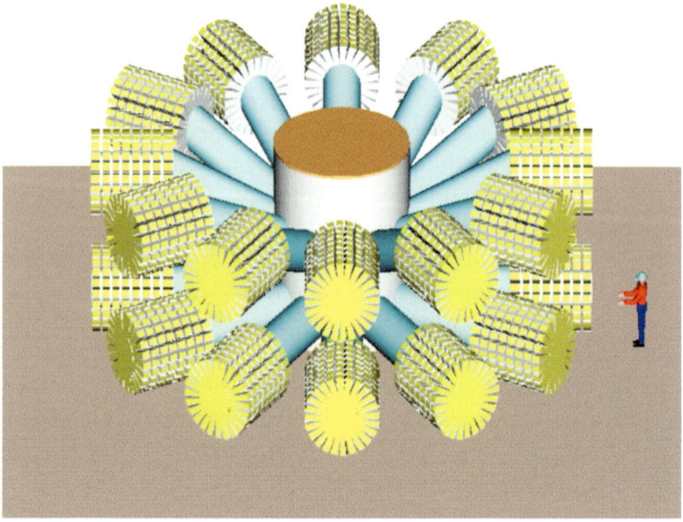

Fig. 3.37 Schematic drawing of the IFE-Z facility [80].

A development of the Z generator at Sandia, USA, is the IFE-Z Project [80], which is substantially no less than a prototype of a pulsed thermonuclear reactor (Fig. 3.37). Use will be made of a double Z-pinch, each pinch generating 8 MJ of X-ray radiation under a 66 MA current pulse. In this case, the planned thermonuclear

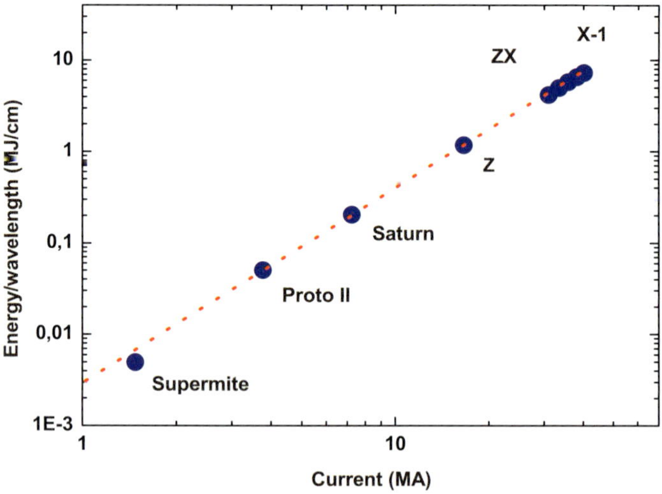


Fig. 3.38 Soft X-ray radiation yield as a function of Z-pinch current.

energy yield is 3 GJ and the gain coefficient $\approx 83$. Figure 3.38 shows the employed scaling of the X-ray energy yield with the pinch current.

The use of chemical explosives for generating intense pulsed currents and magnetic fields relies [55] on the explosive compression of the initial magnetic flux by conducting metallic liners [82, 32, 55] accelerated to velocities of several kilometers per second by the detonation products of condensed explosives. This is precisely how it has been possible to achieve values of electric current of $\approx 300$ MA and magnetic induction of $\approx 29$ MGs, which are the record values achieved under terrestrial conditions [18]. In the latter case, the magnetic field energy density was equal to $3 \times 10^6$ J/cm$^3$, making it possible to carry out interesting physical experiments in magagauss magnetic fields: the quasi-isentropic compression of substances [55, 82], the study of magnetoresistance, magnetooptical effects, Shubnikov–de Haas effect, and many other phenomena at high energy densities [90].

## References

[1] Al'tshuler, L.V.: Use of shock waves in high-pressure physics. Sov. Phys. – Usp. **8**(1), 52–91 (1965). DOI 10.1070/PU1965v008n01ABEH003062. URL http://stacks.iop.org/0038-5670/8/52

[2] Al'tshuler, L.V., Krupnikov, K.K., Fortov, V.E., Funtikov, A.I.: Origins of megabar pressure physics. Herald Russ. Acad. Sci. **74**(6), 613 (2004)

[3] Al'tshuler, L.V., N., K.N., Kuz'mina, L.V., Chekin, B.S.: Shock adiabats for ultrahigh pressures. JETP **45**, 167 (1977)

[4] Al'tshuler, L.V., Trunin, R.F., Krupnikov, K.K., Panov, N.V.: Explosive laboratory devices for shock wave compression studies. Phys. Usp. **39**(5), 539 (1996). DOI 10.1070/PU1996v039n05ABEH000147. URL http://ufn.ru/en/articles/1996/5/f/

[5] Al'tshuler, L.V., Trunin, R.F., Urlin, V.D., et al.: Development of dynamic high-pressure techniques in Russia. Phys. Usp. **42**(3), 261 (1999). DOI 10.1070/PU1999v042n03ABEH000545. URL http://ufn.ru/en/articles/1999/3/c/

[6] Andreev, V.F., Karaev, J.A., Umrihin, N.M., et al. (eds.): Uslovija generacii impul'snogo nejtronnogo izluchenija pri vzryvnom obzhatii termojadernoj plazmy. IAE-5519/7 (Conditions of Generation of Pulse Neutron Radiation by Explosive Compression of Thermonuclear Plasma). IAE, Moscow (1992)

[7] Anisimov, S.I., Prokhorov, A.M., Fortov, V.E.: Application of high-power lasers to study matter at ultrahigh pressures. Sov. Phys. – Usp. **27**(3), 181–205 (1984). DOI 10.1070/PU1984v027n03ABEH004036. URL http://stacks.iop.org/0038-5670/27/181

[8] Atzeni, S., Meyer-ter-Vehn, J.: The Physics of Inertial Fusion. Oxford University Press, Oxford (2004)

[9] Avrorin, E.N., Simonenko, V.A., Shibarshov, L.I.: Physics research during nuclear explosions. Phys. Usp. **49**(4), 432 (2006). DOI

10.1070/PU2006v049n04ABEH005958. URL http://ufn.ru/en/articles/2006/4/j/

[10] Avrorin, E.N., Vodolaga, B.K., Simonenko, V.A., Fortov, V.E.: Intense shock waves and extreme states of matter. Phys. Usp. **36**(5), 337–364 (1993). DOI 10.1070/PU1993v036n05ABEH002158. URL http://stacks.iop.org/1063-7869/36/337

[11] Azizov, E.A., Alexandrov, V.V., Alikhanov, S.G., et al.: Pulse power system development for megajoule X-ray facility BAIKAL. AIP Conf. Proc. **651**(1), 29–32 (2002). DOI 10.1063/1.1531274. URL http://link.aip.org/link/?APC/651/29/1

[12] Bazanov, O.V., Bespalov, V.E., Zharkov, A.P., et al.: Irregular reflection of conically converging shock-waves in Plexiglas and copper. High Temp. **23**(5), 781–787 (1985)

[13] Belov, S.I., Boriskov, G.V., Bykov, A.I., et al.: Shock compression of solid deuterium. JETP Lett. **76**(7), 433–435 (2002)

[14] Betti, R., Anderson, K., Boehly, T.R., et al.: Progress in hydrodynamics theory and experiments for direct-drive and fast ignition inertial confinement fusion. Plasma Phys. Control. Fusion **48**(12B), B153–B163 (2006). DOI 10.1088/0741-3335/48/12B/S15

[15] Boehler, R.: Temperatures in the Earth's core from melting-point measurements of iron at high static pressures. Nature **363**(6429), 534–536 (1993). DOI 10.1038/363534a0

[16] Boehler, R., Forzandonea, D.: The laser heated diamond cell: high P–T phase diagrams. In: G.L. Chiarotti, R.J. Hemley, M. Bernasconi, L. Ulivi (eds.) High Pressure Phenomena, pp. 55–66. IOS Press, Amsterdam (2002)

[17] Boriskov, G.V., Bykov, A.I., Il'kaev, R.I., et al.: Shock compression of liquid deuterium up to 109 GPa. Phys. Rev. B **71**(9), 092104 (2005). DOI 10.1103/PhysRevB.71.092104. URL http://link.aps.org/abstract/PRB/v71/e092104

[18] Boyko, B.A., Bykov, A.I., et al.: More than 20 MG magnetic field generation in the cascade magnetocumulative MC-1 generator. In: H.J. Schneider-Muntau (ed.) Megagauss Magnetic Field Generation, Its Application to Science and Ultra High Pulsed-Power Technology. Proc. VIIIth Int. Conf. Megagauss Magnetic Field Generation and Related Topics, p. 61. World Scientific, Singapore (2004)

[19] Bunkenberg, J., Boles, J., Brown, D., et al.: The omega high-power phosphate-glass system: design and performance. IEEE J. Quantum Electron. **17**(9), 1620–1628 (1981)

[20] Calderola, P., Knopfel, H. (eds.): Physics of High Energy Density. Academic, New York (1971)

[21] Cavailler, C.: Inertial fusion with the LMJ. Plasma Phys. Control. Fusion **47**(12B), B389–B403 (2005). DOI 10.1088/0741-3335/47/12B/S28

[22] Chen, F.F.: Introduction to Plasma Physics and Controlled Fusion, Vol. 1, 2nd edn. Springer, New York (1984)

[23] Chittenden, J.P., Ciardi, A., Jennings, C.A., et al.: Structural evolution and formation of high-pressure plasmas in X-pinches. Phys. Rev. Lett. **98**(2), 025003 (2007). DOI 10.1103/PhysRevLett.98.025003. URL http://link.aps.org/abstract/PRL/v98/e025003

[24] Courant, R., Friedrichs, K.O.: Supersonic Flow and Shock Waves. Interscience, New York (1948)

[25] Cuneo, M.E., Vesey, R.A., Bennett, G.R., et al.: Progress in symmetric ICF capsule implosions and wire-array Z-pinch source physics for double-pinch-driven hohlraums. Plasma Phys. Control. Fusion **48**(2), R1–R35 (2006). DOI 10.1088/0741-3335/48/2/R01

[26] Da Silva, L.B., Celliers, P., Collins, G.W., et al.: Absolute equation of state measurements on shocked liquid deuterium up to 200 GPa (2 Mbar). Phys. Rev. Lett. **78**(3), 483–486 (1997). DOI 10.1103/PhysRevLett.78.483

[27] Ditmire, T., Springate, E., Tisch, J.W., et al.: Explosion of atomic clusters heated by high-intensity femtosecond laser pulses. Phys. Rev. A **57**(1), 369–382 (1998). DOI 10.1103/PhysRevA.57.369. URL http://link.aps.org/abstract/PRA/v57/p369

[28] Drake, R.P.: High-Energy-Density Physics. Springer, Berlin, Heidelberg (2006)

[29] Fortov, V., Iakubov, I., Khrapak, A.: Physics of Strongly Coupled Plasma. Oxford University Press, Oxford (2006)

[30] Fortov, V.E. (ed.): Entsiklopediya nizkotemperaturnoi plazmy (Encyclopedia of Low-Temperature Plasma). Nauka, Moscow (2000)

[31] Fortov, V.E.: Intense Shock Waves and Extreme States of Matter. Bukos, Moscow (2005)

[32] Fortov, V.E. (ed.): Explosive-Driven Generators of Powerful Electrical Current Pulses. Cambridge International Science, Cambridge (2007)

[33] Fortov, V.E.: Intense shock waves and extreme states of matter. Phys. Usp. **50**(4), 333 (2007). DOI 10.1070/PU2007v050n04ABEH006234. URL http://ufn.ru/en/articles/2007/4/c/

[34] Fortov, V.E., Al'tshuler, L.V., Trunin, R.F., Funtikov, A.I.: High-Pressure Shock Compression of Solids VII: Shock Waves and Extreme States of Matter. Springer, New York (2004)

[35] Fortov, V.E., Gryaznov, V.K., Mintsev, V.B., et al.: Thermophysical properties of shock compressed argon and xenon. Contrib. Plasma Phys. **41**(2-3), 215–218 (2001). DOI 10.1002/1521-3986(200103)41:2/3⟨215::AID-CTPP215⟩3.0.CO;2-G

[36] Fortov, V.E., Hoffmann, D.H.H., Sharkov, B.Y.: Intense ion beams for generating extreme states of matter. Phys. Usp. **51**(2), 109 (2008). DOI 10.1070/PU2008v051n02ABEH006420. URL http://ufn.ru/en/articles/2008/2/a/

[37] Fortov, V.E., Ilkaev, R.I., Arinin, V.A., et al.: Phase transition in a strongly nonideal deuterium plasma generated by quasi-isentropical compression at megabar pressures. Phys. Rev. Lett. **99**(18), 185001 (2007). DOI 10.1103/

PhysRevLett.99.185001. URL http://link.aps.org/abstract/PRL/v99/e185001
[38] Fortov, V.E., Khrapak, A.G., Yakubov, I.T.: Fizika neideal'noi plazmy (Physics of Nonideal Plasma). Fizmatlit, Moscow (2004)
[39] Fortov, V.E., Lomonosov, I.V.: Thermodynamics of extreme states of matter. Pure Appl. Chem. **69**(4), 893–904 (1997)
[40] Fortov, V.E., Mintsev, V.B., Ternovoi, V.Y., et al.: Conductivity of nonideal plasma. High Temp. Mater. Processes **8**(3), 447–459 (2004). DOI 10.1615/HighTempMatProc.v8.i3.100
[41] Fortov, V.E., Ternovoi, V.Y., Zhernokletov, M.V., et al.: Pressure-produced ionization of nonideal plasma in a megabar range of dynamic pressures. JETP **97**(2), 259–278 (2003). DOI 10.1134/1.1608993
[42] Fortov, V.E., Ternovoi, V.Y., Zhernokletov, M.V., et al.: Pressure-produced ionization of nonideal plasma in a megabar range of dynamic pressures. JETP **97**(2), 259–278 (2003). DOI 10.1134/1.1608993
[43] Fortov, V.E., Yakushev, V.V., Kagan, K.L., et al.: Anomalous electric conductivity of lithium under quasi-isentropic compression to 60 GPa (0.6 Mbar). Transition into a molecular phase? JETP Lett. **70**(9), 628–632 (1999)
[44] Gasilov, V.A., Zakharov, S.V., Smirnov, V.P.: Generation of intense radiation fluxes and megabar pressures in liner systems. JETP Lett. **53**(2), 85 (1991)
[45] Ginzburg, V.L.: The Physics of a Lifetime: Reflections on the Problems and Personalities of 20th Century Physics. Springer, Berlin, Heidelberg (2001)
[46] Ginzburg, V.L.: On superconductivity and superfluidity (what I have and have not managed to do), as well as on the "physical minimum" at the beginning of the XXI century (December 8, 2003). Phys. Usp. **47**(11), 1155 (2004). DOI 10.1070/PU2004v047n11ABEH001825. URL http://ufn.ru/en/articles/2004/11/d/
[47] Giorla, J., Bastian, J., Bayer, C., et al.: Target design for ignition experiments on the laser Mégajoule facility. Plasma Phys. Control. Fusion **48**(12B), B75–B82 (2006). DOI 10.1088/0741-3335/48/12B/S0
[48] Glidden, S.C., Richter, M., Hammer, D.A., Kalantar, D.H.: 1 kW X-pinch soft X-ray source powered by a 500 kA, 100 ns, 40 pps pulser. In: 9th IEEE Int. Pulsed Power Conf., 1993 Digest of Technical Papers, vol. 1, p. 459 (1993)
[49] Glukhikh, V., Kuchinsky, V., Pechersky, O., et al.: Perspective of kiloterawatt soft X-ray source based on slow inductive storage with energy 1 gigajoule. In: H.V. Horn, S. Ichimaru (eds.) Proc. 12th Int. Conf. on High-Power Particle Beams, BEAMS'98, June 7–12, 1998, Haifa, Israel, p. 71. IEEE, Piscataway, NJ (1998). DOI 10.1109/BEAMS.1998.822392
[50] Grabovskii, E.V., Vorob'ev, O.Y., Dyabilin, K.S., et al.: Excitation of intense shock waves by soft x radiation from a Z-pinch plasma. JETP Lett. **60**(1), 1 (1994)
[51] Grishechkin, S.K., Gruzdev, S.K., Gryaznov, V.K., et al.: Experimental measurements of the compressibility, temperature, and light absorption in dense shock-compressed gaseous deuterium. JETP Lett. **80**(6), 398–404 (2004)

[52] Gryaznov, V.K., Fortov, V.E., Zhernokletov, M.V., et al.: Shock compression and thermodynamics of highly nonideal metallic plasma. JETP **87**(4), 678–690 (1998). DOI 10.1134/1.558710
[53] Gryaznov, V.K., Nikolaev, D.N., Ternovoi, V.Y., et al.: Generation of a nonideal plasma by shock compression of a highly porous $SiO_2$ aerogel. Chem. Phys. Rep. **17**(1-2), 239–245 (1998)
[54] Hammel, B.A., National Ignition Campaign Team: The NIF ignition program: progress and planning. Plasma Phys. Control. Fusion **48**(12B), B497–B506 (2006). DOI 10.1088/0741-3335/48/12B/S47
[55] Hawke, P.S., Burgess, T.J., Duerre, D.E., et al.: Observation of electrical conductivity of isentropically compressed hydrogen at megabar pressures. Phys. Rev. Lett. **41**(14), 994–997 (1978). URL `http://link.aps.org/abstract/PRL/v41/p994`
[56] Hemley, R.J., Ashcroft, N.W.: The revealing role of pressure in the condensed matter sciences. Phys. Today **51**(8), 26–32 (1998). DOI 10.1063/1.882374
[57] Hemley, R.J., Mao, H.K.: Overview of static high pressure science. In: R.J. Hemley, G.L. Chiarotti, M. Bernasconi, L. Ulivi (eds.) High Pressure Phenomena, Proceedings of the International School of Physics "Enrico Fermi" Course CXLVII, p. 3. IOS Press, Amsterdam (2002)
[58] Hogan, W.J. (ed.): Energy from Inertial Fusion. IAEA, Vienna, Austria (1995)
[59] Iosilevskii, I.L., Griaznov, V.K.: Comparative accuracy of thermodynamic description of properties of a gas plasma in the Thomas–Fermi and Saha approximations. High Temp. **19**(6), 799–803 (1982)
[60] Kanel, G.I., Rasorenov, S.V., Fortov, V.E.: Shock-Wave Phenomena and the Properties of Condensed Matter. Springer, New York (2004)
[61] Kirzhnits, D.A.: Extremal states of matter (ultrahigh pressures and temperatures). Sov. Phys. – Usp. **14**(4), 512–523 (1972). DOI 10.1070/PU1972v014n04ABEH004734. URL `http://stacks.iop.org/0038-5670/14/512`
[62] Kirzhnits, D.A., Lozovik, Y.E., Shpatakovskaya, G.V.: Statistical model of matter. Sov. Phys. – Usp. **18**(9), 649–672 (1975). DOI 10.1070/PU1975v018n09ABEH005199. URL `http://stacks.iop.org/0038-5670/18/649`
[63] Knudson, M.D., Hanson, D.L., Bailey, J.E., et al.: Equation of state measurements in liquid deuterium to 70 GPa. Phys. Rev. Lett. **87**(22), 225501 (2001). DOI 10.1103/PhysRevLett.87.225501. URL `http://link.aps.org/abstract/PRL/v87/e225501`
[64] Kruer, W.L.: The Physics of Laser Plasma Interactions. Addison-Wesley, Reading, MA (1988)
[65] Lebedev, S.V., Savvatimskii, A.I.: Metals during rapid heating by dense currents. Sov. Phys. – Usp. **27**(10), 749–771 (1984). DOI 10.1070/PU1984v027n10ABEH004128. URL `http://stacks.iop.org/0038-5670/27/749`

[66] Lee, C.M., Thorsos, E.I.: Properties of matter at high pressures and temperatures. Phys. Rev. A **17**(6), 2073–2076 (1978). DOI 10.1103/PhysRevA.17.2073. URL http://link.aps.org/abstract/PRA/v17/p2073

[67] Lieb, E.H., Simon, B.: Thomas–Fermi theory revisited. Phys. Rev. Lett. **31**(11), 681–683 (1973). URL http://link.aps.org/abstract/PRL/v31/p681

[68] Lindl, J.D.: Inertial Confinement Fusion. Springer, New York (1998)

[69] Loubeyre, P., Celliers, P.M., Hicks, D.G., et al.: Coupling static and dynamic compressions: first measurements in dense hydrogen. High Pressure Res. **24**(1), 25–31 (2004). DOI 10.1080/08957950310001635792

[70] Loubeyre, P., Occelli, F., Le Toulec, R.: Optical studies of solid hydrogen to 320 GPa and evidence for black hydrogen. Nature **416**(6881), 613–617 (2002). DOI 10.1038/416613a

[71] Maksimov, E.G., Magnitskaya, M.V., Fortov, V.E.: Non-simple behavior of simple metals at high pressure. Phys. Usp. **48**(8), 761 (2005). DOI 10.1070/PU2005v048n08ABEH002315. URL http://ufn.ru/en/articles/2005/8/a/

[72] McMahan, A.K., Ross, M.: In: K.D. Timmerhaus, M.S. Barber (eds.) High Pressure Science and Technology, p. 920. Plenum, New York (1979)

[73] Mokhov, V.N.: Formation of the thermonuclear fusion ideas. In: V.D. Selemir, L.N. Plyashkevichu (eds.) Megagauss-IX, p. 665. VNIIEF, Sarov (2004)

[74] More, R.M., Skupsky, S.: Nuclear-motion corrections to the Thomas–Fermi equation of state for high-density matter. Phys. Rev. A **14**(1), 474–479 (1976). DOI 10.1103/PhysRevA.14.474. URL http://link.aps.org/abstract/PRA/v14/p474

[75] Moses, E.I., Bonanno, R.E., Haynam, C.A., et al.: The National Ignition Facility: path to ignition in the laboratory. Eur. Phys. J. D **44**(2), 215–218 (2006). DOI 10.1140/epjd/e2006-00106-3

[76] Mourou, G.A., Tajima, T., Bulanov, S.V.: Optics in the relativistic regime. Rev. Mod. Phys. **78**(2), 1804–1816 (2006). DOI 10.1103/RevModPhys.78.309. URL http://link.aps.org/abstract/RMP/v78/p309

[77] Nabatov, S.S., Dremin, A.M., Postnov, V.I., Yakushev, V.V.: Measurement of the electrical conductivity of sulfur under superhigh dynamic pressures. JETP Lett. **29**(7), 369 (1979)

[78] National Research Council: Frontiers in High Energy Density Physics. National Academies Press, Washington, DC (2003)

[79] Nellis, W.J.: Dynamic compression of materials: metallization of fluid hydrogen at high pressures. Rep. Prog. Phys. **69**(5), 1479–1580 (2006). DOI 10.1088/0034-4885/69/5/R05

[80] Oslon, C., Rochau, G., et al.: Development path for Z-pinch IFE. Fusion Sci. Technol. **47**(3), 633–640 (2005)

[81] Parsons, W., Ballard, E., Bartsch, R., et al.: The atlas project – a new pulsed power facility for high energy density physics experiments. IEEE Trans. Plasma Sci. **25**(2), 205–211 (1997). DOI 10.1109/27.602492

[82] Pavlovski, A.I., Boriskov, G.V., et al.: Isentropic solid hydrogen compression by ultrahigh magnetic field pressure in megabar range. In: C.M. Fowler, R.S. Caird, D.T. Erickson (eds.) Megagauss Technology and Pulsed Power Applications, p. 255. Plenum, New York (1987)

[83] Pukhov, A.: Strong field interaction of laser radiation. Rep. Prog. Phys. **66**(1), 47–101 (2003). DOI 10.1088/0034-4885/66/1/202

[84] Quintenz, J., Sandia's Pulsed Power Team: Pulsed power team. In: Proc. 13th Int. Conf. on High Power Particle Beams. Nagaoka, Japan (2000)

[85] Reinovsky, R.E., Anderson, W.E., Atchison, W.L., et al.: Shock-wave and material properties experiments using the Los Alamos Atlas pulsed power system. AIP Conf. Proc. **706**(1), 1191–1194 (2004). DOI 10.1063/1.1780451. URL `http://link.aip.org/link/?APC/706/1191/1`

[86] Ryutov, D.D., Derzon, M.S., Matzen, M.K.: The physics of fast Z-pinches. Rev. Mod. Phys. **72**(1), 167–223 (2000). DOI 10.1103/RevModPhys.72.167. URL `http://link.aps.org/abstract/RMP/v72/p167`

[87] Sansone, G., Benedetti, E., Calegari, F., et al.: Isolated single-cycle attosecond pulses. Science **314**(5798), 443–446 (2006). DOI 10.1126/science.1132838. URL `http://www.sciencemag.org/cgi/content/abstract/314/5798/443`

[88] Schatz, T., Schramm, U., Habs, D.: Crystalline ion beams. Nature **412**(6848), 717–720 (2001). DOI 10.1038/35089045

[89] Schramm, U., Schatz, T., Bussmann, M., Habs, D.: Cooling and heating of crystalline ion beams. J. Phys. B **36**(3), 561–571 (2003). DOI 10.1088/0953-4075/36/3/314

[90] Selimir, V.D., Tatsenko, O.M., Platonov, V.V.: Investigations in solid state physics in ultra-high magnetic fields – experimental results of Kapitsa series. In: M. von Ortenberg (ed.) Proc. of the Xth Megagauss Conf., Berlin 2004, pp. 219–226. VNIIEF, Sarov, Russia (2005)

[91] Sharkov, B.Y. (ed.): Yadernyi sintez s inertsionnym uderzhaniem (Inertial Confinement Nuclear Fusion). Fizmatlit, Moscow (2005)

[92] Shilkin, N.S., Dudin, S.V., Gryaznov, V.K., et al.: Measurements of the electron concentration and conductivity of a partially ionized inert gas plasma. JETP **97**(5), 922–931 (2003). DOI 10.1134/1.1633948

[93] Sinko, G.V.: Calculation of thermodynamic functions of simple substances on the basis of the equations of the self-coordinated field [in Russian]. Chis. Met. Meh. Spl. Sred. **10**(1), 124 (1979)

[94] Sinko, G.V.: Some results of calculations of thermodynamic functions of aluminum, copper, cadmium and lead. A method of the self-coordinated field [in Russian]. Chis. Met. Meh. Spl. Sred. **12**(1), 121 (1981)

[95] Spielman, R.B., Deeney, C., Chandler, G.A., et al.: Tungsten wire-array Z-pinch experiments at 200 TW and 2 MJ. Phys. Plasmas **5**(5), 2105–2111 (1998). DOI 10.1063/1.872881

[96] Trunin, R.F.: Shock compressibility of condensed materials in strong shock waves generated by underground nuclear explosions. Phys. Usp. **37**(11), 1123

(1994). DOI 10.1070/PU1994v037n11ABEH000055. URL http://ufn.ru/en/articles/1994/11/d/

[97] Trunin, R.F., Podurets, M.A., Simakov, G.V., et al.: An experimental verification of the Thomas–Fermi model for metals under high pressure. Sov. Phys. – JETP **35**, 550 (1972)

[98] Turchi, P.J., Baker, W.L.: Generation of high-energy plasmas by electromagnetic implosion. J. Appl. Phys. **44**(11), 4936–4945 (1973). DOI 10.1063/1.1662066. URL http://link.aip.org/link/?JAPIAU/44/4936/1

[99] Vladimirov, A.S., Voloshin, N.P., Nogin, V.N., et al.: Shock compressibility of aluminum at $p > 1$ Gbar. JETP Lett. **39**(2), 82 (1984)

[100] Wikipedia: ISKRA lasers. URL http://en.wikipedia.org/wiki/ISKRA_lasers

[101] Winterberg, F.: The magnetic booster target inertial confinement fusion driver. Z. Naturforsch. A **39A**, 325 (1984)

[102] Zababahin, E.I., Zababahin, I.E.: Yavleniya neogranichennoj kumulyacii (The phenomena of unlimited cumulating). Nauka, Moscow (1988)

[103] Zasov, A.V., Postnov, K.A.: Obshchaya astrofizika (General Astrophysics). Vek 2, Fryazino (2006)

[104] Zel'dovich, Y.B., Raizer, Y.P.: Fizika udarnykh voln i vysokotemperaturnykh gidrodinamicheskikh yavlenii, 2nd edn. Nauka, Moscow (1966). [English Transl.: Physics of Shock Waves and High-Temperature Hydrodynamic Phenomena. Dover, Mineola, NY (2002)]

[105] Zhernokletov, M.V.: Shock compression and isentropic expansion of natural uranium. High Temp. **36**(2), 214–221 (1998)

[106] Zhernokletov, M.V., Zubarev, V.N., Trunin, R.F., Fortov, V.E.: Eksperimental'nye dannye po udarnoi szhimaemosti i adiabaticheskomu rasshireniju kondensirovannyh vewestv pri vysokih plotnostjah energii (Experimental data on shock compressibility and adiabatic expansion of condensed matter at high energy density). IHF RAN, Chernogolovka (1996)

# Chapter 4
# High-Power Lasers in High-Energy-Density Physics

The rapid progress of laser technology has opened up the possibility of generating ultrashort laser pulses of the nano–pico–femto[1]–atto-second range and of bringing (see Tables 3.2, 3.3; Fig. 3.2) the existing and projected laser complexes into the petawatt–zettawatt power range (Figs 4.1, 4.2), making it possible to span a wide range of power densities up to the highest values achievable today, $q \approx 10^{22}$–$10^{23}$ W/cm$^2$ [70, 69, 6, 14], which will undoubtedly rise with time. The action of these enormous intensities on targets leads to diverse new physical effects in hot plasmas [7, 70, 14], such as multiphoton ionization, self-focusing and filamentation of different types, the generation of superintense electric and magnetic fields, electron and ion acceleration to relativistic velocities, nuclear reactions caused by these fast particles, relativistic plasma "transparentization", nonlinear modulation and multiple generation of harmonics, ponderomotive effects in hydrodynamics, and many other effects, which are the subject of vigorous research today (see reviews [70, 80] and references therein). Some of these effects will be briefly considered in this chapter.

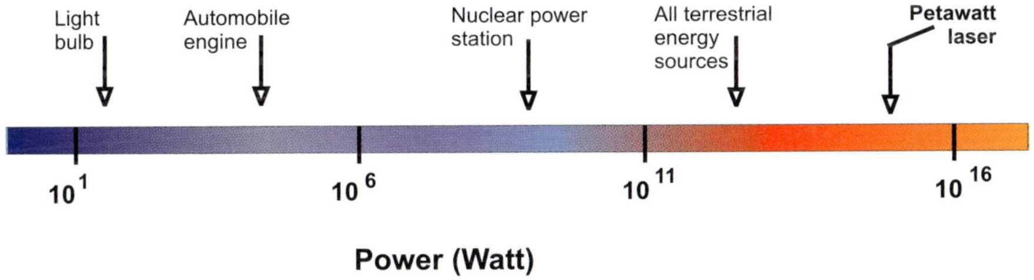


**Fig. 4.1** Power scale.

[1] For comparison, the ratio between 10 femtoseconds and 1 minute is equal to the ratio between 1 minute and the age of the universe.

V.E. Fortov, *Extreme States of Matter*, The Frontiers Collection,
DOI 10.1007/978-3-642-16464-4_4, © Springer-Verlag Berlin Heidelberg 2011

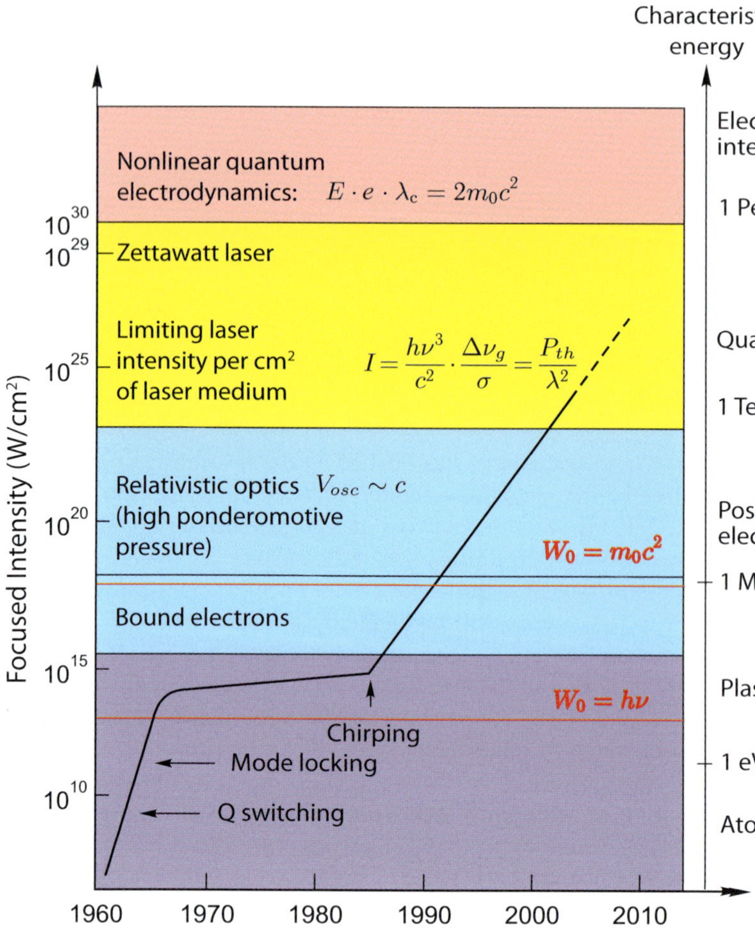


**Fig. 4.2** Growth of laser radiation intensity with time [74, 70]. The rapid increase of radiation intensity in the 1960s resulted in the discovery of numerous nonlinear effects induced by bound electrons (characteristic energies on the order of an electronvolt). Contemporary swift intensity growth permits processes at relativistic electron energies ($W \sim m_ec^2 \simeq 0.5$ MeV) to be studied.

The further growth of power density will be accompanied by qualitatively new phenomena [74], such as spontaneous electron–positron pair production (the "boiling" of the vacuum and the loss of its transparency [74, 1, 49, 3, 54, 48, 30, 80, 22, 61, 55, 84, 70, 57, 18, 7]), the emergence of microscopic quantities of relativistic matter, the generation of relativistic plasma streams and shock waves, solitons, jets, and $\gamma$-bursts similar to the astrophysical ones, and – in the distant future – the realization of quantum gravitation conditions [100, 86, 20]. This new intensity level may see the advent of new schemes of controlled thermonuclear fusion (Sect. 6.1), a new way of producing short-lived isotopes, as well as extraordinary schemes of high-efficiency compact accelerators (Sect. 6.3).

Figure 4.2 shows the progress in increasing the irradiation intensity and the possibilities that open up for studying the processes in high-energy density physics [70, 74].

With the invention of lasers it became possible to record numerous nonlinear optical effects, such as the deformation of intraatomic and molecular fields by laser radiation; stimulated Raman, Brillouin, and Thomson scattering; multiphoton ionization; and several other nonlinear effects related to bound electrons. Since the advent of the first laser, raising the peak power of laser radiation has been and still is one of the principal goals of quantum electronics. The very notion of high peak power is permanently changing, and today means a power of no less than 1 PW ($10^{15}$ W). The rapid progress of output laser power in the 1960s and 1970s was based on the principles of Q-switching and mode locking, which enabled the laser pulse duration to be reduced from microseconds to picoseconds over 40 years [70]. Further advancement in this direction was limited by the large dimensions and cost of lasers and the need to operate at the limit of radiation resistance of the optical elements.

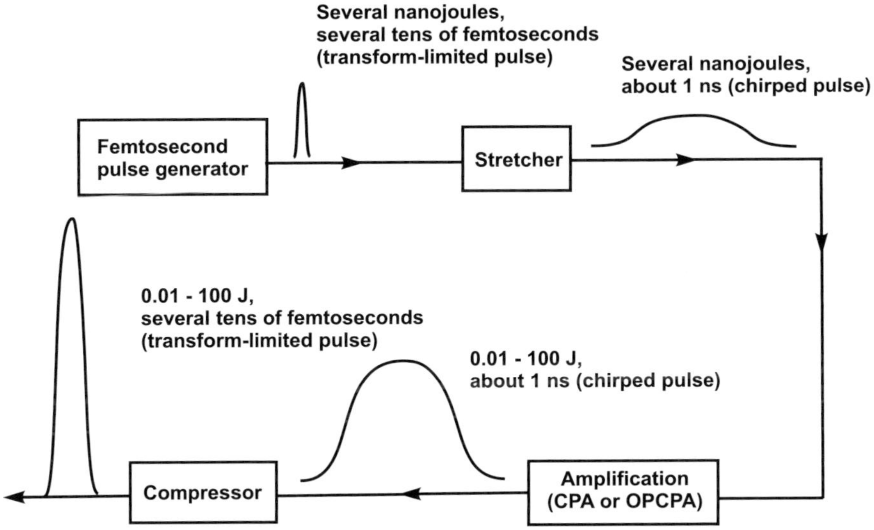


**Fig. 4.3** General schematic of a femtosecond laser (CPA: chirped pulse amplification; OPCPA: optical parametric chirped pulse amplification). Reprinted, with permission, from [45].

The present-day "renaissance" in laser physics is related to the invention of the chirped optical pulse technique (Fig. 4.2) in 1985 [90, 91, 58, 59, 17], which opened up the way for multiterawatt, petawatt, and even exawatt laser systems to raise the intensities on target to $q \approx 10^{22}$ W/cm$^2$, with the theoretical limit equal to $\approx 3 \times 10^{23}$ W/cm$^2$. In this technique [90, 91, 58, 59, 17], an initially short laser pulse is stretched in time (Fig. 4.3, [45]) on passing through dispersion elements (for instance, diffraction gratings). The pulse is decomposed into spectral components, each of which traces a slightly different path, depending on its wavelength, and the pulse stretches in time. Spectral clipping ("chirping") also occurs – the frequency continuously varies from the beginning to the end of the pulse. The stretched pulse possesses a lower intensity and is amplified in the ordinary way by the laser-active medium and arrives at the other nonlinear element for optical compression in the compressor. While conventional techniques enable the radiation to be focused by lenses in two mutually perpendicular directions, the new technique does this simultaneously in three dimensions and sharply increases the resultant intensity at the target.

The chirp principle has made it possible to raise the attainable intensity of laser radiation by 5–6 orders of magnitude and to radically lower the cost and dimensions of lasers, which have become "table-top" devices affordable by university laboratories. Furthermore, these lasers combine well with big facilities for controlled laser fusion ("fast" ignition, see Sect. 6.1.3) and charged-particle accelerators (Sect. 6.3, Fig. 5.8) and have provided a way of recording nonlinear quantum-electrodynamic effects such as pair production in vacuum [18, 7, 13] as well as intense radiation for studying photon–photon collisions [96].

The advent of this laser technology has enabled a substantial advancement along the intensity scale and hence the energy density scale, and has made it possible in laboratory research to go over from atomic and solid-state processes to studies of hot plasmas, to nuclear physics, high-energy physics, relativism, and cosmology beyond the standard model [70, 38, 77].

At the present time there are about 20 petawatt lasers, which are either in operation (see Fig. 3.2 and Table 3.3) or will soon be commissioned, and the technologies developed to date give promise that intensities of $\approx 10^{28}$ W/cm$^2$ will be attained in the future [70, 80].

Of the most interesting laser facilities, mention should be made of several petawatt laser systems [45]. First and foremost, there is the Russian project [9] executed in the Russian Federal Nuclear Center (Sarov) and in the Institute of Applied Physics of the Russian Academy of Sciences, where another parametric amplifier was made involving a DKDP [$K(H_{1-x}D_x)_2PO_4$] crystal 200 mm in diameter. One of the beams of the "Luch" facility (with an energy of 1 kJ of the second harmonic pulse and a pulse duration of 2 ns) was employed for pumping. The highest energy of the chirped pulse after the final amplifier stage was equal to about 100 J. The four-grating compressor efficiency is 68%.

Projected for 2011 is the completion of construction of a 10 PW optical parametric chirped pulse amplification (OPCPA) laser at the Rutherford Laboratory (Great Britain) [19]. Two channels of the Vulcan neodymium glass laser with an energy of 600 J each are employed to pump two final amplifiers. Advantage is taken of super-broadband phase matching in a DKDP crystal at the chirped pulse wavelength near 910 nm. A special feature of this project is a very long (3 ns) chirped pulse.

Recently, two major pan-European projects have started up: High Power Laser Energy Research (HiPER) [41] and Extreme Light Infrastructure (ELI) [26]. The

HiPER project is aimed at the investigation of controlled laser thermonuclear fusion for a relatively modest energy of target-compressing radiation: less than 0.4 MJ in the second harmonic as compared to 1.8 MJ in the third harmonic for the National Ignition Facility (NIF). This energy "economy" is achieved by using, along with nanosecond pulses, shorter (of the order of 1 ps) pulses with a power ranging from 150 to 2000 PW for the ignition of the thermonuclear target. The goal of the ELI project is to make a superhigh-power (50–1000 PW) femtosecond laser for the pursuance of unique research.

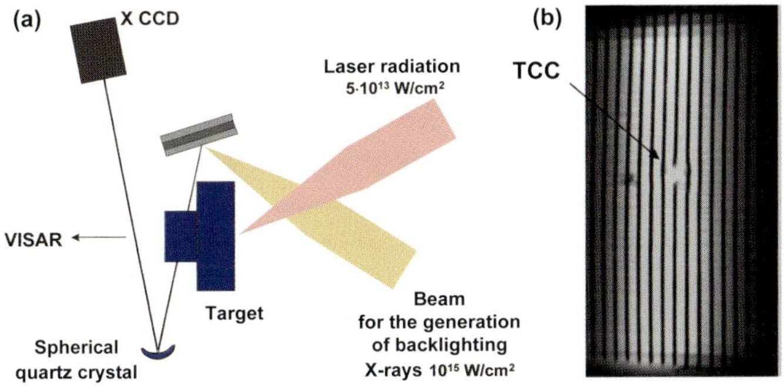


**Fig. 4.4** Experiment (a) with laser-driven shock waves [11] involving the measurement (b) of the density of shock-compressed plasma from the absorption of 5 keV X-ray radiation (VISAR: velocity interferometry system for any reflector; CCD: charge-coupled device; TCC: target chamber center).

The action of high-power laser radiation leads to new and strongly nonlinear physical phenomena in relativistic plasmas with pressures up to gigabars, with electric fields of teravolts per centimeter, and gigagauss magnetic fields [70, 80] (Fig. 4.2).

In addition to the well-known effects of self-focusing, stimulated scattering, and front steeping there emerge new effects: light filamentation, relativistic and ponderomotive effects in hydrodynamics, as well as fully developed generation of nonthermal gigavolt electrons and multimegavolt ions in laser plasmas, which result in nuclear reactions [74, 80, 70]. Here, we are dealing with extremely short – femtosecond – durations of a laser pulse, during which the electromagnetic wave executes only a few cycles.

Let us begin to ascend the curve in Fig. 4.2. Beginning with $q > 10^{14}$ W/cm$^2$ (for $\lambda = 1$ $\mu$m), the amplitude pressures of laser-driven shock waves pass into the megabar range [1, 80, 70, 49, 3], in accordance with the scaling law

$$P\,[\mathrm{TPa}] = 0.87(q\,[\mathrm{W/cm^2}])^{2/3}(\lambda\,[\mu\mathrm{m}])^{-2/3}.$$

Beginning with $q > 3.4 \times 10^{18}$ W/cm$^2$, the electric intensity $E = \sqrt{4\pi q/c}$ in the laser wave is comparable with the intensity of the electric field of the nucleus $E_a = e/a_B^2 \approx 5 \times 10^9$ V/cm in the first Bohr orbit of hydrogen. To ionize the energy level $U_i$ [eV] requires an intensity

$$q\left[\mathrm{W/cm^2}\right] = \frac{4 \times 10^9 U_i^4\,[\mathrm{eV}]}{Z^2}.$$

Under these conditions, laser radiation ionizes the medium, which turns into a heated plasma. Interesting experiments in the generation of intense shock waves by laser radiation [1, 11] (Fig. 4.4) and in the production of fast charged particles in laser plasmas [7, 80, 22, 61, 55, 81] (Fig. 4.5) can be conducted in this parameter domain. Beginning with roughly the same intensities $q > 10^{17}$ W/cm$^2$, an appreciable number of nonthermal electrons and ions of the multimegaelectronvolt range are generated in the absorption region [22, 61, 55, 81]. Beginning with $10^{18}$ W/cm$^2$, the ponderomotive light pressure is comparable with the hydrodynamic plasma pressure [74, 49, 3, 80, 70].

Relativistic effects become significant when the kinetic energy of an electron accelerated in the laser wave field is of the order of its rest energy $m_e c^2$, which leads to the condition

$$q_{\mathrm{pe}}\lambda^2 \approx 1.37 \times 10^{18}\,\frac{\mathrm{W}\,\mu\mathrm{m}^2}{\mathrm{cm}^2}.$$

For $\lambda = 1\ \mu$m this gives $10^{18}$ W/cm$^2$.

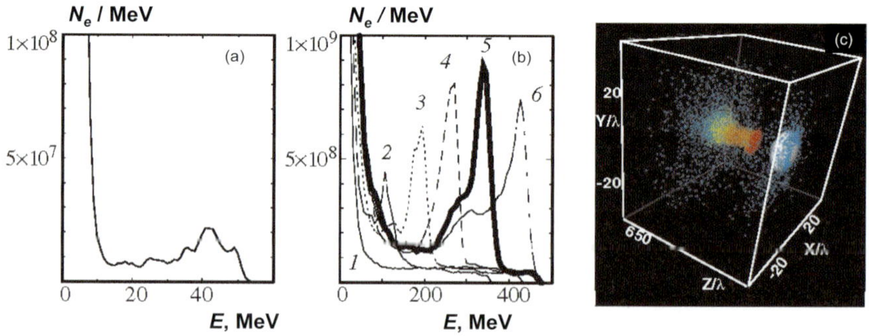


**Fig. 4.5** Energy spectrum of electrons [80, 81] accelerated by laser radiation: a) 20 mJ, 6.6 fs. b) 12 J, 33 fs. Temporal evolution of the spectrum: 1–6) $ct/\lambda = 350, 450, 550, 650, 750, 850$, respectively. c) Propagation of the laser pulse 12 J, 33 fs, $z/\lambda = 690$ through a plasma with a density of $10^{19}$ cm$^{-3}$. Three-dimensional picture of nonthermal electron energy distribution for $q \approx 10^{19}$ W/cm$^2$. The electrons that differ by 10 keV above 10 MeV are indicated by different colors.

Therefore, for the first time it has been possible to obtain microscopic quantities of matter with relativistic energies in terrestrial conditions [80, 70], with relativistic electron mass on the order of 100 times the mass of the rest energy.

The proton motion will become relativistic at intensities

$$q_{\mathrm{pp}} = \left(\frac{M_{\mathrm{p}}}{m_{\mathrm{e}}}\right)^2 q_{\mathrm{pe}} \approx 5 \times 10^{24}\ \mathrm{W/cm^2},$$

which will hopefully be realized in experiments soon.

The transition to "relativistic" intensities of laser radiation [70] has already brought several interesting physical results [69]. The case in point is the generation of X-ray [8] and gamma-ray [76] radiation, relativistic self-focusing (Fig. 4.6) [85], high-order harmonic generation [16, 47], acceleration of electrons [7, 80, 64], protons [60], and ions [22, 61, 55, 28], neutron [25] and positron [34] generation, electron vortex and soliton generation [29, 70, 80], multimegagauss magnetic field generation [68], as well as the manifestations of quantum electrodynamics [18, 7] (Fig. 4.6).

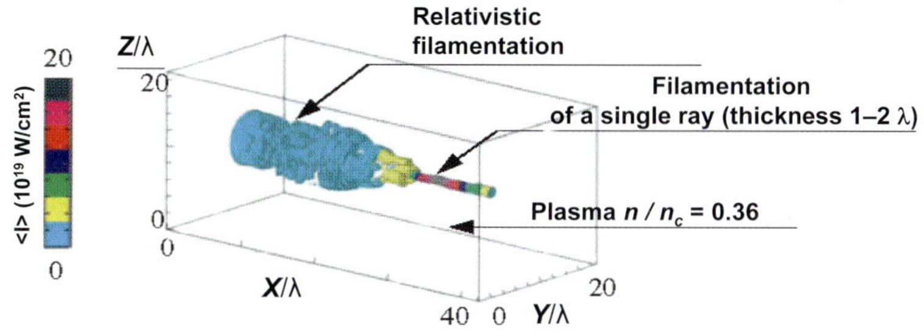


**Fig. 4.6** Self-focusing of a laser pulse [80].

Under high-intensity irradiation there occurs relativistic plasma "transparentization" [80, 70, 74], which is related to the relativistic growth of the electron mass and the corresponding lowering of the critical plasma frequency $\omega_{\mathrm{P}} = \sqrt{4\pi e^2 n_{\mathrm{e}}/\gamma m}$ (where $\gamma$ is the relativistic Lorentz factorand $n_{\mathrm{e}}$ the number density of electrons), the plasma density modification by ponderomotive forces, as well as the frequency transformation of a laser pulse itself [34].

Along with the effect of relativistic plasma "transparentization", of considerable interest is the effect of relativistic self-focusing of laser radiation [70] caused by plasma permittivity variation due to the relativistic growth of the electron mass in the transverse direction relative to the beam propagation direction and the spatial plasma density redistribution under the action of ponderomotive forces. The critical power for self-focusing was derived by Norreys et al. [76] and Sarkisov et al. [85]:

$$W_{\mathrm{cr}} = \frac{m_{\mathrm{e}} c^5 \omega^2}{e^2 \omega_{\mathrm{pe}}^2} \approx 17 \left(\frac{\omega}{\omega_{\mathrm{pe}}}\right)^2\ \mathrm{GW}.$$

Experiments carried out to observe this effect are reported in [16, 64]. Particle-in-cell simulations [80] of laser beam filamentation are presented in Fig. 4.7. This multiray structure of the laser pulse occurs due to the filamentation instability and the buildup of transverse beam nonuniformity, which resembles the effect of self-focusing. The self-focusing and filamentation of laser radiation in plasma result in the formation of electron vortices [60], high-intensity compensative currents, and the generation of high magnetic fields. Measurements [25, 95, 65] for an intensity of $9 \times 10^{19}$ W/cm$^2$ yielded a magnetic strength of 340 MGs, which is an order of magnitude higher than the record fields obtained employing explosion magnetic generators [12, 79, 39, 31].

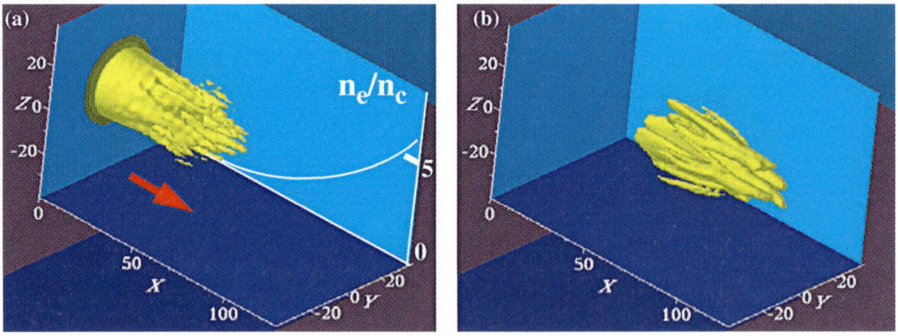


**Fig. 4.7** Multiple filamentation of a wide laser beam of petawatt power [80]. There emerge several channels, which propagate quasi-independently due to their plasma screening.

Superhigh magnetic field generation in laser-produced plasma is among the interesting and vigorously advancing lines of research (for a review see [10], which we shall follow in the subsequent discussion).

Different mechanisms of magnetic field generation in the interaction of high-intensity laser radiation with solid targets give rise to magnetic fields with induction up to 1 GGs in the resultant dense plasma. These fields are localized near the critical surface, where the bulk of the laser energy is absorbed. Several main mechanisms of quasi-static magnetic field generation were put forward: (i) different directions of the plasma temperature and density gradients; (ii) the flow of fast electrons accelerated by longitudinal ponderomotive forces transverse to the laser pulse direction; (iii) the Weibel instability in a collisionless plasma.

Sudan [92] predicted the generation of magnetic fields with an induction up to $\approx$ 1 GGs in a relativistic dense plasma. According to the theory outlined in that paper, the ponderomotive force acting on the electrons is the source of the quasi-stationary magnetic field. This force gives rise to the electron current directed from the laser beam axis to the periphery, until there sets in a cooperative oscillatory motion of ions and electrons caused by the electroneutrality requirement.

The Weibel instability mechanism is related to the anisotropy of electron velocities in the plasma. This anisotropy emerges in the course of ionization of atoms and atomic ions by a superstrong laser field. The majority of electrons escape in the direction of the electric vector of the linearly polarized laser wave. A substantially lower number of electrons escape in the transverse directions. Both the longitudinal and transverse velocities obey the energy–time uncertainty relations. Weibel [104] was the first to show that in the Maxwell equations in the presence of electron current anisotropy there emerges instability relative to the spontaneous buildup of the quasi-static magnetic field.

The thermoelectric mechanism [10], unlike the previous one, is realized in collisional plasmas which exhibit electron density and temperature gradients, these gradients forming an angle with each other. The density gradient is aligned with the radius of the electron beam. It arises from the nonuniformity of laser radiation intensity over the area of the focal spot. As a result, there are substantially more electrons on the axis of the laser beam than at the beam periphery, owing to a strong difference in the degree of ionization of the atoms of the medium. The temperature gradient is evidently directed along the normal to the target surface. The increment of spontaneous magnetic field growth is proportional both to the temperature gradient and to the velocity gradient. In this case, the magnetic field possesses toroidal symmetry: its annular lines of force encompass the laser beam.

In the passage of a laser pulse of relativistic intensity, the magnetic part of the Lorentz force accelerates the plasma electrons along the direction of laser pulse propagation. This gives rise to a magnetic field, which is also inherently annular.

There exist even more sophisticated methods of magnetic field generation in laser plasma, some of them are discussed in [10].

Najmudin et al. [71] measured the magnetic field generated in subcritical laser plasmas. A 30-fs long laser pulse at an irradiation intensity of $4.2 \times 10^{18}$ W/cm$^2$ was shown to give rise to an azimuthal magnetic field with an induction of 2–8 MGs on a scale length of 200 $\mu$m; for an intensity of $8 \times 10^{18}$ W/cm$^2$, the field amounted to 7 MGs. Both magnetic fields existed for several picoseconds. Their formation is attributed to fast electron streams generated in the laser plasma.

The measurements made by Tatarakis et al. [94] allow one to determine the range of magnetic fields generated on the critical surface in a dense plasma, which corresponds to magnitudes of 340–460 MGs (Fig. 4.8). The magnetic pressure of these fields exceeds $10^9$ atm.

Wagner et al. [103] reported experiments involving solid-state target irradiation by a laser pulse with an intensity of about $10^{20}$ W/cm$^2$. By means of polarization measurements of the yield of radiation harmonics, they found that in the laser plasma there emerged magnetic fields with an induction of about 700 MGs. These magnetic fields exist in layers, which are much thicker than the skin layer. This is attributable to the ultrarelativistic electron motion in so strong a laser field as well as to the deep penetration of high-order harmonics of laser radiation into the target material. Eliezer et al. [27] estimated the magnetic field in laser experiments at $\approx 1$ GGs.

Similar interesting effects of relativistic nature [3, 80] are related to strongly nonlinear plasma waves, which form vacuum channels and "bubbles" (Fig. 4.9 [80]) in the plasma, produce plasma lenses for charged particles, and give rise to intense

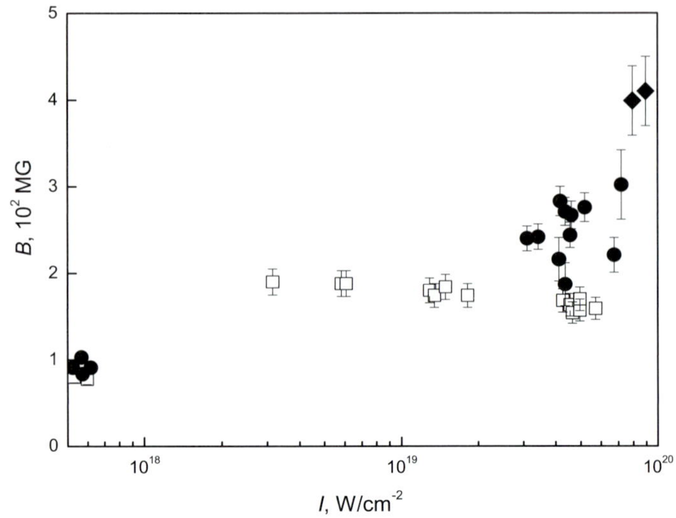


**Fig. 4.8** Magnetic induction as a function of laser radiation intensity [94], which was measured from harmonics $5\omega$ (♦), $4\omega$ (•) and $3\,\omega$(□).

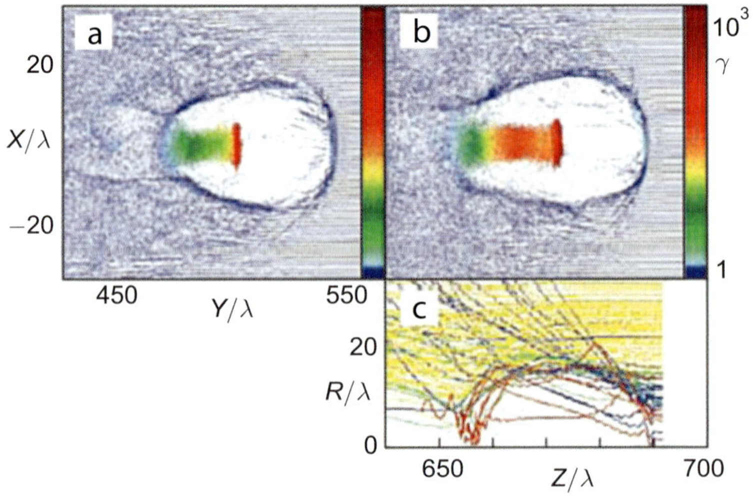


**Fig. 4.9** Soliton "bubbles" in plasma under the irradiation of a 33-fs laser pulse with an energy of 12 J: a) $ct/\lambda = 500$; b) $ct/\lambda = 700$. c) Electron trajectories in the frame of reference co-moving with the laser pulse. The trajectories of electrons initially located at different distances from the axis are indicated in color [80].

electromagnetic radiation in the frequency range from terahertz to X-rays, as well as excite collisionless shock waves [70].

The propagation of two collinear laser beams of relativistic intensity through a plasma [14] leads to the generation of intense electromagnetic wake waves. These wake waves enable electrons to be accelerated (Fig. 4.5) with acceleration rates up to 100 MV/cm, which is thousands of times greater than ordinary acceleration gradients of $\approx 5$ kV/cm (see Sect. 6.3). Gigantic longitudinal electric fields are generated in this case [7]: for an intensity $q \approx 10^{18}$ W/cm$^2$ the electric intensity is equal to $\approx 2$ TV/m, for $10^{23}$ W/cm$^2$ it amounts to $\approx 0.1$ PV/m. These acceleration rates imply that the laser version of the ordinary 50 GeV SLAC accelerator would be only 100 $\mu$m long [7]. Successful experiments have been carried out on laser-driven acceleration of electrons to energies of 10–170 MeV at laser radiation intensities of $10^{18}$–$10^{19}$ W/cm$^2$ [72, 63]. There are reasons to believe that the advent of multi-petawatt and exawatt lasers in the future would lead to the implementation of acceleration rates of the order of a teravolt per centimeter for a total kinetic energy ranging into the gigavolts – for more details, see Sect. 6.3.

Among other interesting manifestations of nonlinearities in laser plasma, we mention the nonlinear steepening of the optical front (similar to the formation of a shock wave in hydrodynamics [15]), plasma jet formation [43, 47], and high-order harmonic generation [16, 47], which is of practical interest for lithography, holography, medicine, etc.

Under the action of a circularly polarized electric field, the plasma electrons are set in motion to generate synchrotron radiation [37, 70], relativistic effects being significant in its description at high intensities. For $\lambda = 1$ $\mu$m this radiation-dominant regime is realized, beginning at intensities $q \approx 3 \times 10^{23}$ W/cm$^2$, when a substantial fraction of laser energy is radiated in the form of hard X-rays. Quantum optical effects come into play at $q \approx 1.4 \times 10^{26}$ W/cm$^2$ [7], the kinetic electron energy being equal to $\approx 50$ TeV in this case. For $q \approx 10^{21}$ W/cm$^2$, the light pressure is equal to $\approx 300$ Gbar, which is close to the pressure at the center of the Sun and is much higher than the pressure in the near-source zone of a nuclear explosion [5, 102, 4, 98].

High-intensity lasers make it possible to achieve superstrong acceleration $a_e = a_0 \omega \cdot c \approx 10^{30}$g (for a dimensionless radiation amplitude $a_0 = eA/m_e c \approx 10^5$), which is close to the accelerations in the vicinity of the Schwarzschild radius of a black hole [106, 44] – see Sect. 7.2. This permits the conditions in the neighborhood of black holes and wormholes to be modeled, and therefore the predictions of general relativity theory to be verified.

In particular, for $q \approx 10^{26}$ W/cm$^2$ the electron acceleration amounts to $a = 10^{27} g$, which is close to the conditions of the black-hole event horizon [74, 38, 106, 101]. If such acceleration is realized, an opportunity is expected to open up to study the specific electromagnetic Unruh radiation [100, 86, 20], which is similar to the Hawking radiation caused by gravitational effects. In this case, additional (in comparison with calculations by Maxwell's equations) electromagnetic radiation with an effective temperature $kT = \hbar a/c$ is bound to occur. The ratio between the power of this radiation and the synchrotron radiation power is equal to $10^{-6}$ at an intensity

$q \approx 10^{18}$ W/cm$^2$ and increases proportionally with intensity, which raises expectations that it would be possible to observe it at high intensities of laser radiation.

At higher laser radiation intensities $q \approx 3 \times 10^{29}$ W/cm$^2$ there is a good chance to verify the predictions of modern quantum gravity theories [38, 83, 2, 36, 82] about the change of space–time dimensionality at short distances. According to Arkani-Hamed et al. [2], this distance $r_n \approx 10^{32/n-17}$ cm, where $n$ is the number of dimensions greater than 4. In this case, the electron wave function will be reflective of a different law of gravitation for $n$ up to 3 at distances less than $10^{-6}$ cm.

The effects of quantum electrodynamics, polarization, breakdown of vacuum, and spontaneous electron–positron pair production, and then of the emergence of quark–gluon plasma become significant at high optical radiation intensities $q \approx 3 \times 10^{29}$ W/cm$^2$.

The problem of spontaneous electron–positron pair production in vacuum touches upon many interesting situations, such as the collisions of heavy nuclei (with $Z_1 + Z_2 > 135$) [107], evaporation of black holes [40], and particle production in the universe [78]. The characteristic electric field scale [74] for the manifestation of the breakdown effect in quantum electrodynamics is the Schwinger intensity

$$E_{\mathrm{S}} = \frac{m^2c^3}{e\hbar} \approx 10^{16}\ \mathrm{V/cm},$$

which is sufficient to accelerate an electron to relativistic velocities over the Compton wavelength $\lambda_{\mathrm{C}} = 2\pi\hbar/(mc)$ and corresponds to ultrahigh intensities of laser radiation

$$q_{\mathrm{QED}} = q_{\mathrm{pe}}\frac{\lambda^2}{\lambda_{\mathrm{C}}^2} \approx 8.1 \times 10^{30}\ \mathrm{W/cm^2}.$$

To realize these ultraextreme conditions, the focusing of $\lambda \approx 1\ \mu$m laser radiation in a 1 mm$^3$ volume requires the release of $\approx 1$ MJ of energy! This intensity threshold for electron–positron pair production is substantially lowered (to $10^{22}$ W/cm$^2$) in the case of scattering by nuclei [89, 67, 53]. The composition of two counterpropagating laser beams yields a critical intensity of $\approx 10^{26}$ W/cm$^2$ for pair production, which is two orders of magnitude below the limit for a single beam [73].

Thoma [97] analyzed the feasibility of producing an ultrarelativistic $kT \gg m_{\mathrm{e}}c^2$ electron–positron plasma with a temperature of $\approx 10$ MeV by way of double-sided irradiation of a gold foil by 330-fs long laser pulses at an intensity of $\approx 7 \times 10^{21}$ W/cm$^2$. He took advantage of quantum electrodynamics (QED) techniques and analogies with quark–gluon plasma.

The high energy densities supplied to laser plasmas turn out to be sufficient not only for diverse electronic processes, but also for nuclear transformations such as activation, fission and fusion, and transmutation [7, 35, 87]. Apart from low-energy nuclear reactions caused by gamma-ray, electron, and proton radiation fluxes, which are generated by plasma fields, more sophisticated mechanisms of nuclear transformations are also possible. For instance, at intensities $\approx 10^{17}Z^6$ W/cm$^2$ there occurs a $Z$-fold ionization of atoms, which destabilizes nuclei of charge $Z$. Thus knocking out inner-shell electrons leads to disintegration of heavy elements [42, 88].

One of the directions in this research [23, 24, 35, 88] involves harnessing laser-accelerated electrons with energies above 200 MeV that decelerate in heavy targets to generate multimegavolt gamma-ray bremsstrahlung, which gives rise to photonuclear reactions. The photonuclear fission of uranium was successfully demonstrated and theoretically substantiated in [51, 99, 87, 62]. Recently Magill et al. [56] realized (as a result of the $(\gamma, \mathrm{n})$-reaction) the laser-induced transmutation of $^{129}$I – the main radioactive waste component with a half-life of 15.7 million years – into $^{128}$I with a half-life of 25 min. Of considerable importance in experiments of this kind is the mechanism of the reactions of electron–positron pair production in the Coulomb field of heavy nuclei, the cross sections for these reactions being substantially (two orders of magnitude) greater than the cross section for the photonuclear reactions of $^{238}\mathrm{U}(\gamma, \mathrm{f})$.

Among the possible uses of laser-induced photonuclear reactions, mention should be made [35] of the production of radioactive isotopes for medicine, the transmutation of long-lived isotopes, determination of nuclear constants, and radioactive material detection for antiterrorist purposes.

Owing to a high ( $\approx 12\%$ [51]) efficiency of laser radiation energy transformation to high-energy protons they may be used for the development of pulsed neutron sources by means of the nuclear reactions $(\mathrm{p}, x\mathrm{n})$ and $(\mathrm{p}, \mathrm{f})$. In any case their intensity may [35] far exceed the intensity ($10^5$ neutrons per joule) of Coulomb explosions of deuterium clusters. In the experiments reported in [66, 105], about $1.2 \times 10^9$ bismuth atoms were recorded, which requires via the $(\mathrm{p}, x\mathrm{n})$-reaction more than $2 \times 10^9$ neutrons arising from the $(\mathrm{p}, x\mathrm{n})$ reaction. In this case, the neutron yield per joule of laser energy is equal to $\approx 5 \times 10^6$. Estimates for the conditions of the Vulcan laser system made in [75] predict a neutron yield of $2 \times 10^9$ for the $\mathrm{Zn}(\mathrm{p}, x\mathrm{n})$ reaction and of order $10^8$–$10^{10}$ for the $\mathrm{Li}(\mathrm{p}, \mathrm{n})$ reaction. We emphasize that the currently available compact neutron generators or spontaneous fission sources exhibit characteristic intensities on the order of $10^8$–$10^{10}$ neutron/s and $10^{10}$ neutron/(s cm$^2$), respectively, while stationary and pulsed reactors possess intensities of $10^7$–$10^{13}$ neutron/s and $10^{14}$ neutron/(s cm$^2$).

As noted in the foregoing, in the field of high-intensity laser radiation there occurs production of positrons, which annihilate with target electrons to emit two photons with an energy of $\approx 0.511$ MeV. An analysis of the spectra of this radiation yields valuable information about the electronic properties of the target. Therefore positron sources are presently employed in annihilation positron spectroscopy in materials science, in positron spectroscopy in basic research, etc. Low-energy positrons find use in electron–positron plasma research, atomic and molecular physics, antihydrogen production, simulations of astrophysical phenomena, and materials science.

Estimates and experiments in positron production in laser plasmas made to date [35] suggest that the number of positrons per plasma electron may amount to $10^{-4}$–$10^{-2}$ for a total number on the order of $10^9$ positrons per laser pulse. For comparison, a linear electron accelerator yields on the order of $10^8$ positrons per second, which in turn is 1000 times the yield of a radioisotope source. Laser sources of positrons therefore turn out to be comparable to the traditional ones when the pulse repetition rate is increased to 10 Hz or higher.

Electrons and ions accelerated in laser plasmas may be employed to produce isotopes by means of $(\gamma, \mathrm{n})$ and $(\mathrm{p}, \mathrm{n})$ nuclear reactions. Such devices are more compact than the existing cyclotrons and are located directly in clinics, for instance, for positron-emission tomography [32, 50, 21, 52]. In particular, in the Lawrence Berkeley National Laboratory a laser-plasma electron accelerator on the basis of a 10-TW solid-state laser was employed to produce the $^{62}$Cu and $^{61}$Cu isotopes in the $(\gamma, \mathrm{n})$ reaction of $^{63}$Cu. Considerable recent attention in the USA, Europe, and Japan has been paid to the idea of developing laser-plasma accelerators of protons with energies on the order of 200 MeV for different medical applications.

Proton tomography shows considerable promise for the diagnostics of fast processes in shock-wave physics and defense engineering, making it possible to obtain high-contrast pictures with a high temporal resolution [46]. Along with proton beams, the high-intensity 20–100 keV X-ray radiation of laser plasmas has been validly used for the X-ray diagnostics of pulsed plasma processes (Fig. 4.4) [11]. An important step in this area was made by French scientists [11], who accomplished a direct X-ray measurement of the density of a shock-compressed plasma in experiments involving intense laser-driven shock waves.

Electron–positron pair production under laser irradiation was reported in [18, 7, 89, 67, 53, 73, 24], where observations were made of other interesting effects as well, such as multiphoton interference and polarization of the vacuum. Nonlinear Compton scattering and pair production were recorded in the interaction of 46.6- and 49.1-GeV electrons accelerated by SLAC accelerators with laser radiation, $q \approx 5 \times 10^{18}$ W/cm$^2$ [74].

Tajima [93] proposed employing a high-energy (8 GeV) electron accelerator (SPring-8) and high-power lasers for experiments in QED by generating high-intensity $\gamma$-ray radiation for photonuclear reactions. In this case, a study can be made of the interplay of strong and weak interactions. Combining accelerators with lasers may hold much promise for the development of new-generation optical sources, such as femtosecond synchrotrons and coherent X-ray radiation sources [74, 33].

Further advancement along the laser intensity scale (Fig. 4.2) is hard to predict, for it is limited by our knowledge of the structure of matter in the immediate spatiotemporal neighborhood of the big bang at ultrahigh energy densities.

## References

[1] Anisimov, S.I., Prokhorov, A.M., Fortov, V.E.: Application of high-power lasers to study matter at ultrahigh pressures. Sov. Phys. – Usp. **27**(3), 181–205 (1984). DOI 10.1070/PU1984v027n03ABEH004036. URL `http://stacks.iop.org/0038-5670/27/181`

[2] Arkani-Hamed, N., Dimopoulos, S., Dvali, G.: Phenomenology, astrophysics, and cosmology of theories with submillimeter dimensions and TeV scale

quantum gravity. Phys. Rev. D **59**(8), 086004 (1999). DOI 10.1103/PhysRevD.59.086004. URL http://link.aps.org/abstract/PRD/v59/e086004

[3] Atzeni, S., Meyer-ter-Vehn, J.: The Physics of Inertial Fusion. Oxford University Press, Oxford (2004)

[4] Avrorin, E.N., Simonenko, V.A., Shibarshov, L.I.: Physics research during nuclear explosions. Phys. Usp. **49**(4), 432 (2006). DOI 10.1070/PU2006v049n04ABEH005958. URL http://ufn.ru/en/articles/2006/4/j/

[5] Avrorin, E.N., Vodolaga, B.K., Simonenko, V.A., Fortov, V.E.: Intense shock waves and extreme states of matter. Phys. Usp. **36**(5), 337–364 (1993). DOI 10.1070/PU1993v036n05ABEH002158. URL http://stacks.iop.org/1063-7869/36/337

[6] Bahk, S.W., Rousseau, P., Planchon, T.A., et al.: Generation and characterization of the highest laser intensities ($10^{22}$ W/cm$^2$). Opt. Lett. **29**(24), 2837–2839 (2004). DOI 10.1364/OL.29.002837. URL http://www.opticsinfobase.org/abstract.cfm?URI=ol-29-24-2837

[7] Bamber, C., Boege, S.J., Koffas, T., et al.: Studies of nonlinear QED in collisions of 46.6 GeV electrons with intense laser pulses. Phys. Rev. D **60**(9), 092004 (1999). DOI 10.1103/PhysRevD.60.092004. URL http://link.aps.org/abstract/PRD/v60/e092004

[8] Beg, F.N., Bell, A.R., Dangor, A.E., et al.: A study of picosecond laser–solid interactions up to $10^{19}$ W cm$^{-2}$. Phys. Plasmas **4**(2), 447–457 (1997). DOI 0.1063/1.872103

[9] Belov, I.A., et al.: In: Int. conf. "X Kharitonov's thematic scientific readings", p. 145. RPhNZ-VNIIEPh, Sarov (2008)

[10] Belyaev, V.S., Krainov, V.P., Lisitsa, V.S., Matafonov, A.P.: Generation of fast charged particles and superstrong magnetic fields in the interaction of ultrashort high-intensity laser pulses with solid targets. Phys. Usp. **51**(8), 793 (2008). DOI 10.1070/PU2008v051n08ABEH006541. URL http://ufn.ru/en/articles/2008/8/b/

[11] Benuzzi-Mounaix, A., Koenig, M., Ravasio, A., et al.: Laser-driven shock waves for the study of extreme matter states. Plasma Phys. Control. Fusion **48**(12B), B347–B358 (2006). DOI 10.1088/0741-3335/48/12B/S32

[12] Boyko, B.A., Bykov, A.I., et al.: More than 20 MG magnetic field generation in the cascade magnetocumulative MC-1 generator. In: H.J. Schneider-Muntau (ed.) Megagauss Magnetic Field Generation, Its Application to Science and Ultra-High Pulsed-Power Technology. Proc. VIIIth Int. Conf. Megagauss Magnetic Field Generation and Related Topics, p. 61. World Scientific, Singapore (2004)

[13] Bula, C., McDonald, K.T., Prebys, E.J., et al.: Observation of nonlinear effects in Compton scattering. Phys. Rev. Lett. **76**(17), 3116–3119 (1996). DOI 10.1103/PhysRevLett.76.3116. URL http://link.aps.org/abstract/PRL/v76/p3116

[14] Bulanov, S.V.: New epoch in the charged particle acceleration by relativistically intense laser radiation. Plasma Phys. Control. Fusion **48**(12B), B29–B37 (2006). DOI 10.1088/0741-3335/48/12B/S03

[15] Bulanov, S.V., Inovenkov, I.N., Kirsanov, V.I., et al.: Nonlinear depletion of ultrashort and relativistically strong laser pulses in an underdense plasma. Phys. Fluids B **4**(7), 1935–1942 (1992). DOI 10.1063/1.860046. URL `http://dx.doi.org/10.1063/1.860046`

[16] Bulanov, S.V., Naumova, N.M., Pegoraro, F.: Interaction of an ultrashort, relativistically strong laser pulse with an overdense plasma. Phys. Plasmas **1**(3), 745–757 (1994). DOI 10.1063/1.870766

[17] Bunkenberg, J., Boles, J., Brown, D., et al.: The omega high-power phosphate-glass system: design and performance. IEEE J. Quantum Electron. **17**(9), 1620–1628 (1981)

[18] Burke, D.L., Field, R.C., Horton-Smith, G., et al.: Positron production in multiphoton light-by-light scattering. Phys. Rev. Lett. **79**(9), 1626–1629 (1997). DOI 10.1103/PhysRevLett.79.1626. URL `http://link.aps.org/abstract/PRL/v79/p1626`

[19] Checkhlov, O., Divall, E.J., Ertel, K., et al.: Development of petawatt laser amplification systems at the central laser facility. Proc. SPIE **6735**(1), 67350J (2007). DOI 10.1117/12.753222. URL `http://link.aip.org/link/?PSI/6735/67350J/1`

[20] Chen, P., Tajima, T.: Testing Unruh radiation with ultraintense lasers. Phys. Rev. Lett. **83**(2), 256–259 (1999). DOI 10.1103/PhysRevLett.83.256. URL `http://link.aps.org/abstract/PRL/v83/p256`

[21] Chiu, C., Fomytskyi, M., Grigsby, F., et al.: Laser electron accelerators for radiation medicine: A feasibility study. Med. Phys. **31**(7), 2042–2052 (2004). DOI 10.1118/1.1739301. URL `http://dx.doi.org/10.1118/1.1739301`

[22] Clark, E.L., Krushelnick, K., Davies, J.R., et al.: Measurements of energetic proton transport through magnetized plasma from intense laser interactions with solids. Phys. Rev. Lett. **84**(4), 670–673 (2000). DOI 10.1103/PhysRevLett.84.670. URL `http://link.aps.org/abstract/PRL/v84/p670`

[23] Cowan, T.E., Hunt, A.W., Johnson, J., et al.: High energy electrons, positrons and photonuclear reactions in petawatt laser–solid experiments. In: H.B. I. Tajima K. Mima (ed.) High Field Science, p. 145. Kluwer/Plenum, New York (2000)

[24] Cowan, T.E., Hunt, A.W., Phillips, T.W., et al.: Photonuclear fission from high energy electrons from ultraintense laser-solid interactions. Phys. Rev. Lett. **84**(5), 903–906 (2000). DOI 10.1103/PhysRevLett.84.903. URL `http://link.aps.org/abstract/PRL/v84/p903`

[25] Disdier, L., Garconnet, J.P., Malka, G., Miquel, J.L.: Fast neutron emission from a high-energy ion beam produced by a high-intensity subpicosecond laser pulse. Phys. Rev. Lett. **82**(7), 1454–1457 (1999). DOI 10.1103/

PhysRevLett.82.1454. URL http://link.aps.org/abstract/PRL/v82/p1454

[26] ELI: The Extreme Light Infrastructure European Project: ELI homepage. URL http://www.extreme-light-infrastructure.eu/

[27] Eliezer, S., Mendonca, J.T., Bingham, R., Norreys, P.: A new diagnostic for very high magnetic fields in expanding plasmas. Phys. Lett. A **336**(4-5), 390–395 (2005). DOI doi:10.1016/j.physleta.2005.01.040

[28] Esirkepov, T., Yamagiwa, M., Tajima, T.: Laser ion-acceleration scaling laws seen in multiparametric particle-in-cell simulations. Phys. Rev. Lett. **96**(10), 105001 (2006). DOI 10.1103/PhysRevLett.96.105001. URL http://link.aps.org/abstract/PRL/v96/e105001

[29] Esirkepov, T.Z., Bulanov, S.V., Nishihara, K., et al.: Proposed double-layer target for the generation of high-quality laser-accelerated ion beams. Phys. Rev. Lett. **89**(17), 175003 (2002). DOI 10.1103/PhysRevLett.89.175003. URL http://link.aps.org/abstract/PRL/v89/e175003

[30] Fortov, V.E.: Intense Shock Waves and Extreme States of Matter. Bukos, Moscow (2005)

[31] Fortov, V.E. (ed.): Explosive-Driven Generators of Powerful Electrical Current Pulses. Cambridge International Science, Cambridge (2007)

[32] Fortov, V.E., Hoffmann, D.H.H., Sharkov, B.Y.: Intense ion beams for generating extreme states of matter. Phys. Usp. **51**(2), 109 (2008). DOI 10.1070/PU2008v051n02ABEH006420. URL http://ufn.ru/en/articles/2008/2/a/

[33] Fujiwara, M., Kawase, K., Titov, A.T.: Parity non-conservation measurements with photons at SPring-8. AIP Conf. Proc. **802**(1), 246–249 (2005). DOI 10.1063/1.2140661. URL http://link.aip.org/link/?APC/802/246/1

[34] Gahn, C., Tsakiris, G.D., Pretzler, G., et al.: Generating positrons with femtosecond-laser pulses. Appl. Phys. Lett. **77**(17), 2662–2664 (2000). DOI 10.1063/1.1319526. URL http://link.aip.org/link/?APPLAB/77/2662/1

[35] Galy, J., Maucec, M., Hamilton, D.J., et al.: Bremsstrahlung production with high-intensity laser matter interactions and applications. New J. Phys. **9**(2), 23 (2007). DOI 10.1088/1367-2630/9/2/023

[36] Giddings, S.B., Thomas, S.: High energy colliders as black hole factories: The end of short distance physics. Phys. Rev. D **65**(5), 056010 (2002). DOI 10.1103/PhysRevD.65.056010. URL http://link.aps.org/abstract/PRD/v65/e056010

[37] Ginzburg, V.L.: Applications of Electrodynamics in Theoretical Physics and Astrophysics. Gordon and Breach, New York (1989)

[38] Ginzburg, V.L.: The Physics of a Lifetime: Reflections on the Problems and Personalities of 20th Century Physics. Springer, Berlin, Heidelberg (2001)

[39] Hawke, P.S., Burgess, T.J., Duerre, D.E., et al.: Observation of electrical conductivity of isentropically compressed hydrogen at megabar pressures.

Phys. Rev. Lett. **41**(14), 994–997 (1978). URL http://link.aps.org/abstract/PRL/v41/p994
[40] Hawking, S.W.: Particle creation by black holes. Commun. Math. Phys. **43**(3), 199–220 (1975). DOI 10.1007/BF02345020
[41] HiPER: High Power Laser Energy Research Project: HiPER homepage. URL http://www.hiper-laser.org/
[42] Jung, I.D., Kartner, F.X., Matuschek, N., et al.: Self-starting 6.5-fs pulses from a Ti:sapphire laser. Opt. Lett. **22**(13), 1009–1011 (1997). DOI 10.1364/OL.22.001009. URL http://www.opticsinfobase.org/abstract.cfm?URI=ol-22-13-1009
[43] Kando, M., Nakajima, K., Arinaga, M., et al.: Interaction of terawatt laser with plasma. J. Nucl. Mater. **248**(1), 405–407 (1997). DOI 10.1016/S0022-3115(97)00177-3
[44] Kanel, G.I., Rasorenov, S.V., Fortov, V.E.: Shock-Wave Phenomena and the Properties of Condensed Matter. Springer, New York (2004)
[45] Khazanov, E.A., Sergeev, A.M.: Petawatt laser based on optical parametric amplifiers: their state and prospects. Phys. Usp. **51**(9), 969 (2008). DOI 10.1070/PU2008v051n09ABEH006612. URL http://ufn.ru/en/articles/2008/9/h/
[46] King, N.S.P., Ables, E., Adams, K., et al.: An 800-MeV proton radiography facility for dynamic experiments. Nucl. Instrum. Meth. Phys. Res. A **424**(1), 84–91 (1999). DOI 10.1016/S0168-9002(98)01241-8
[47] Kodama, R., Tanaka, K.A., Sentoku, Y., et al.: Observation of ultrahigh gradient electron acceleration by a self-modulated intense short laser pulse. Phys. Rev. Lett. **84**(4), 674–677 (2000). DOI 10.1103/PhysRevLett.84.674. URL http://link.aps.org/abstract/PRL/v84/p674
[48] Konyukhov, A.V., Likhachev, A.P., Oparin, A.M., et al.: Numerical modeling of shock-wave instability in thermodynamically nonideal media. JETP **98**(4), 811–819 (2004). DOI 10.1134/1.1757680
[49] Kruer, W.L.: The Physics of Laser Plasma Interactions. Addison-Wesley, Reading, MA (1988)
[50] Ledingham, K.W.D., McKenna, P., Singhal, R.P.: Applications for nuclear phenomena generated by ultra-intense lasers. Science **300**(5622), 1107–1111 (2003). DOI 10.1126/science.1080552. URL http://www.sciencemag.org/cgi/content/abstract/300/5622/1107
[51] Ledingham, K.W.D., Spencer, I., McCanny, T., et al.: Photonuclear physics when a multiterawatt laser pulse interacts with solid targets. Phys. Rev. Lett. **84**(5), 899–902 (2000). DOI 10.1103/PhysRevLett.84.899. URL http://link.aps.org/abstract/PRL/v84/p899
[52] Leemans, W.P., Rodgers, D., Catravas, P.E., et al.: Interaction of an ultrashort, relativistically strong laser pulse with an overdense plasma. Phys. Plasmas **8**(5), 2510–2516 (2001). DOI 10.1063/1.1352617
[53] Liang, E.P., Wilks, S.C., Tabak, M.: Pair production by ultraintense lasers. Phys. Rev. Lett. **81**(22), 4887–4890 (1998). DOI 10.1103/PhysRevLett.81.4887. URL http://link.aps.org/abstract/PRL/v81/p4887

[54] Lindl, J.D.: Inertial Confinement Fusion. Springer, New York (1998)
[55] Mackinnon, A.J., Borghesi, M., Hatchett, S., et al.: Effect of plasma scale length on multi-MeV proton production by intense laser pulses. Phys. Rev. Lett. **86**(9), 1769–1772 (2001). DOI 10.1103/PhysRevLett.86.1769. URL http://link.aps.org/abstract/PRL/v86/p1769
[56] Magill, J., Schwoerer, H., Ewald, F., et al.: Laser transmutation of iodine-129. Appl. Phys. B **77**(4), 387–390 (2003). DOI 10.1007/s00340-003-1306-4
[57] MAGPIE Project: MAGPIE Project Home Page. URL http://dorland.pp.ph.ic.ac.uk/magpie/
[58] Maine, P., Mourou, G.: Amplification of 1-nsec pulses in Nd:glass followed by compression to 1 psec. Opt. Lett. **13**(3), 467–469 (1988). URL http://www.opticsinfobase.org/abstract.cfm?URI=ol-13-6-467
[59] Maine, P., Strickland, D., Bado, P., et al.: Generation of ultrahigh peak power pulses by chirped pulse amplification. IEEE J. Quantum Electron. **24**(2), 398–403 (1988). DOI 10.1109/3.137
[60] Maksimchuk, A., Flippo, K., Krause, H., et al.: Plasma phase transition in dense hydrogen and electron—hole plasmas. Plasma Phys. Rep. **30**(6), 473–495 (2004). DOI 10.1134/1.1768582
[61] Maksimchuk, A., Gu, S., Flippo, K., et al.: Forward ion acceleration in thin films driven by a high-intensity laser. Phys. Rev. Lett. **84**(18), 4108–4111 (2000). DOI 10.1103/PhysRevLett.84.4108. URL http://link.aps.org/abstract/PRL/v84/p4108
[62] Malka, G., Aleonard, M.M., Chemin, J.F., et al.: Relativistic electron generation in interactions of a 30 TW laser pulse with a thin foil target. Phys. Rev. E **66**(6), 066402 (2002). DOI 10.1103/PhysRevE.66.066402. URL http://link.aps.org/abstract/PRE/v66/e066402
[63] Malka, V., Fritzler, S., Lefebvre, E., et al.: Electron acceleration by a wake field forced by an intense ultrashort laser pulse. Science **298**(5598), 1596–1600 (2002). DOI 10.1126/science.1076782. URL http://www.sciencemag.org/cgi/content/abstract/298/5598/1596
[64] Mangles, S.P.D., Murphy, C.D., Najmudin, Z., et al.: Monoenergetic beams of relativistic electrons from intense laser—plasma interactions. Nature **431**(7008), 535–538 (2004). DOI 10.1038/nature02939
[65] Mason, R.J., Tabak, M.: Magnetic field generation in high-intensity-laser—matter interactions. Phys. Rev. Lett. **80**(3), 524–527 (1998). DOI 10.1103/PhysRevLett.80.524. URL http://link.aps.org/abstract/PRL/v80/p524
[66] McKenna, P., Ledingham, K.W., Shimizu, S., et al.: Broad energy spectrum of laser-accelerated protons for spallation-related physics. Phys. Rev. Lett. **94**(8), 084801 (2005). DOI 10.1103/PhysRevLett.94.084801. URL http://link.aps.org/abstract/PRL/v94/e084801
[67] Mima, K., Fast Ignition Research Group: Present status and future prospects of laser fusion and related high energy density plasma research. AIP Conf.

Proc. **740**(1), 387–397 (2004). DOI 10.1063/1.1843522. URL http://link.aip.org/link/?APC/740/387/1

[68] Mima, K., Ohsuga, T., Takabe, H., et al.: Wakeless triple-soliton accelerator. Phys. Rev. Lett. **57**(12), 1421–1424 (1986). URL http://link.aps.org/abstract/PRL/v57/p1421

[69] Mourou, G.A., Barry, C.P.J., Perry, M.D.: Ultrahigh-intensity lasers: physics of the extreme on a tabletop. Phys. Today **51**(1), 22–28 (1998). DOI 10.1063/1.882131

[70] Mourou, G.A., Tajima, T., Bulanov, S.V.: Optics in the relativistic regime. Rev. Mod. Phys. **78**(2), 1804–1816 (2006). DOI 10.1103/RevModPhys.78.309. URL http://link.aps.org/abstract/RMP/v78/p309

[71] Najmudin, Z., Walton, B.R., Mangles, S.P.D., et al.: Measurements of magnetic fields generated in underdense plasmas by intense lasers. AIP Conf. Proc. **827**(1), 53–64 (2006). DOI 10.1063/1.2195197. URL http://link.aip.org/link/?APC/827/53/1

[72] Nakajima, K., Fisher, D., Kawakubo, T., et al.: Observation of ultrahigh gradient electron acceleration by a self-modulated intense short laser pulse. Phys. Rev. Lett. **74**(22), 4428–4431 (1995). DOI 10.1103/PhysRevLett.74.4428. URL http://link.aps.org/abstract/PRL/v74/p4428

[73] Narozhny, N.B., Bulanov, S.S., Mur, V.D., Popov, V.S.: $e^+e^-$ – pair production by a focused laser pulse in vacuum. Phys. Lett. A **330**(1-2), 1–6 (2004). DOI 10.1016/j.physleta.2004.07.013

[74] National Research Council: Frontiers in High Energy Density Physics. National Academies Press, Washington, DC (2003)

[75] Nellis, W.J.: Shock compression of hydrogen and other small molecules. In: G.L. Chiarotti, R.J. Hemley, M. Bernasconi, L. Ulivi (eds.) High Pressure Phenomena, Proceedings of the International School of Physics "Enrico Fermi" Course CXLVII, p. 607. IOS Press, Amsterdam (2002)

[76] Norreys, P.A., Santala, M., Clark, E., et al.: Observation of a highly directional $\gamma$-ray beam from ultrashort, ultraintense laser pulse interactions with solids. Phys. Plasmas **6**(5), 2150–2156 (1999). DOI 10.1063/1.873466

[77] Okun', L.B.: Leptony i kvarki, 2nd edn. Nauka, Moscow (1990) [English Transl.: Leptons and Quarks. North-Holland, Amsterdam (1982)]

[78] Parker, L.: Quantized fields and particle creation in expanding universes. I. Phys. Rev. **183**(5), 1057–1068 (1969). DOI 10.1103/PhysRev.183.1057. URL http://link.aps.org/abstract/PR/v183/p1057

[79] Pavlovski, A.I., Boriskov, G.V., et al.: Isentropic solid hydrogen compression by ultrahigh magnetic field pressure in megabar range. In: C.M. Fowler, R.S. Caird, D.T. Erickson (eds.) Megagauss Technology and Pulsed Power Applications, p. 255. Plenum, New York (1987)

[80] Pukhov, A.: Strong field interaction of laser radiation. Rep. Prog. Phys. **66**(1), 47–101 (2003). DOI 10.1088/0034-4885/66/1/202

[81] Pukhov, A., Meyer-ter-Vehn, J.: Laser wake field acceleration: the highly non-linear broken-wave regime. Appl. Phys. B **74**(4-5), 355–361 (2002). DOI 10.1007/s003400200795

[82] Rubakov, V.A.: Multidimensional models of particle physics. Phys. Usp. **46**(2), 211 (2003). DOI 10.1070/PU2003v046n02ABEH001355. URL http://ufn.ru/en/articles/2003/2/e/

[83] Rubakov, V.A., Shaposhnikov, M.E.: Do we live inside a domain wall? Phys. Lett. B **125**(2-3), 136–138 (1983). DOI 10.1016/0370-2693(83)91253-4

[84] Ryutov, D.D., Remington, B.A., Robey, H.F., Drake, R.P.: Magnetodynamic scaling: from astrophysics to the laboratory. Phys. Plasmas **8**(5), 1804–1816 (2001). DOI 10.1063/1.1344562

[85] Sarkisov, G.S., Bychenkov, V.Y., Novikov, V.N., et al.: Self-focusing, channel formation, and high-energy ion generation in interaction of an intense short laser pulse with a He jet. Phys. Rev. E **59**(6), 7042–7054 (1999). DOI 10.1103/PhysRevE.59.7042. URL http://link.aps.org/abstract/PRE/v59/p7042

[86] Schutzhold, R., Schaller, G., Habs, D.: Signatures of the Unruh effect from electrons accelerated by ultrastrong laser fields. Phys. Rev. Lett. **97**(12), 121302 (2006). DOI 10.1103/PhysRevLett.97.121302. URL http://link.aps.org/abstract/PRL/v97/e121302

[87] Schwoerer, H., Ewald, F., Sauerbrey, R., et al.: Fission of actinides using a tabletop laser. Europhys. Lett. **91**(1), 47–52 (2003). DOI 10.1209/epl/i2003-00243-1

[88] Schwoerer, H., Magill, J., Beleites, B. (eds.): Lasers and Nuclei: Applications of Ultrahigh Intensity Lasers in Nuclear Science, *Lecture Notes in Physics*, vol. 694. Springer, Berlin (2006)

[89] Shearer, J.W., Garrison, J., Wong, J., Swain, J.E.: Pair production by relativistic electrons from an intense laser focus. Phys. Rcv. A **8**(3), 1582–1588 (1972). DOI 10.1103/PhysRevA.8.1582. URL http://link.aps.org/abstract/PRA/v8/p1582

[90] Strickland, D., Mourou, G.: Compression of amplified chirped optical pulses. Opt. Commun. **56**(3), 219–221 (1985). DOI 10.1016/0030-4018(85)90120-8

[91] Strickland, D., Mourou, G.: Compression of amplified chirped optical pulses. Opt. Commun. **55**(6), 447–449 (1985). DOI 10.1016/0030-4018(85)90151-8

[92] Sudan, R.N.: Mechanism for the generation of $10^9$ G magnetic fields in the interaction of ultraintense short laser pulse with an overdense plasma target. Phys. Rev. Lett. **70**(20), 3075–3078 (1993). DOI 10.1103/PhysRevLett.70.3075

[93] Tajima, T.: Summary of Working Group 7 on "Exotic acceleration schemes". AIP Conf. Proc. **569**(1), 77–81 (2001). DOI 10.1063/1.1384337. URL http://link.aip.org/link/?APC/569/77/1

[94] Tatarakis, M., Gopal, A., Watts, I., et al.: Measurements of ultrastrong magnetic fields during relativistic laser–plasma interactions. Phys. Plasmas **9**(5), 2244–2250 (2002). DOI 10.1063/1.1469027. URL http://link.aip.org/link/?PHP/9/2244/1

[95] Tatarakis, M., Watts, I., Beg, F.N., et al.: Laser technology: Measuring huge magnetic fields. Nature **415**(6869), 280 (2002). DOI 10.1038/415280a
[96] Telnov, V.: Photon collider at TESLA. Nucl. Instrum. Methods Phys. Res. A **472**(1-2), 43–60 (2001). DOI 110.1016/S0168-9002(01)01161-5
[97] Thoma, M.H.: Field theoretic description of ultrarelativistic electron–positron plasmas (2008). URL `http://arxiv.org/abs/0801.0956`
[98] Trunin, R.F.: Shock compressibility of condensed materials in strong shock waves generated by underground nuclear explosions. Phys. Usp. **37**(11), 1123 (1994). DOI 10.1070/PU1994v037n11ABEH000055. URL `http://ufn.ru/en/articles/1994/11/d/`
[99] Umstadter, D.: Photonuclear physics: Laser light splits atom. Nature **404**(6775), 239 (2000). DOI 10.1038/35005202
[100] Unruh, W.G.: Notes on black-hole evaporation. Phys. Rev. D **14**(4), 870–892 (1976). DOI 10.1103/PhysRevD.14.870. URL `http://link.aps.org/abstract/PRD/v14/p870`
[101] Vacca, J.R. (ed.): The World's 20 Greatest Unsolved Problems. Prentice Hall PTR, Englewood Cliffs, NJ (2004)
[102] Vladimirov, A.S., Voloshin, N.P., Nogin, V.N., et al.: Shock compressibility of aluminum at $p > 1$ Gbar. JETP Lett. **39**(2), 82 (1984)
[103] Wagner, U., Tatarakis, M., Gopal, A., et al.: Laboratory measurements of 0.7 *gg* magnetic fields generated during high-intensity laser interactions with dense plasmas. Phys. Rev. E **70**(2), 026401 (2004). DOI 10.1103/PhysRevE.70.026401
[104] Weibel, E.S.: Spontaneously growing transverse waves in a plasma due to an anisotropic velocity distribution. Phys. Rev. Lett. **2**(3), 83–84 (1959). DOI 10.1103/PhysRevLett.2.83
[105] Zagar, T., Galy, J., Magill, J., Kellett, M.: Laser-generated nanosecond pulsed neutron sources: scaling from VULCAN to table-top. New J. Phys. **7**, 253 (2005). DOI 10.1088/1367-2630/7/1/253. URL `http://www.iop.org/EJ/abstract/1367-2630/7/1/253`
[106] Zasov, A.V., Postnov, K.A.: Obshchaya astrofizika (General Astrophysics). Vek 2, Fryazino (2006)
[107] Zeldovich, Y.B., Popov, V.S.: Electronic structure of superheavy atoms. Sov. Phys. – Usp. **14**, 673 (1972)

# Chapter 5
# Relativistic Charged Particle Beams

The highest energy densities attainable under terrestrial conditions are generated in relativistic heavy-ion collisions. Accelerators [23] required for this purpose operate in several laboratories throughout the world and are well known as the principal experimental tool in nuclear physics, elementary particle physics, quantum chromodynamics, and superdense nuclear matter physics research [56, 29, 54, 65, 41], i.e., in the areas which have always been at the forefront of the natural sciences. There exists a demand for constant advancement into the domain of higher energies and higher phase densities of accelerated particle beams.

Accelerator science and technology have come a long way from the first 1.2 MeV proton cyclotron invented by E. Lawrence in 1932 to the Large Hadron Collider (LHC) (Fig. 5.1), just beginning operation at CERN, with energies of 7 000 000 MeV in either of the counter-propagating beams, which is 7000 times the proton rest energy $m_\mathrm{p}c^2$. During this period, the world has seen the construction of thousands of accelerators of different types, which are giant electrical facilities incorporating cutting-edge engineering ideas and exhibiting a high degree of reliability. Today they hold the record in high-energy-density physics.

It is planned to collide two proton beams with an energy of $7 \times 7$ TeV in the LHC accelerator complex, which is the largest man-made experimental installation, to reach the new domain of distances of $10^{-16}$ cm and energies on the 1 TeV scale, which is sufficient, in particular, for the production of the particles of dark matter (their mass $m_\mathrm{DM} \approx 10$ GeV–1 TeV), the Higgs boson, quark–gluon plasma, perhaps for discovering new dimensions, and for the solution of other intriguing problems of high energy physics [66, 64]. The main purpose of the experiments at LHC is to check the so-called Standard Model of particle physics and to see if there is any new physics in the teraelectronvolt range of energies [43].

The Standard Model (SM), which describes strong and electroweak interactions of elementary particles, is based on several main principles: renormalizability, gauge invariance, and spontaneous gauge-symmetry breaking (see [43]). Despite the amazing achievements of the SM in describing the experimental data, there are many reasons why it cannot be considered the final theory. In this model,

V.E. Fortov, *Extreme States of Matter*, The Frontiers Collection,
DOI 10.1007/978-3-642-16464-4_5, © Springer-Verlag Berlin Heidelberg 2011

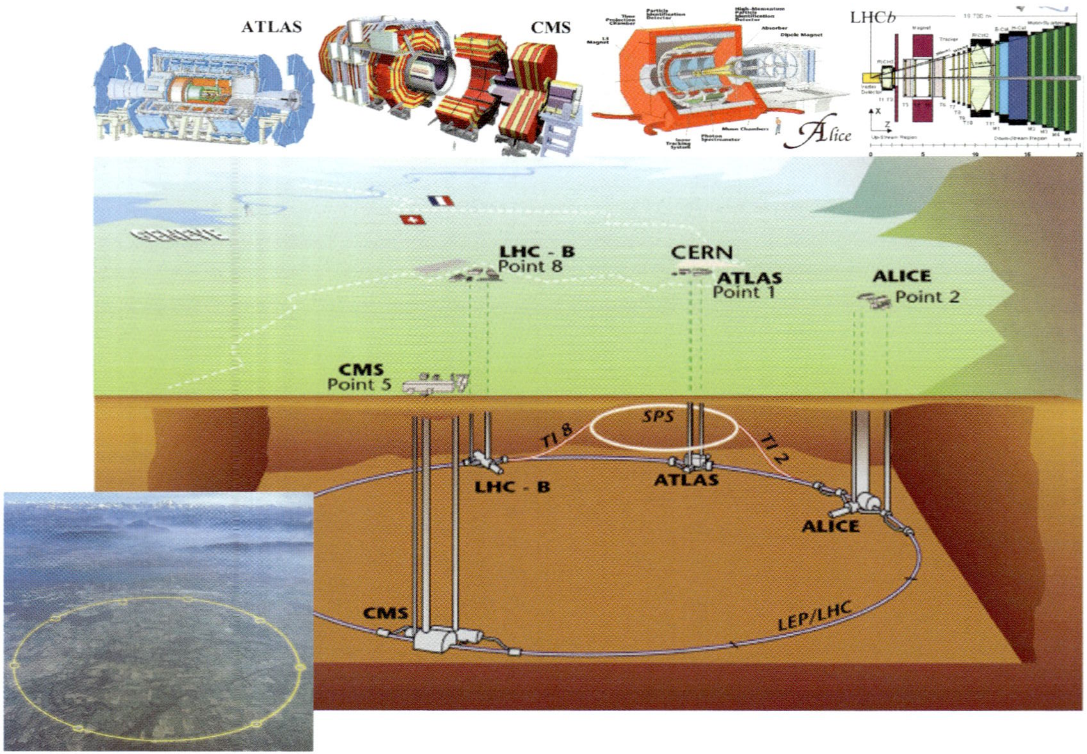


**Fig. 5.1** Schematic of the large hadron collider (LHC) at the European Center for Nuclear Research (CERN). Its underground tunnel measures about 27 km in diameter. Shown at the top are the main detectors: ALICE, ATLAS, CMS, and LHC-B.

neutrinos are massless particles, with the result that there are no neutrino oscillations. But there presently exist clear indications that neutrino oscillations indeed occur; this follows from the detection of neutrinos born in the atmosphere and from the deficit of solar electron-neutrino flux.

Krasnikov and Matveev [43] analyzed the main investigation areas at the LHC:

- the search for the Higgs boson;
- the search for supersymmetry;
- the search for new physics outside the scope of the SM and MSSM (Minimal Supersymmetric Standard Model);
- B-physics;
- heavy-ion physics;
- top-quark physics; and
- standard physics (QCD and electroweak interactions).

In the LHC, protons will move in the form of 3000 bunches distributed over the entire 27 km circumference of the collider. Each bunch of up to 100 billion protons measures, at the points of collision, several centimeters in length and only 16 $\mu$m in diameter – like a thin human hair. A total of 2808 bunches of 100 billion protons each will collide in the zones where detectors are located to cause over 600 million proton collisions of particles per second (up to 20 collisions at intersection). These collisions will take place between the particles that constitute protons – quarks and gluons. At the maximum particle energy, the energy release will be equal to approximately one seventh of the energy contained in the initial neutrons, or about 2 TeV.

During the low-luminosity stage (the first two to three years of operation), the luminosity is planned to be $L_{\text{low}} = 10^{33}$ cm$^{-2}$ s$^{-1}$ with the integrated luminosity $L_{\text{t}} = 10$ fb$^{-1}$ per annum. During the high-luminosity stage, the luminosity is planned to be $L_{\text{high}} = 10^{34}$ cm$^{-2}$ s$^{-1}$ with the integrated luminosity $L_{\text{t}} = 100$ fb$^{-1}$ per annum.

Proton bunches will smash into each other at four points, where detectors will be located. There are two multipurpose detectors – the Compact Muon Solenoid (CMS) and A Toroidal LHC Apparatus (ATLAS) – as well as the ALICE detector for studying the physics of heavy ions and the LHC-B detector for studying B-physics.

The detectors, which are impressive giant engineering structures – the biggest one having a volume half that of Notre Dame cathedral in Paris and the heaviest one containing more iron than the Eiffel Tower – will measure the parameters of the thousands of particles escaping at each collision event. Despite the huge size of the detectors, their individual elements should be mounted with a precision of 50 $\mu$m. Later on it will be possible to study the processes in the collision of highly ionized lead ions ($Pb^{82+}$) with energies of up to 155 GeV per nucleon.

The Stanford Linear Accelerator at SLAC, Menlo Park, California, generates a 5 ps pulse of 10 electrons with a kinetic energy of 50 GeV, which is focused to a 3-$\mu$m-sized spot to provide a power density of $10^{20}$ W/cm$^2$.

The actively operating Relativistic Heavy Ion Collider (RHIC) (Fig. 5.2) of the Brookhaven National Laboratory (Long Island, NY, USA) affords an energy of up

**Fig. 5.2** Relativistic Heavy Ion Collider (RHIC) of the Brookhaven National Laboratory (USA).

to 500 GeV per nucleon in the center-of-mass system for colliding gold ions (5–200 GeV for A+A), 39 TeV for Au+Au, and 13 TeV for Cu+Cu [62]. The new experimental data obtained on this accelerator are discussed in [32]. In Darmstadt (Germany), in November 2007 a start was made on the construction of the unique ion and antiproton accelerator facility FAIR (Facility for Antiproton and Ion Research) with an energy of 1.5–34 GeV per nucleon for $\approx 5 \times 10^{11}$ and $\approx 4 \times 10^{13}$ accelerated $U^{92+}$ ions and antiprotons, respectively. The construction cost of each of these major ultrarelativistic hadron accelerator complexes amounts to several billion dollars and is close to the limit of the economic potentials of the world's wealthiest countries and even international communities such as the European Union.

The scientific programs at these complexes comprise experimental research into the basic problems of high energy physics in hadron collisions, which are accompanied by the production of superdense nuclear substance – quark–gluon plasma. According to contemporary notions, this was precisely the state of the universe's substance during the first microseconds after the big bang (Fig. 7.45), and this is precisely the state of the substance of astrophysical objects such as gamma-ray bursts, neutron stars, and black holes. It is no coincidence that during the preparation of these experiments there emerged concerns about their safety because of the possibility that a black hole would emerge in laboratory conditions with the subsequent absorption of our civilization. However, calculations showed that the state of a mini black hole in accelerator experiments would exist for a very short time ($\approx 10^{-26}$ s), which is insufficient for catastrophic consequences. Furthermore, the absence of catastrophic events in the transit of superhigh-energy cosmic particles is supposedly an additional argument for the safety of the acceleration experiments in this energy density range. It is pertinent to note that numerical experiments in this area are carried out employing the world's highest-power supercomputers with recourse to seven- and more-dimensional mathematical models [54, 65].

For our consideration it is significant that these acceleration experiments are aimed not only at the production of particle beams with ultrarelativistic energies and the investigation of individual hadron collision events (Fig. 7.46) [45], but also for macroscopic substance heating [68, 23, 36].

## 5.1 Production of Macroscopic Hot Plasma Volumes

The methods of gas-dynamic acceleration of condensed strikers as described in Sects. 3.4 and 3.5 have a fundamental disadvantage arising from the limited value of the velocity of sound $c_s$ in the pusher gas, so that the acceleration efficiency decreases sharply (exponentially) when the striker reaches the velocity of sound. The techniques of high-energy-density generation relying on the use of high-intensity fluxes of charged particles – electrons, light or heavy ions – as well as on electrodynamic acceleration techniques, where the velocity of light fulfills the role of the velocity of sound, are devoid of these limitations. An important positive feature of charged particle beams is the volume nature of their energy release [68]. This distinguishes them from laser irradiation, where the bulk of the energy for radiation of frequency $\omega_l$ is released in a narrow critical zone [6, 44, 7, 46] $\omega_l \approx \omega_p \approx \sqrt{4\pi e^2 n_e / m_e}$ and is then transferred to the target interior by electron thermal conduction [6, 44, 7, 46].

As a result of deceleration of the charged particles there emerges a layer of isochorically (at constant-volume) heated plasma, which subsequently broadens to generate a shock wave traveling into the target or a cylindrical shock diverging from the beam axis. Modern research in the area of high-energy-density physics takes advantage of both of these techniques – isochoric heating and compression by shocks generated by particle beams.

For particle beam generators use is made of either cyclotrons developed for high-energy and nuclear physics research [36], or high-current diode systems [9, 49]. Involved in the latter case are subnanosecond current pulses of the megampere range with kinetic particle energies of 1–20 MeV [58, 67].

Fortov et al. [21] employed relativistic MeV electron beams to excite shock waves in aluminum targets for the purpose of studying the features of electron absorption in a dense plasma and elucidating the effect of intrinsic magnetic fields of the beams on the stopping power of magnetized plasma (the magnetic "stopping" effect).

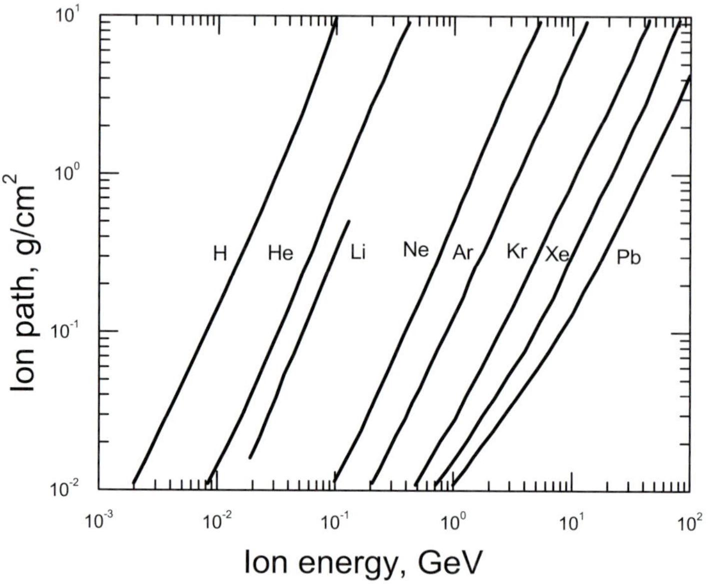


**Fig. 5.3** Characteristic ion paths in a cold material [68].

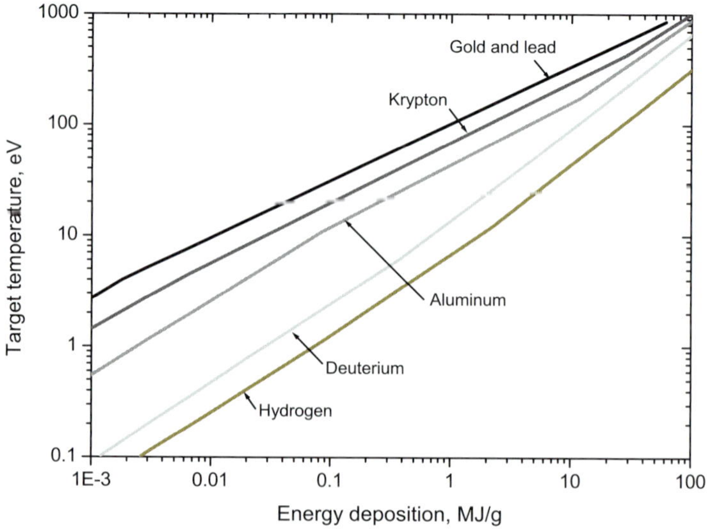


**Fig. 5.4** Typical plasma temperature as a function of energy deposited by charged particle beams [68].

Owing to the substantially shorter paths of ions in comparison with electrons (Fig. 5.3), ion beams make it possible to obtain higher specific energy densities in plasmas (Fig. 5.4). In a series of investigations by Baumung et al. [9] performed on the pulsed high-current KALIF accelerator, the ≈2 MeV proton beam with a current of ≈400 kA produced a power density on the order of $10^{12}$ W/cm$^2$. This enabled thin (50–100 m) flyers to be accelerated to velocities of 12–14 km/s and informative measurements of the dense-plasma stopping power to be performed for fast protons, the thermodynamic parameters and viscosity of the shock-compressed plasma to be recorded, and the spalling resistance of metals at ultrahigh rates of straining to be found. It turned out (Fig. 5.5) that the spalling resistance of metals significantly rises (by 1–2 orders of magnitude) with increasing rate of strain to approach its theoretical limit, which is related to the propagation kinetics of dislocations and cracks in a pulsed stress field [9, 39].

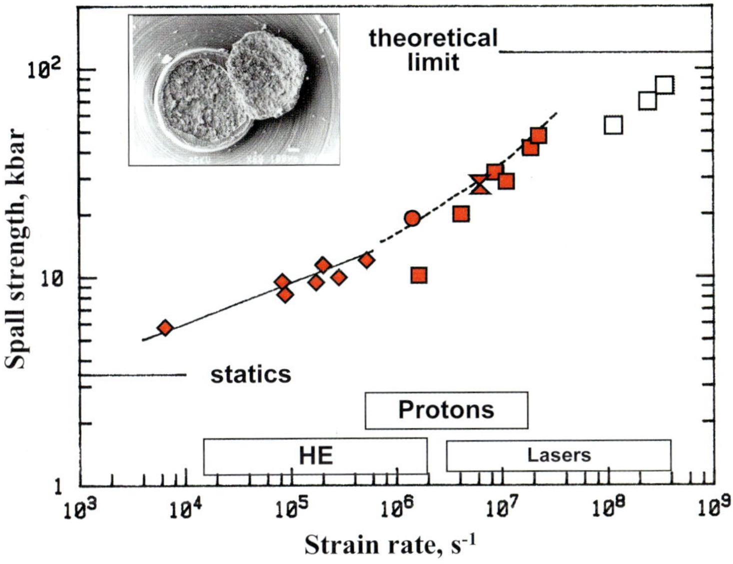


**Fig. 5.5** Spalling strength of an aluminum-magnesium alloy for high strain rates (HE: high explosive).

The pulsed high-current BPFA-II generator (Sandia National Laboratories, USA, Fig. 5.6) was earlier used to advantage for producing megavolt lithium ion beams with an intensity on the order of $10^{12}$ W/cm$^2$ for the inertial confinement fusion (ICF) program [15, 47]. At present this facility is employed in the high-current Z-pinch mode for the production of soft X-rays and electrodynamic generation of shocks and adiabatic compression waves [42, 79, 16, 60].

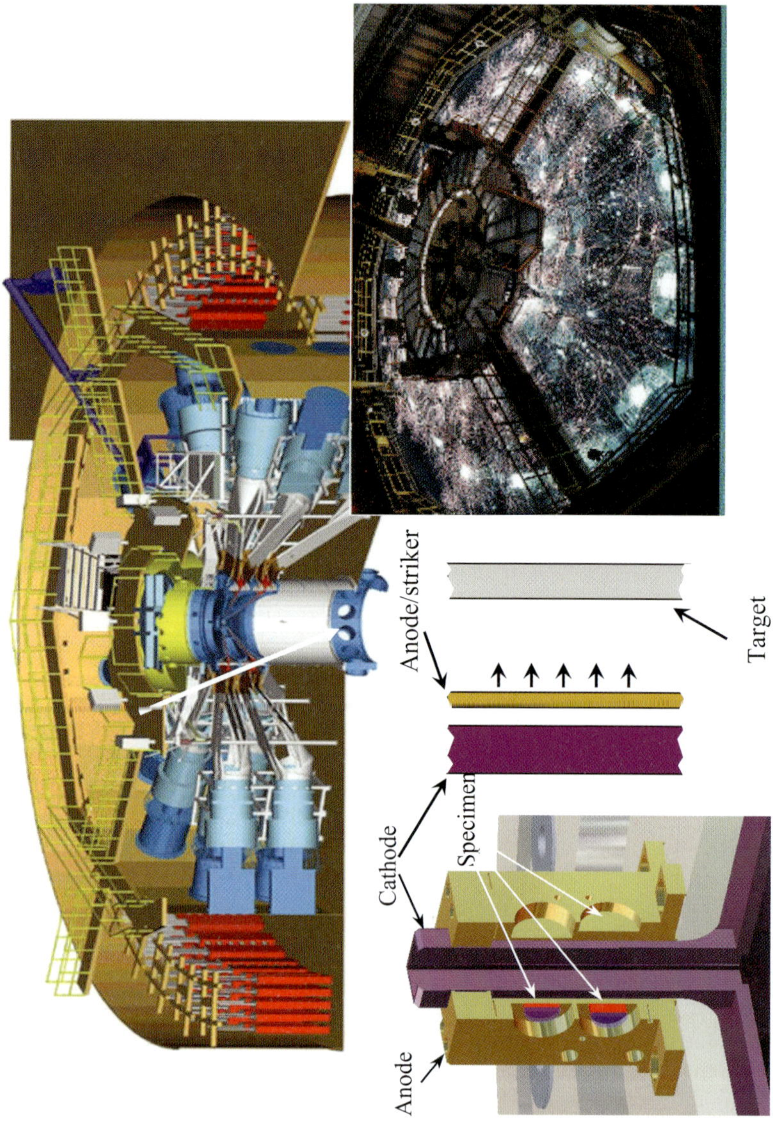


**Fig. 5.6** Schematic of electrodynamic acceleration of metal liners at the Z-pinch facility of the Sandia National Laboratories, USA [42].

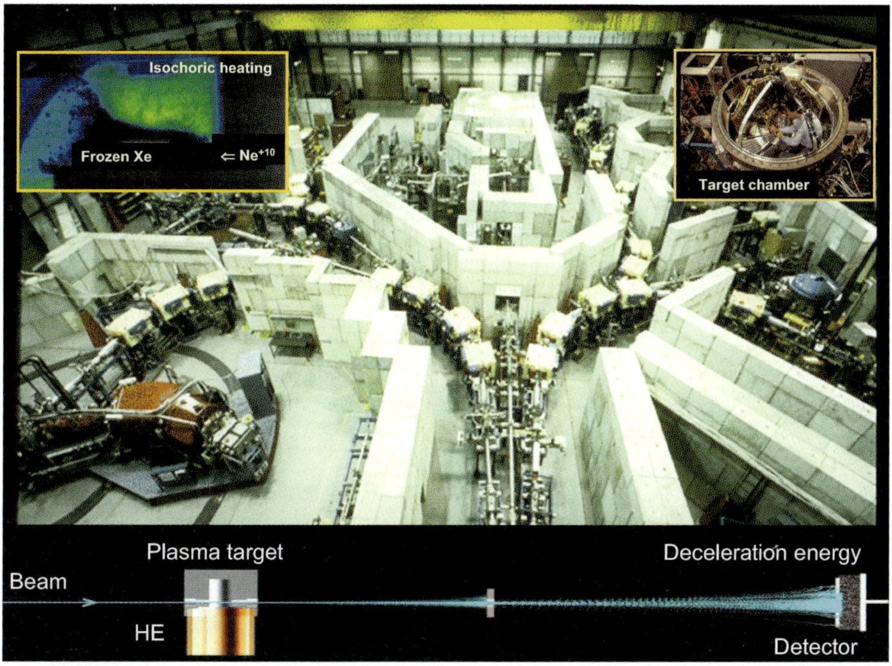


**Fig. 5.7** Relativistic heavy-ion beam interaction experiments at the GSI accelerator (Darmstadt, Germany) [36, 23].

The relativistic heavy-ion accelerators constructed for high-energy physics experiments turned out to be candidates (Table 3.2) for controlled nuclear fusion with inertial plasma confinement and for experiments on the compression and heating of dense plasmas (Fig. 5.7) [68, 23].

The LHC (CERN) (Fig. 5.1), constructed to study the collisions of two proton beams with an energy of 7 TeV each, generates 2808 bunches, each 0.5 ns long, which contain $1.1\times10^{11}$ protons each and are spaced at 25 ns, so that the total beam duration is 89 s and the energy $\approx$350 MJ, which is sufficient to melt 500 kg of copper. The energy density in one beam is $10^{10}$ J/cm$^3$. The characteristic kinetic energy of one heavy relativistic ion is comparable to the kinetic energy of a metal liner accelerated by the high explosive (HE) detonation products of the launching system described in Sect. 3.5.

Heavy-ion beams with a kinetic energy of 3–300 MeV per nucleon were employed in experiments on the heating of condensed and porous targets, the measurement of plasma stopping power for ions, and the interaction of charged beams with shock-compressed plasma produced by miniexplosion-driven shock tubes [23, 63, 18, 50, 82].

Of special interest is the possibility of employing the GSI (Darmstadt, Germany) heavy-ion accelerator in combination with the high-power petawatt PHELIX laser system (Fig. 5.8), which would qualitatively enhance the experimental capabilities

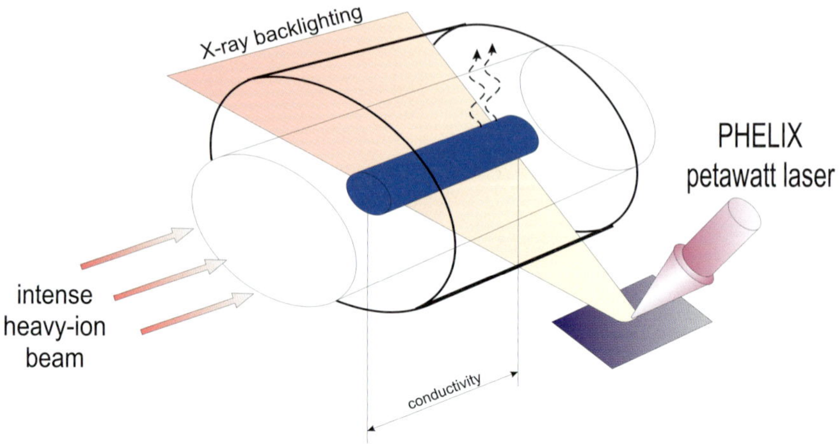


**Fig. 5.8** Setup of the experiment using a relativistic heavy-ion beam and a petawatt laser [23, 36, 82, 81].

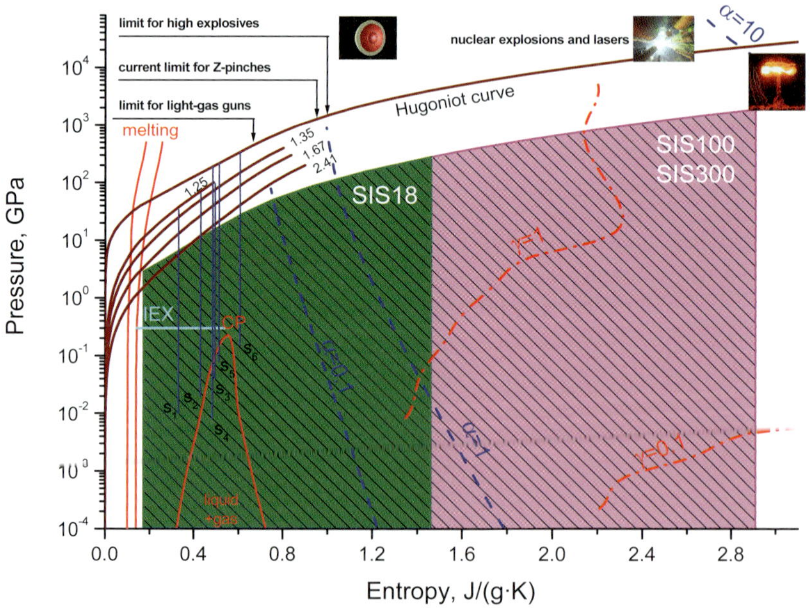


**Fig. 5.9** Capabilities of the SIS18, SIS100, and SIS300 heavy-ion accelerators (see Table 3.2) for producing high energy densities in lead [23, 36, 82, 81].

of this device. The potential and application prospects of the acceleration complexes at the GSI heavy-ion center are depicted in Figs. 5.9 and 5.10. One can see that high-

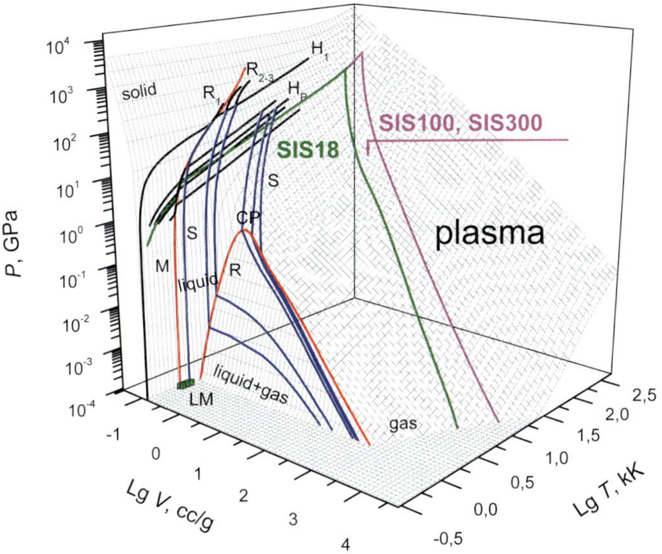


**Fig. 5.10** Parts of the phase diagram of zinc attainable with heavy-ion generators [23, 36, 82, 81].

intensity relativistic heavy-ion beams are interesting candidates for the generation of high-energy-density plasmas as well as for pulsed nuclear fusion [36, 23].

## 5.2 The Nuclear Matter Phase Diagram and Quark–Gluon Plasma

As noted in the foregoing, record high energy densities today are attained in head-on collisions of heavy nuclei (Fig. 5.14) accelerated to subluminal velocities in synchrotrons. These experiments are aimed at the search for new particles (Fig. 5.11) and the experimental investigation of the basic problems of high energy physics in hadron collisions, which are accompanied by the production of superdense nuclear matter – the quark–gluon plasma (QGP) (Figs. 5.15 and 5.20). In particular, unique experiments on the generation of baryonic matter in an ultraextreme superdense heated state with a density of $\approx 10^{15}$ g/cm$^3$, a pressure of $10^{30}$ bar, and a temperature of $\approx$ 200 MeV (Figs 2.1 and 2.2) in Cu–Cu and Au–Au heavy-nuclei individual collision events were carried out using the accelerators at CERN and Brookhaven. According to contemporary notions, this was precisely the state of matter in the universe during the first few microseconds after the big bang (Fig. 7.105), and this is the state of matter in astrophysical objects such as gamma-ray bursts, neutron stars, and black holes.

Theoretical and experimental investigations of the quantum chromodynamics (QCD) phase diagram are among the most prominent research directions in contemporary physics [71]. For more than a quarter of a century, the international scientific community, in the area of high energy physics, has undertaken great efforts in order to reveal new phases of strongly interacting matter in extreme conditions, as produced in ultrarelativistic heavy ion collisions.

Investigation of hot and dense baryonic matter is indeed a challenging task in fundamental physics. It provides information on the in-medium properties of the hadrons and on the nuclear matter equation of state; moreover, it allows a search for manifestations of de-confinement and/or chiral symmetry restoration. Similarly, it might indicate phase transitions, the appearance of mixed phases and critical points, if any, so that it might shed light on the evolution of the early universe and on the formation process of neutron stars.

A vast research program in this branch of fundamental science has been carried out over the last 30 years. The main research centers equipped with high energy ion accelerators are:

- GSI (Germany), $E_{\rm lab} = 1$ to 2 $A$GeV, equivalent to a nucleon–nucleon center-of-mass energy $\sqrt{S_{NN}}$ of 2–2.5 GeV;
- CERN, SPS (Switzerland/France), $E_{\rm lab} = 20$ to 158 $A$GeV, equivalent to $\sqrt{S_{NN}}$ of 6–17 GeV;
- BNL, RHIC (USA), $\sqrt{S_{NN}}$ of 20–200 GeV.
- JINR, Nuclotron (Russia), $E_{\rm lab} = 1$ to 6 $A$GeV, equivalent to $\sqrt{S_{NN}}$ of 2.5–3 GeV.

The most intensive exploration has been recently performed at $\sqrt{S_{NN}}$ of 20–200 GeV in the center-of-mass system.

As we discuss later, a number of new phenomena were discovered in relativistic nuclei collisions:

- Strong stopping power of the colliding nuclei.
- Collective flow of secondary particles.
- Unexpectedly large broadening of the transverse momentum distributions.
- Irregularities in the beam-energy behavior of the K/$\pi$ ratio.
- Dramatic enhancement of the multistrange hyperon production.
- Essential broadening of the spectral function maximum for $\rho$-mesons in experiments with leptons, instead of the expected shift towards lower lepton-pair invariant mass.
- Strong in-medium modification of the produced fast hadrons.
- Jet quenching during propagation through the excited nuclear medium.

Thus, recent observations bear witness to the creation of a new kind of QCD matter, which is named the strongly interacting quark-gluon plasma (QGP).

In the course of time, experiments with heavy ions displayed the tendency of continuous increase of the energy transferred by the accelerators to the colliding particles. In the quest for higher and higher energies more and more powerful machines were built, which obviously happened to be financially more and more expensive. Quite recently a new reverse tendency has showed up, namely the tendency

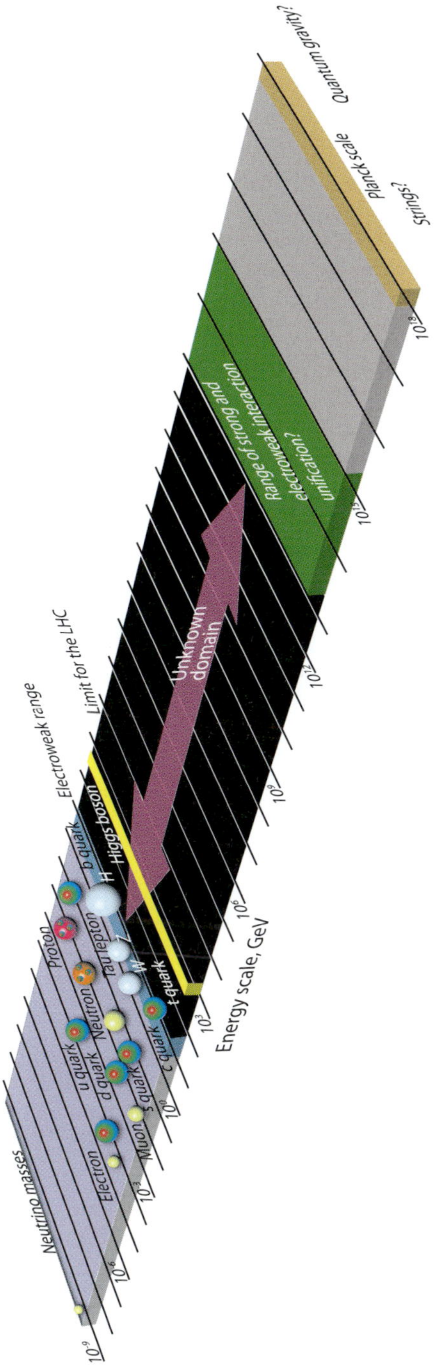


**Fig. 5.11** Particle hierarchy problem. Reprinted, with permission, from [59].

to reduce the accelerator energies in order to search for evidence of the existence of new states of strongly interacting nuclear matter. At the same time one would like to investigate the corresponding phase transitions and critical phenomena, if any. At present, it appears that to fulfill these challenging investigations, the required energy range is quite modest, corresponding to 8–40 GeV per nucleon in the laboratory coordinate system [76, 77, 75]. Indeed it is obviously reasonable that, before plunging deeply into supposedly existing new phases of nuclear matter, one should begin by establishing their existence and this is better done by searching around the phase boundaries [76, 77, 75, 73]. Here is where the signatures of phase transitions should manifest themselves, which is the best evidence for the existence of new phases. Such an optimization is clear both from the point of view of physics and from the point of view of financial costs.

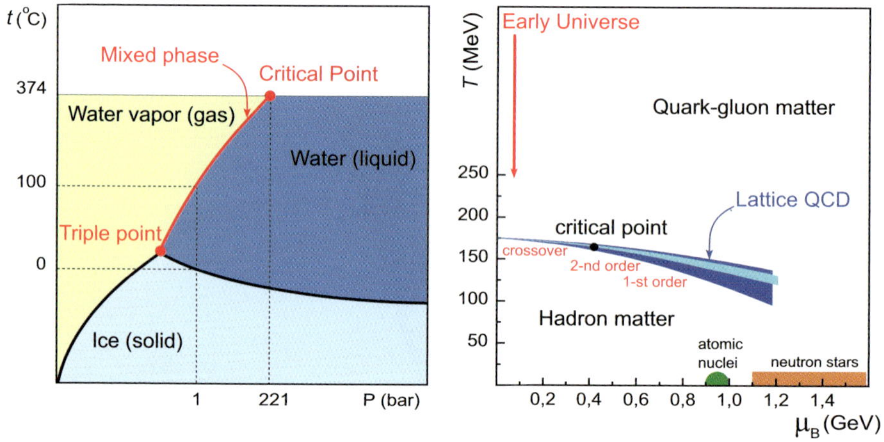


**Fig. 5.12** Left panel: Water–steam transition (first-order phase transition with latent heat) ends at a critical point (second order). There is no difference between steam and water above the critical point. Right panel: According to modern theoretical models, quark–hadron phase transitions have a structure similar to that of the the water–steam transition, with a crossover above the critical point ($\mu_B$ is the baryon chemical potential related to the baryon density, $T$ is temperature).

According to current views of theoretical physics, nuclear matter can undergo a series of first-order phase transitions when the temperature and/or the baryonic density is increased (see the right-hand panel of Fig. 5.12 and Fig. 5.13). One of these transitions is the first-order one corresponding to chiral symmetry restoration. Strong interactions break chiral symmetry stronly at low temperatures and/or low baryon densities, but they conserve it at high values of the same parameters. As a consequence, one can predict the existence of a mixed state where restored sSymmetry and broken symmetry phases coexist [77]. Such a mixed state is reminiscent of the vapor–liquid mixed phase that we observe in our daily experience when we boil water in a kettle (see the left panel of Fig. 5.12). At the boiling point, bubbles of vapor (the gaseous phase, corresponding to the restored symmetry state of nuclear matter in our analogy) are created and "swim" through the bulk of hot water (the liquid phase, corresponding to the broken symmetry state of nuclear matter in the same analogy).

At least one more phase transition is expected. It should correspond to the deconfinement of nuclear matter into the (hypothetical) QGP.

Hence, the central goal in heavy ion experimental studies consists in the exploration of the nuclear matter phase diagram. The two main phases are the *hadronic* and *quark–gluon* ones. They should be separated, according to the above discussion, by a first-order phase transition at high baryon densities. In fact we do not know whether the two above advocated phase transitions coincide, and this ignorance introduces various different possible scenarios for the detailed structure of the phase diagram. If we plot data in the temperature–baryon density plane, the first-order transitions manifest themselves through finite width bands where two or more phases coexist. These strips could be separated, fully coincident, or build up into a bifurcation strip. These are the hypothetical mixed phase regions [77, 17]. (Have a look at the yellow strips in Fig. 5.13).

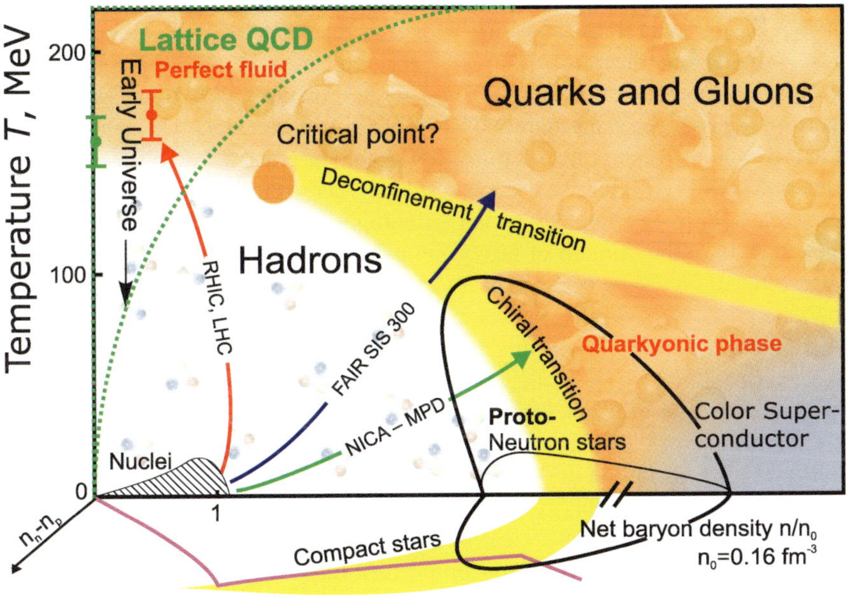


**Fig. 5.13** The phase diagram of strongly interacting QCD matter as presented on the NICA homepage [53, 17].

In recent years, our ideas about the structure of the QCD phase diagram have developed considerably and the several hypothetical scenarios that have been conceived now demand experimental confirmation or falsification. Specifically, ideas on the structure of phases at high baryon density have been developed.

At very high baryonic density and very low temperature, the emergence of a color superconductive phase is expected. This hypothesis triggered an avalanche of papers suggesting new complicated phenomena that might be relevant in neutron star physics.

The classification in terms of large $N_c$ (number of colors) behavior suggests the existence of a confined baryonic phase where thermodynamical quantities should scale proportionally to $N_c$ (this resembles the behavior of the Skyrme crystal). Such a hypothetical phase of strongly interacting matter that is thought to occur at high baryonic density and moderate temperatures was recently dubbed the quarkyonic phase.

At moderate baryon density and high temperature the confined and de-confined phases should be separated by a first-order phase transition, corresponding in the phase diagram to a strip that is expected to terminate at the chiral critical point (see the yellow strip in Fig. 5.13). Typically this strip might be bifurcate, as we mentioned before, since the two phase transitions (chiral and de-confinement) can coincide in one region of the plane and can be separated in another.

Let us now give a short description of the present state of the art at the experimental level, highlighting the main facts that have so far been established. The search for the QGP at Brookhaven RHIC and CERN SPS has revealed first indications of a rich physics, going beyond the naive picture of a gas of quarks and gluons. Namely, indications for the existence of a strongly coupled QGP (sQGP) were found. However, results from the energy-scan program of the pilot NA49 experiment suggest that the onset of the hadron-to-quark matter transition should be expected at lower center-of-mass energies of the colliding heavy-ion systems. Yet the theoretical comprehension of these data is far from complete, which encourages different laboratories to undertake new efforts in the domain of heavy-ion physics. Recently it was realized that the picture is not complete because of a lack of data from the pilot experiments in the low energy domain with $\sqrt{S_{NN}}$ of 2–10 GeV. According to recent theoretical considerations, it is precisely in this energy range that the excited, strongly interacting system formed by the collision enters into a new phase, and the transition between different phases of nuclear matter might be observed. In response to such a quest, GSI announced the construction of a large accelerator complex, named FAIR. The heavy-ion beams that in the future could be extracted from this new, forthcoming machine will have an energy of $E_{\rm lab}$ of 5–35 AGeV in the laboratory system, corresponding to $\sqrt{S_{NN}}$ of 3–8.5 GeV in the center-of-mass system. A large international collaboration project, named Compressed Baryon Matter (CBM), has been launched at FAIR.

Recently also the BNL-RHIC Laboratory planned to reduce the collider energy from $\sqrt{S_{NN}} = 200$ GeV to $\sqrt{S_{NN}} = 5$ GeV. Unfortunately, implementing such an energy downgrading of a high energy machine decreases the luminosity of the beam by a few orders of magnitude. This implies a corresponding decrease in the statistics and the need to run the experiment for much longer times in order to improve them. Clearly this leads to the conclusion that only some of the conclusive measurements can be done at this experimental facility.

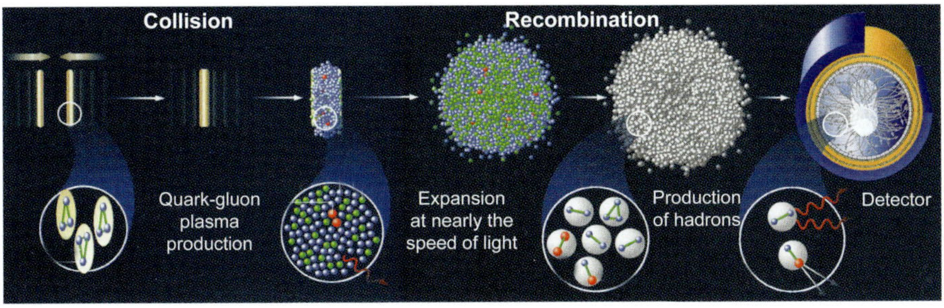


**Fig. 5.14** Dynamics of relativistic heavy-nuclei collisions in accelerators. Reprinted, with permission, from [62]

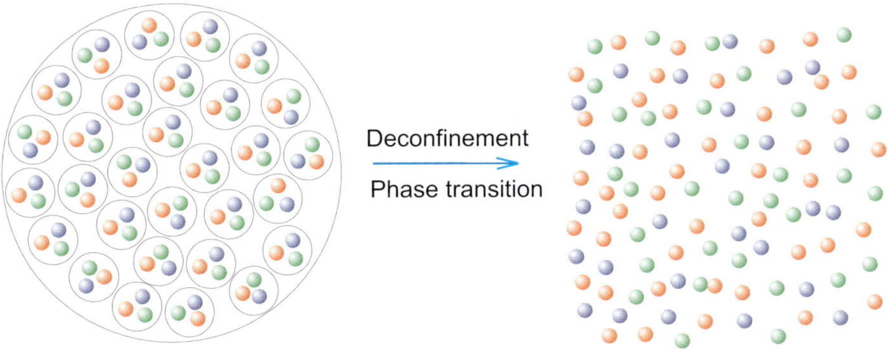


**Fig. 5.15** Quark–gluon plasma production. Under ordinary conditions (at the left) quarks (shaded balls) are bound in hadrons. At high temperatures $T > T_c$ there occurs deconfinement of quarks and they cease to be bound in hadrons and form a QGP.

Among the host of interesting physical results obtained using accelerator devices, here we enlarge on the production of a QGP, which emerges in quark deconfinement – at energies $\geqslant 200$ MeV [32, 51]. The experimental scheme is such that the kinetic energy of motion in the collision of two nuclei (Fig. 5.14 [62]) is transformed into the internal energy of nucleons. In accordance with the predictions of QCD [51], this gives rise to the so-called color glass condensate and subsequently – as it experiences thermalization – to a new state of matter – a QGP, or a "quark soup" [32, 51] (Fig. 5.15). It is assumed in this case that there is enough time for the matter to thermalize, so that the kinetic energy manages (this is the subject of special consideration) to be transformed into the internal energy of the resultant plasma.

The QGP resulting from these collisions consists of quarks, antiquarks, and gluons [35, 54, 65, 30]. This plasma is sometimes referred to as the "oldest" form of matter, because it existed even during the first microseconds after the big bang (Sect. 7.6). Hadrons were formed from this plasma as it underwent expansion and cooling. The QGP possesses a maximal density of $9\rho_0$–$10\rho_0$ (where $\rho_0 = 2.5 \times 10^{14}$ g/cm$^3$ is the nuclear density), and may emerge at the center of neutron stars and black holes, or in the collapse of ordinary stars. In the quest for QGP, large-scale experimental programs have been launched to create ultrarelativistic ion collisions in the accelerators HERA and RHIC in Brookhaven, GSI in Darmstadt, and SPS and LHC at CERN. The case in point is the collisions of heavy nuclei with energies of $\approx 100$ GeV per nucleus and higher in the center-of-mass frame or with energies of 20 TeV per nucleus in the laboratory frame of reference. The conditions attainable with contemporary accelerators are shown in the phase diagram of nuclear matter (Fig. 5.16). The domain of low temperatures and baryon densities is occupied by hadrons (nuclei and mesons) [52, 56, 54, 65, 30]. The limiting case of high densities (5–10 times the density of large nuclei: $\approx 0.17$ particles per cubic femtometer, $\approx 2.5 \times 10^{14}$ g/cm$^3$) and high temperatures ($T > 200$ MeV $\approx 10^{12}$ K) corresponds to quarks and gluons, which are not bound in hadrons under these conditions and form a QGP. The transition between these states may by smeared or abrupt, like a phase transition of the first kind with a critical point (Fig. 5.13). To describe the behavior of compressed baryon matter in the corresponding domain of the phase diagram, advantage is taken of the methods of QCD, which are in themselves the object of experimental verification.

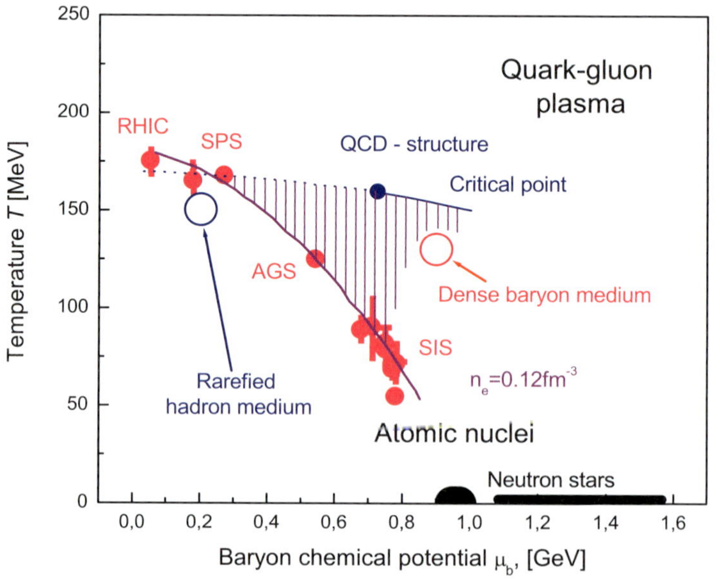


**Fig. 5.16** Phase diagram of nuclear matter (RHIC at BNL, USA; SPS at CERN, Switzerland; AGS at BNL, USA; SIS at GSI, Germany

.

Quark-gluon plasma is an essential experiment in understanding the transformation of matter immediately after the beginning of our universe (see Sect. 7.6). During the first microseconds [54, 65] after the big bang, the temperature decreased

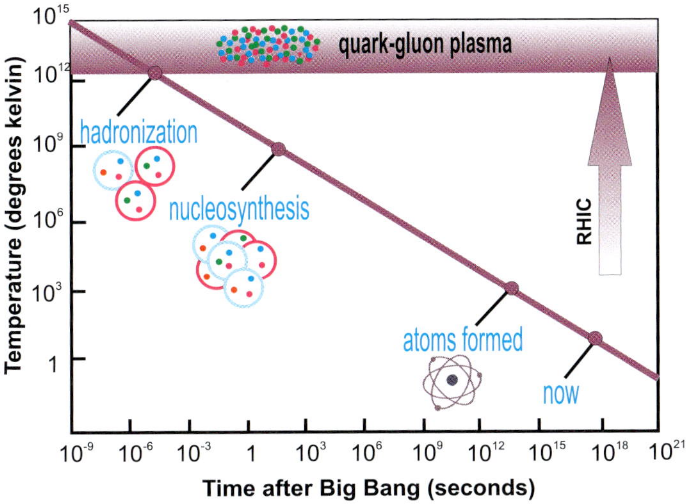


**Fig. 5.17** Transformation of matter after the big bang. The arrow indicates the RHIC accelerator parameters.

as $T[\text{MeV}] \approx 1/t^{1/2}$, where $t$ is time [s], so that a QGP with a temperature of several hundred MeV could exist during the first 5–10 $\mu$s after the big bang (for more details, see Fig. 5.17). At that time, the baryon density was not so high. As the universe expanded, the plasma became cooler, which resulted in the "hadronization" of matter and the subsequent production of pions. If a first-order transition occurred, "bubbles" of hadrons (neutrons, protons, and pions) would form inside the plasma. QGP [70, 69, 38, 40] is a superhot and superdense form of nuclear matter with unbound quarks and gluons, which are bound inside hadrons at lower energies (Fig. 5.15).

The characteristics of quarks (which possess a fractional charge), which make up hadrons, are collected in Table 5.1. For instance:

$$\begin{aligned}\text{proton} &= u + u + d\\ \text{neutron} &= u + d + d\\ \pi^+ &= u + \bar{d}, \text{ etc.}\end{aligned}$$

The existence of QGP follows from the property of asymptotic freedom of QCD [13, 11, 27, 12], which yields a value of 1–10 GeV/fm for the corresponding transition energy density, which is close to the energy density inside a proton and exceeds the nuclear energy density by an order of magnitude. Detailed numerical calculations give the critical conditions for the emergence of QGP: $T_c \approx$ 150–200 MeV $\approx (1.8\text{–}2.4) \times 10^{12}$ K (Fig. 5.18).

The emergence of this plasma manifests itself as an increase in the number of degrees of freedom – from 3 inherent to hadrons to 8 for gluons multiplied by 2 helical ones, plus 2–3 due to the light flavors of quarks, which in turn possess 2 spins and 3 colors. Therefore, according to QCD quarks possess 24–26 degrees of free-

**Table 5.1** Characteristics of quarks.

| Flavor | Charge [$\|e\|$] | Mass [MeV] |
|---|---|---|
| $u$ | $2/3$ | 5 |
| $d$ | $-1/3$ | 10 |
| $s$ | $-1/3$ | 150 |
| $c$ | $2/3$ | 1 300 |
| $b$ | $-1/3$ | 4 200 |
| $t$ | $11/3$ | 175 000 |

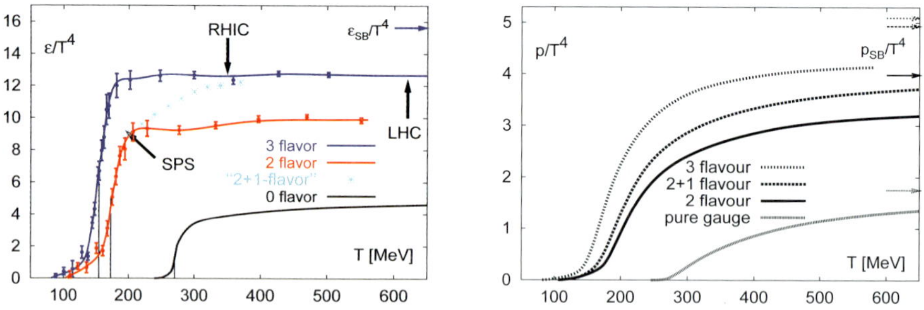


**Fig. 5.18** QCD calculations of the temperature dependence of the energy density (left) and pressure (right) [32]. The most realistic case corresponds to flavor 2+1. (The subscript SB stands for Stefan–Boltzmann.)

dom, and in a QGP for $T \approx (1\text{–}3)T_c$ 40–50 degrees of freedom are excited, versus 3 in a low-temperature ($T < T_c$) pion gas. Since the energy density, the pressure, and the entropy are approximately proportional to the excited degrees of freedom of the system, a sharp variation of these thermodynamic parameters in a narrow temperature range near $T_c$ accounts for a large (up to roughly tenfold in Fig. 5.18) energy difference between ordinary nuclear matter and the QGP.

In view of the aforesaid, the equation of an ideal, $g \to 0$, QGP is of the form [31]

$$P_{\text{SB}}^{\text{QCD}}(T) = (\underbrace{2_s \times 8_c}_{\text{gluons}} + \underbrace{\frac{7}{8} \times 2_c \times 3_c \times 2_{q\bar{q}} \times n_f}_{\text{quarks}})\frac{\pi^2 T^4}{90} - \underbrace{B}_{\text{vac}}.$$

For hadronic matter [31]:

$$P^{\text{H}}(T) = (\underbrace{3_{\text{ISO}}}_{\text{pions}} + \underbrace{O(e^{-M/T})}_{\rho,\omega,\ldots})\frac{\pi^2 T^4}{90}.$$

A comparison of these expressions defines the critical QGP transition temperature $T_c$.

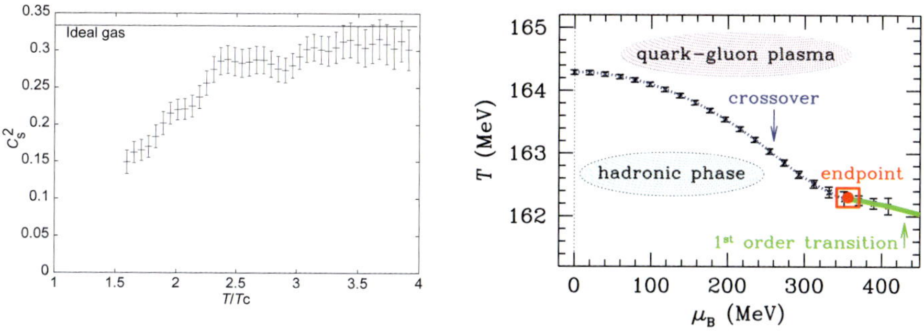


**Fig. 5.19** Features of the equation of state of a QGP [32]. Left: Temperature dependence of the speed of sound. Right: Phase boundary and critical point according to [19, 14].

Like our customary "electromagnetic" plasma (EMP), the QGP may be ideal for $T \gg T_c$ and nonideal for $T \approx (1\text{–}3)T_c$. The corresponding nonideality parameter – the ratio between the interparticle interaction energy and the kinetic energy – in this case is of the form $\Gamma = 2Cg^2/4\pi aT = 1.5\text{–}5$, where $C$ is the Casimir invariant ($C = 4/3$ for quarks and $C = 3$ for gluons), the interparticle distance $a \approx 1/T \approx 0.5$ fm, $T = 200$ eV, and the strong interaction constant $g \approx 2$. The factor 2 in the numerator takes into account magnetic interaction, which is of the same order of magnitude as the Coulomb interaction in the relativistic case.

At present it is hard to tell unambiguously whether the transition to a QGP is a true thermodynamic phase transition with an energy density jump or a sharp and

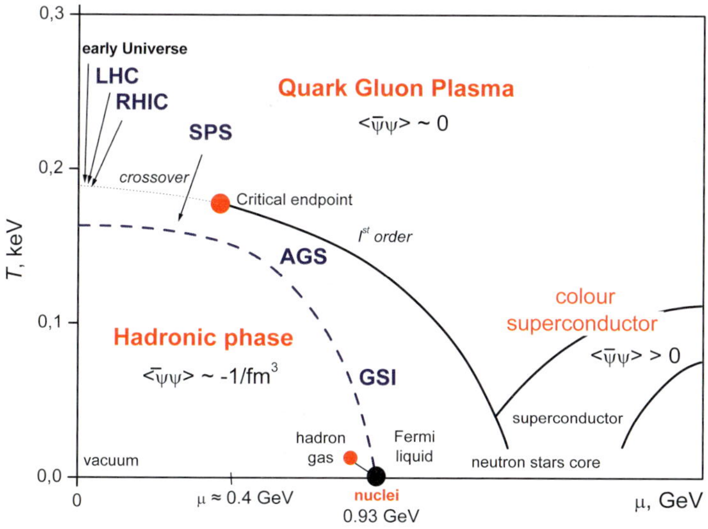


**Fig. 5.20** Phase diagram of QGP [31, 10].

yet continuous transition [32]. Conceivably (Fig. 5.19 [32]) for low baryon densities $\mu_B$ this might be a continuous transition and for high $\mu_B$ a phase transition of the first kind (Fig. 5.20). In any case, the theory [32] predicts a low value for the velocity of sound in the transition region (Figs. 5.19 and 5.18), which is reflected in the hydrodynamic anomalies accessible to observations in the relativistic collisions of heavy nuclei. The specified features of adiabatic compressibility of the QGP testify to a "softer" equation of state for $T \approx T_c$ and a "stiffer" one at high temperatures as well as for $T \leqslant T_c$. In the limit $T \ll T_c$, the equation of state of hadronic matter becomes "softer", although in this case the uncertainty is quite high and the description of this matter by quantum electrodynamics (QED) techniques encounters serious problems.

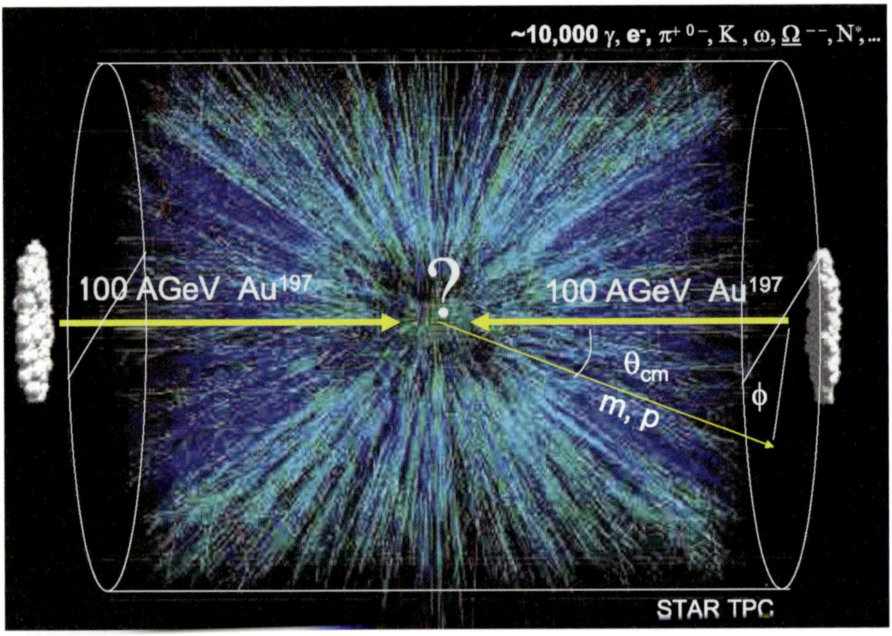


**Fig. 5.21** A relativistic hadron collision: "burning of the vacuum" [45].

A picture of a relativistic heavy nuclei collision is shown in Fig. 5.21. Under the conditions of the RHIC experiment, the longitudinal Lorentzian compression of the size of colliding nuclei is on the order of 100. The characteristic volume of the U–U collision domain ($\approx 3000$ fm$^3$) contains $\approx 10000$ quarks and gluons, while the characteristic collision time is $\tau_0 \approx 0.2$–$2$ fm/$c \approx (5$–$50)\times 10^{-25}$ s. That is why some of high-energy processes supposedly take place in the expanding substance, when the nuclear bunches have already passed through each other.

Gyulassy and McLerran [32] called attention to the fact that the production of fast particles in the expanding plasma after nuclear collisions is similar to the production of new forms of matter after the big bang (Sect. 7.6). However, the difference is that the expansion in nuclear collisions is one-dimensional rather than three-dimensional as in cosmology. The spatiotemporal evolution of matter following a relativistic collision is shown in Fig. 5.22.

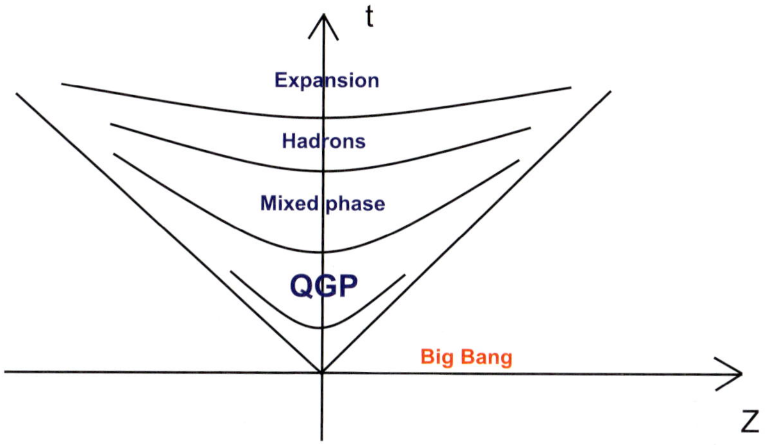


**Fig. 5.22** Spatiotemporal matter evolution after a collision [51].

During the collision, as the nuclear substance expands and cools down, the emergent quarks and gluons are thermalized (the time $\tau_{eq} \leqslant 1$ fm/$c \approx 3 \times 10^{-24}$ s) and may reach local thermodynamic equilibrium during the plasma lifetime $\tau_0 \approx$ (1–2)$R/c \approx 10$ fm/$c$. In this case, the medium will be set in hydrodynamic motion; recording this may yield experimental information about the properties of hadronic or quark–gluon matter as well as about the limit of their back and forth transition, which should occur at an energy density of $\approx 1$ GeV/fm$^3$ according to QED.

Figure 5.23 [32] serves to illustrate the characteristic energy density in nuclear collisions as a function of time. An analysis of the collision and expansion dynamics shows [32] that the transition from a relatively slow one-dimensional expansion to a faster three-dimensional expansion occurs in a characteristic time $\approx 0.3$ fm/$c$. The upper bound of the dark band in Fig. 5.23 corresponds to the assumption that the system is in thermodynamic equilibrium and is an ideal massless gas, while the lower one corresponds to nonequilibrium "frozen" conditions. By the point in time 3 fm/$c$ the plasma is a mixture of quarks, gluons, and hadrons, and for 10 fm/$c$ the quarks and the gluons recombine in hadrons. The lower bound of attainable energy density is realized for $t \approx 1$ fm/$c$ and the upper limit (massless gas) for 0.3 fm/$c$. The general estimate of energy density is of the form [32]

$$(2\text{–}3)\ \text{GeV/fm}^3 \leq E \leq (20\text{–}30)\ \text{GeV/fm}^3.$$

For comparison, the energy density in neutron stars (Sect. 7.2) amounts to $\approx$ 1 GeV/fm$^3$.

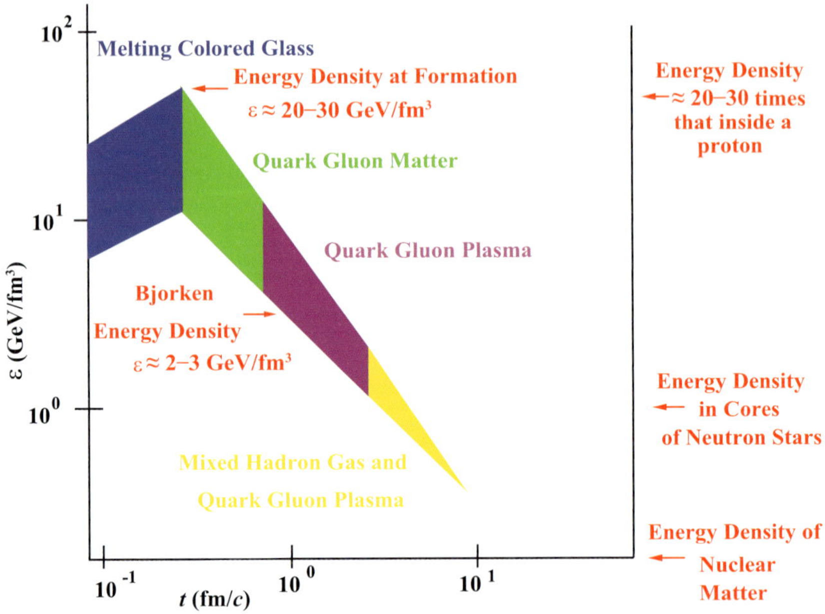


**Fig. 5.23** Characteristic energy density in the collision of nuclei as a function of interaction time [32].

The emergence of a new form of matter – QGP – should be accompanied by qualitatively new physical phenomena, which should manifest themselves in experiments. First, the emergence of new degrees of freedom in the plasma must be reflected in the collision and expansion relativistic hydrodynamics, which in turn is described by the equations of motion of a viscous fluid in the conditions of local thermodynamic equilibrium. This formalism is simplified for a nonviscous fluid (the Euler equation), while the experimental manifestation of collective (viscous) effects may be indicative of plasma effects.

The second manifestation of QGP is the difference between experiment and the parameters of hydrodynamic effects calculated from a given equation of state. The results of such a comparison for the azimuthal flow components for $\pi$, K, $p$ and $\Lambda$ in 200*A* GeV Au–Au collisions are shown in Fig. 5.24 [1, 78, 2, 4, 5]. One can see that for energies of up to $\approx 1$ GeV/fm there is good agreement between the calculated and measured data, which is not the case at higher energy values. This disagreement is attributed to the emergence of QGP. Taking into account the lowering of the sound velocity in the vicinity of $T \approx T_c$, which is caused by the emergence of this plasma, and the corresponding "softening" of the equation of state would improve the correlation between the data of calculations and experiments.

The emergence of QGP may show up not only in the properties of the equation of state, but also in the behavior of viscosity in the hydrodynamic motion.

It is precisely these effects that are thought to be responsible for the lower – in comparison with calculations (Fig. 5.25) – elliptic expansion velocities measured

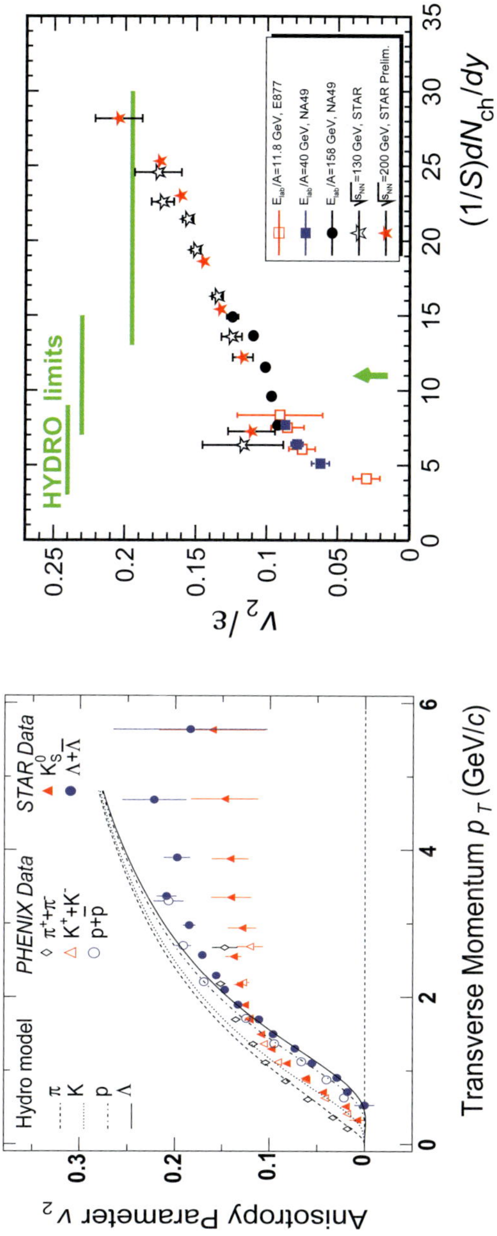


**Fig. 5.24** Experimental manifestation of a QGP [32]. Left: STAR [1, 78, 2] and PHENIX [4] measurement data. Right: Data from [5].

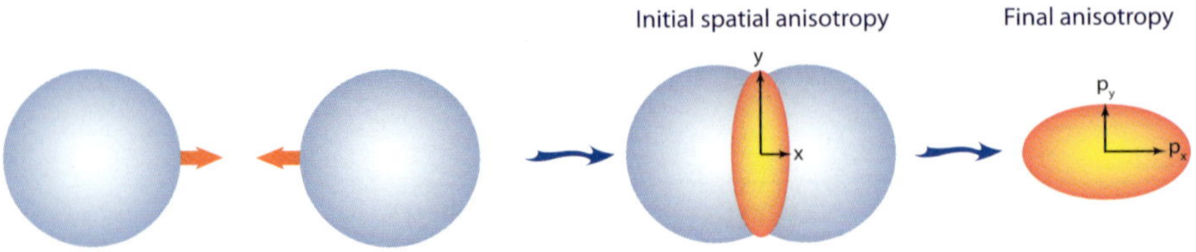


**Fig. 5.25** Relativistic nuclear collision, generation of elliptical flows. The high-energy domain is elliptical in shape, so that the spatial anisotropy generates the anisotropy of momenta of the expanding medium [57].

at the SPS accelerator (this discrepancy is unavoidable in the framework of three-dimensional transient (3+1)D hydrodynamics) and for the reduction of this discrepancy with increase in impact parameter [32], which is due to the lower efficiency of pressure transfer to the hydrodynamic flow by hadrons as against plasma. In any case, ideal nonviscous flow calculations by the Euler equations [10] are in better agreement (Fig. 5.26) with experiments than are the calculations by equations with viscous dissipation.

Interestingly, for moderate energies the RHIC experiments display anomalies in QGP dissipation and yield effective viscosity values that are up to 10 times lower than one would expect from the models of weakly nonideal Debye plasma. This is thought to be [32] due to plasma nonideality effects.

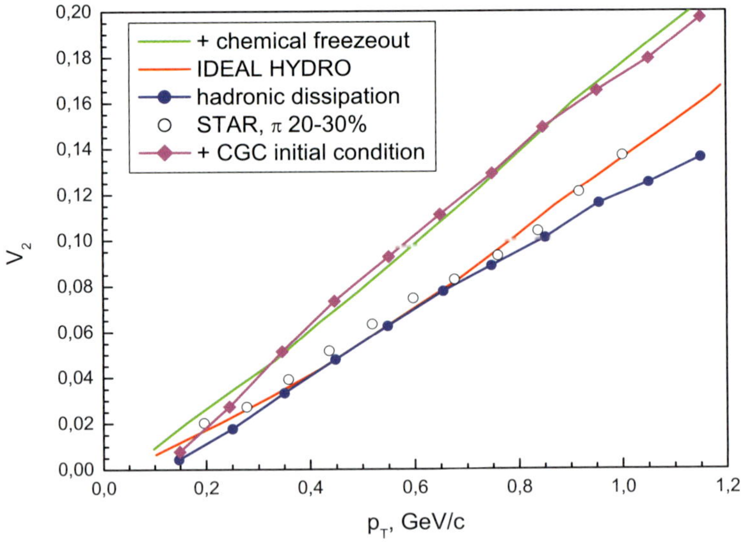


**Fig. 5.26** Comparison of measured (circles) and calculated expansion velocities in nuclear collisions [10].

The quenching of the jets produced in relativistic nuclear collisions also contains information about the properties of shock-compressed matter [33, 34, 85] and the emergence of QGP. From the order of magnitude this quenching is determined by the radiative losses of gluons, while the contribution of elastic losses is relatively low.

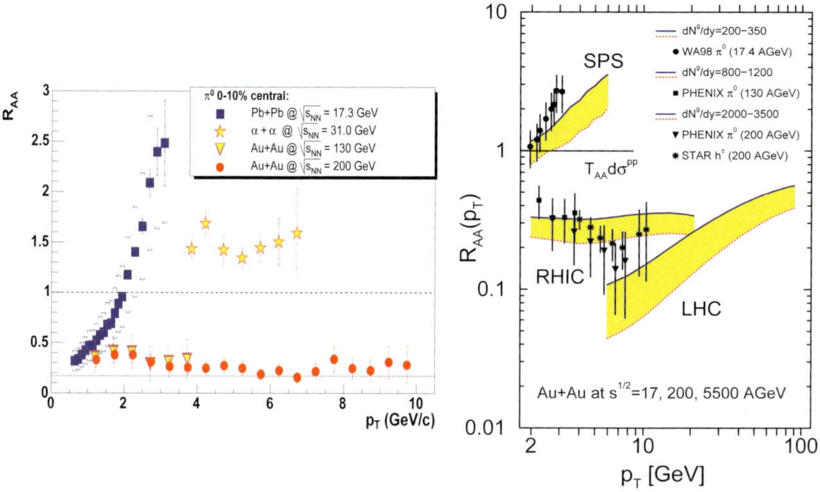


**Fig. 5.27** Left: $\pi^0$ flux quenching data recorded at PHENIX and RHIC in comparison with different ISR and SPS observations (both at CERN). Right: Factor $R_{AA} = dN_{AA}/T_{aa}(b)d\sigma_{pp}$ measured at SPS, RHIC, and LHC in comparison with radiative energy loss models [32, 83].

The results of such "tomography" for PHENIX experiments are given in the left-hand panel of Fig. 5.27 [32, 83]. They show that the initial reduced gluon density should be $dN^g/dy \approx 1000 \pm 200$ to account for the observed jet quenching. These values are in reasonable agreement with other independent measurements [32]:

- with initial entropy values determined from the plasma expansion after a collision,
- with initial plasma parameters following from hydrodynamic calculations of "elliptical" flows,
- with density variation rates calculated by techniques of QED.

These four data sets permit the initial energy density in relativistic collisions to be determined:

$$E_0 = E(1/\rho_0) \approx \rho_0^2/\pi R^2 \times dN^g/dy \approx 20\ \text{GeV/fm}^3 \approx 100\ E_a,$$

where $\rho_0 \approx Q_{\text{sat}} \approx (1.0\text{–}1.4)$ GeV is the characteristic gluon momentum. This energy density in turn determines the formation time $\hbar/\rho_0 \approx 0.2$ fm/$c$ of the primary nonequilibrium QGP. Under these conditions, the local thermodynamic equilibrium required for the application of hydrodynamics sets in for $\tau_{eq} \approx (1\text{–}3)B/\rho_0 < 0.6$ fm/$c$. By this time the temperature will be $T(\tau_{eq}) \sim (\varepsilon_0/(1\text{–}3)\times 12)^{1/4} \approx 2T_c$. According to one of the models [32], for $\rho_0 \approx (2\text{–}2.2)$ GeV the number of minijets should be on the order of 1000.

Experimental data on correlated double jets in nuclear collisions carry more detailed information about the plasma properties (Fig. 5.28). The correspondence of the characteristics of these jets obtained under different conditions (central and peripheral collisions, protons and gold) is regarded as a substantial argument for the applicability of QCD techniques and for QGP production.

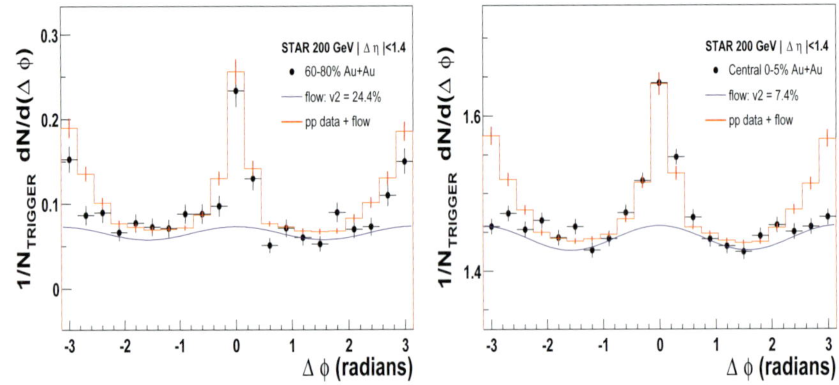


**Fig. 5.28** Strongly correlated backward jets in STAR and RHIC experiments [32, 37, 3] in Au+Au collisions (at the left) are compared with p+p collisions and with off-center collisions with monojets in head-on Au+Au collisions (at the right).

Therefore, the jet quenching effects observed in nuclear collisions allow the energy density of nuclear matter to be determined and conclusions to be drawn about a strong collective interaction (nonideality) of this plasma, proceeding from an analysis of the energy losses of the jets in their motion through the QGP.

Today, the observational manifestations of QGP [31] that are being actively analyzed include

- "Barometric" effects: from the parameters of collective flows (elliptic, longitudinal, radial)
- "Thermodynamic" effects: photons, leptonic pairs, vector mesons
- "Critical" effects: hadronic density fluctuations
- "Tomography": short jets, heavy quark fluxes
- "Exotics": multiquark states, femtometer-dimensional "fullerenes"

Among the interesting hydrodynamic phenomena, special mention should be made of Stöcker's elegant and beautiful idea [80] of the production of conic Mach shock waves (Fig. 5.29), whose properties permit the characteristics of compressed nuclear matter to be judged.

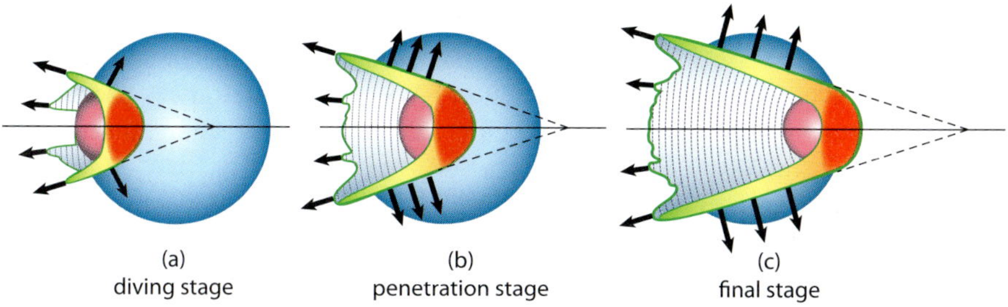


**Fig. 5.29** Formation of Mach shock waves in nuclear matter [80] in the collision of a light nucleus (at the left) with a "heavy" one.

The quest for QGP and the study of its physical properties are currently the subject of intensive theoretical and experimental investigations in many laboratories throughout the world. This research will undoubtedly receive a new impetus when the Large Hadron Collider is fully operational.

The QGP and the ordinary plasma, which constitutes 98–99% of the visible universe and is referred to as the "electromagnetic" plasma (EMP), have, on the one hand, many differences and, on the other, much in common [51]. Unlike the EMP, the QGP is relativistic or ultrarelativistic. This distinction in relativism shows up not only in the kinematics of motion, but also in that the nonrelativistic case sees only particles, whose number is a conserved quantity, while the relativistic case also sees antiparticles, and the number of leptons is a conserved quantity. The particle density is no longer an adequate characteristic: the density of baryons and "strange" particles (strangeness) takes its place.

In EMPs, the large difference in the mass of electrons and ions leads to the difference in their dynamics and kinetics and underlies, in particular, the difference of electron and ion temperatures in relaxation processes. In a QGP, there also are heavy (charmed, top, and bottom) particles, which, however, are fewer in number than the light quarks and gluons, and their lifetime is shorter. That is why the contribution of heavy quarks in QGP dynamics is small. QGPs are described by QCD and EMPs by QED. The latter theory is Abelian, unlike chromodynamics. In QCD, gluons carry color charge, determining the quark–quark and quark–antiquark interaction, and interact with each other. Unlike photons, gluons make a contribution to the color charge density and the color current.

A feature that the QGP and EMP share is the collective nature of interparticle interaction [51]. Despite screening, the radius of effective electromagnetic interaction is normally much longer than the interparticle distance, so that there are many particles in the Debye sphere and their motion is strongly correlated. Quantum electrodynamics gives a solution, which corresponds to the Debye one: $\Phi(r) = (q/r) \times \exp(-m_{\mathrm{D}}^2)$, in which the Debye mass (which plays the part of the inverse Debye radius in the atomic system of units) $m_{\mathrm{D}}^2 = e^2T/3$ is of the order of $(qT)^2$, where $q$ is the constant of QED. Since the number of particles in this density theory $\sim T^3$, the number of particles in the Debye sphere is $\sim 1/q^3$ in the weak-compression limit $(1/q \gg 1)$. Interestingly, the pseudopotential of the interparticle interaction of like charges in QGP may become attractive in some cases [51]. The long-range interparticle interaction inherent in EMP and QGP has the consequence that an important role is played by collective effects such as screening, plasma oscillations, and instability.

Unlike experiments with EMPs, where use is made of external electromagnetic or gravitational fields, for QGPs the fields of requisite intensity are inconceivably high, and only self-induced fields are of significance in relativistic collisions.

The description of an EMP is also very different from that of the QGP [51]. For electromagnetic plasmas, under the corresponding time limitations extensive use is made $(m_{\rm i} \gg m_{\rm e})$ of a two-fluid (electron–ion) model with different electron and ion temperatures. The local electroneutrality condition hinders appreciable charge separation, which leads to equations of magnetohydrodynamics where the plasma motion takes place under the action of pressure gradients and a magnetic field.

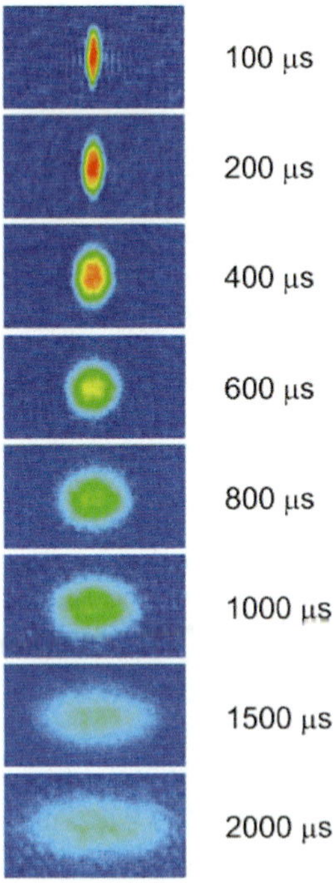


**Fig. 5.30** Asymmetric "elliptical" dynamics of nonideal cryogenic $^6$Li lithium plasma expansion upon deenergizing an electrostatic trap [55]. An analogy with an "elliptical" flow in nuclear collisions.

In QGP there is no analog to magnetohydrodynamics, because every quark or gluon may carry charges of different color. On attaining local thermodynamic equilibrium, different color components will therefore possess the same temperature and velocity. Furthermore, a quark–gluon system becomes neutral in color, and in a QGP the hydrodynamics of a neutral fluid without chromodynamic fields is realized. Naturally, in the absence of local thermodynamic equilibrium use is made of more complex kinetic equations with a "collisional" term of one form or another [32].

To close the equations of motion, which express the mass, momentum, and energy conservation laws, necessitates invoking the equations of state. In the simplest case of an ideal ultrarelativistic gas of massless particles it is assumed that $E_{(x)} = 3P(X)$.

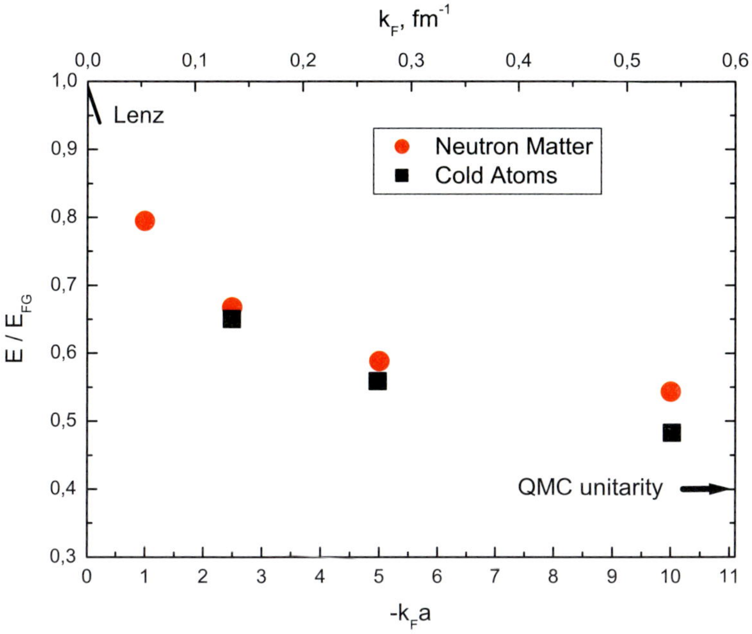


**Fig. 5.31** Reduced energy of neutron matter (circles) and ultracold atoms (squares) [28]. QMC: Quantum Monte Carlo.

Many hydrodynamic and kinetic plasma instabilities typical of EMPs are thought to show up in the QGP [51], although they are extremely hard to observe there. However, the development of these perturbations is associated with the short ($\leqslant 1$fm/$c$) measured QGP thermalization time and the effect of jet quenching in relativistic nuclear collisions. The experimentally observed [51] fast matter thermalization, the parameters of elliptical flows, the spectrum of outgoing particles, jet quenching, and low viscosity are attributed [51] to the nonideality of the QGP close to the deconfinement threshold. The nonideality parameter estimate $\Gamma \approx 1.5$–$5.0$ given above may increase by an order of magnitude when higher-order terms in the interaction potentials are taken into account [51]. This in turn may give rise to a "plasma" phase transition similar to that observed in a strongly nonideal nonrelativistic plasma [22, 20, 26, 25, 24].

In [51], the effects of plasma nonideality revealed for compressed EMP are employed in the analysis of the behavior of the viscosity, cross sections, collisions, and stopping power of QGP. In QGP the ratio between the Landau length $\Gamma \approx q^2/E$ and

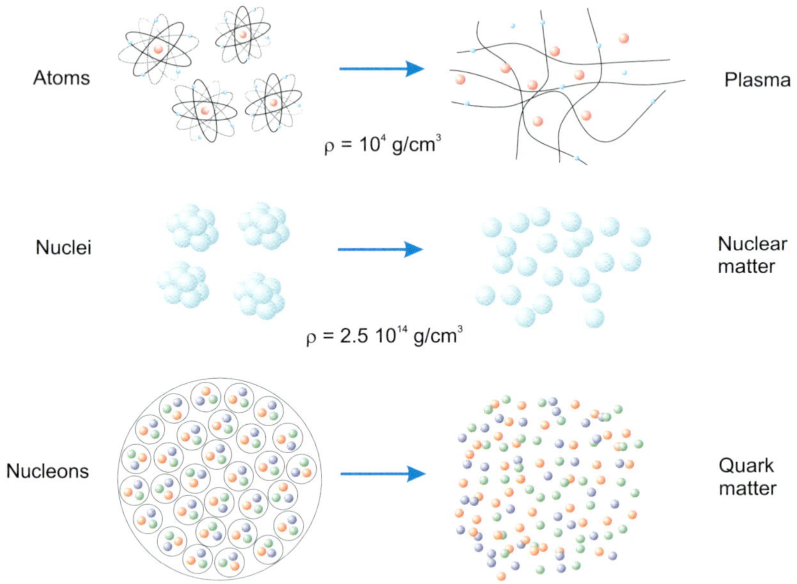


**Fig. 5.32** Matter transformation at high energy densities.

the Debye radius is equal to 1–5; this increases the scattering cross section by a factor of 2–9, which shortens the free path $\lambda$ and hence the viscosity ($\eta \sim \lambda$) also by an order of magnitude. This is consistent with the measured parameters of elliptical flows and particle spectra in nuclear collisions as well as with the recorded increase of radiation losses.

Interesting analogies [51] are drawn between strongly nonideal QGPs and strongly nonideal "dust" plasmas [20, 26, 25]. In both cases we assume to be dealing with a Newtonian liquid, in which the shear viscosity depends on the velocity of motion. Furthermore, the QGP possesses features of a nanoliquid [51]. In particular, the initial QGP size immediately after a collision is $\approx$ 10 fm ($\approx$ 20 interparticle distances), which distinguishes it from a continuous medium. This is also typical for a nonideal "dust" plasma.

In [10], an analysis was made of analogies between a QGP and a strongly nonideal EMP that consists of ions in electrostatic traps cooled by laser and evaporative techniques to ultralow temperatures. In both cases we are dealing with strongly interacting systems with a limited volume and a large number (of order $10^4$–$10^7$) degrees of freedom. Figure 5.30 shows the expansion of a cryogenic ion EMP cloud [55], which exhibits pronounced features of an "elliptical" flow typical of the expansion of superdense nuclear matter in relativistic collision experiments (Fig. 5.25). Similar energy characteristics of neutral matter and ultracold Fermionic atoms (Fig. 5.31) [28] demonstrate a striking similarity at temperatures that differ by $\approx$20 orders of magnitude.

These analogies may be a helpful aid in the elucidation of several fundamental properties of the QGP, especially those which are hard to derive directly from relativistic ion collision experiments:

- kinetic energy distribution in a compressed medium,
- collective motions and their measures,
- screening,
- collisions and expansion,
- flows, hydrodynamics,
- thermalization,
- correlations,
- interaction with fast particles, stopping power,
- viscosity, dissipation.

Further analogies of this kind can be easily added.

To conclude this section we give a scheme of matter transformation at high energy densities (Fig. 5.32), which in a sense extends the conclusions of Sect. 3.6 (Fig. 3.29) about substance simplification as you move toward extremely high pressures and temperatures.

## 5.3 Low-Energy Scan Experiments with Heavy Ions at the NICA Collider

The first accelerator in Russia that allowed beams of high-energy nuclei to be generated was the Synchrophasatron of the Joint Institute for Nuclear Research (JINR) in Dubna [71]. Nowadays this accelerator has been replaced by the Nuclotron, which belongs to a new generation of machines based on the use of superconducting magnets. In the years following the first pioneering Dubna experiments in 1971 on the relativistic acceleration of deuterons, at the Synchrophasatron, and the discovery of the *cumulative effect*[1] in the reactions with nuclei, a new challenging direction in high-energy physics dubbed relativistic nuclear physics [8], a few generations of heavy ion accelerators succeeded each other. Through such technical advances and through the construction of new machines, the laboratory equivalent accessible energy scale grew constantly in time, from a few GeV per nucleon to some tens of TeV per nucleon.

The collider project named Nuclotron-based Ion Collider fAcility (NICA) [75, 74, 48], which was initiated in 2006 at JINR, is similarly devoted to exploring heavy ion collisions in the energy range $\sqrt{S_{NN}} = 4$–$11$ GeV. Such a center-of-mass energy

[1] Cumulative effect means the following. In the collision of elementary particles there are kinematical forbidden regions for secondary particles. If one instead collides an elementary particle, for example a proton, with a compound object, such as a nucleus, secondary particles emerging from the collision can appear also in the kinematically forbidden region as a result of the internal dynamical structure of the compound object. Secondary particles of this type have been named cumulative particles (see e.g. [8] and references therein).

corresponds to about 8–60 GeV per nucleon in the laboratory frame. NICA, with its Multipurpose Particle Detector (MPD), has been defined as the new flagship of the Joint Institute for Nuclear Research in Dubna (JINR). With this instrument, JINR plans to provide a collider facility aimed at the exploration of the high-density region in the QCD phase diagram, thus providing unique opportunities. Specifically NICA will provide collisions of ions over a wide range of atomic masses: nucleon–nucleon, nucleon–nucleus, and nucleus–nucleus collisions are planned. For the time being the maximal foreseen atomic mass is $Au^{79+}$. Hence, altogether, the actual measurements will range from $Au^{79+} + Au^{79+}$ collisions at a center-of-mass energy up to $\sqrt{S_{NN}} = 11$ GeV with an average luminosity of $L = 10^{27}$ cm$^{-2}$ s$^{-1}$, to proton–proton collisions up to $\sqrt{S_{NN}} = 20$ GeV and an average luminosity of $L \approx 10^{30}$ cm$^{-2}$ s$^{-1}$.

The mission of the ambitious NICA project is to investigate the properties of hot and highly compressed nuclear matter, which emerges from collisions of heavy ions in the vicinity of the maximal baryonic density achievable in accelerator experiments (see Fig 5.35). The goal of such investigations is to reveal possible phase transitions, new states of nuclear matter, and signatures of formation of the quark–hadron mixed phase.

The NICA project also foresees the possibility of electron–ion collisions. This adds a unique feature to this facility: the possibility of measurements of electromagnetic form factors at large momenta for the nucleon and nuclei. This includes the investigation of the spatial distribution of charge, magnetization both in nuclei and in the nucleon, as well as the determination of the valence quark generalized parton distribution. These measurements would reveal correlations of spatial and momentum distributions. Finally, through these additional measurements one can study color transparency, which provides a new insight into the behavior of QCD matter at absolute zero.

Summarizing, we have now a collection of new experimental programs that are being developed throughout the world, aimed at the investigation of the QCD phase diagram in its central region (namely at moderate temperatures and at reasonably high baryon densities), where all the most dramatic new features of QCD collective behavior are expected to be located. These are all listed here:

- Low-energy RHIC at Brookhaven (USA).
- NA61 (SHINE) at SPS-CERN (Switzerland).
- CBM (Compressed Baryonic Matter) at FAIR Darmstadt (Germany).
- MPD at JINR-NICA in Dubna (Russian Federation).

RHIC will work in the collider mode, similarly to what is planned for NICA, whereas SPS-CERN and FAIR will operate in the fixed target mode.

Let us now give more detailed information about the forthcoming NICA facility. An essential part of this scientific program arose out of a host of preliminary discussions held since 2005, in view of the necessary upgrade of the JINR Nuclotron. Since its beginnings, this machine had been dedicated to the study of hot and dense baryonic matter, so that its conversion into the NICA mission is a natural development in view of the new worldwide scientific perspectives [76, 77, 75, 74, 73].

Two interaction points are foreseen at the NICA collider (Fig. 5.33). Hence there is room for the installation of two detectors, which will operate simultaneously. One such detectors, named the Multipurpose Particle Detector (MPD; Fig. 5.34), which we already mentioned, will be optimized for to study the properties of hot and dense nuclear matter in heavy-ion collisions. In particular, it will be designed to fulfill the following tasks by means of careful scanning in beam energy and centrality of excitation functions [72]:

- Search for signatures of the hypothetical deconfinement and/or chiral symmetry restoration phase transitions.
- Search for critical points.
- Discovery of the mixed quark–hadron phase.

The second detector, named Spin Particle Detector (SPD) has a different mission, namely the investigation of spin physics both in heavy-ion collisions and in the collisions of polarized nucleon and deuteron beams.

The new JINR facility will make it possible to study in-medium properties of the hadrons. For instance, data will be accessible that can be used to determine the equation of state of nuclear matter, which is very important missing information in astrophysical studies. Neutron stars, whose astrophysical relevance does not need to be emphasized, are macroscopic realizations of nuclear matter and the study of their properties needs an accurate determination of the underlying equation of state.

In the first stage of the NICA project the following measurements are planned:

- Multiplicity and spectral characteristics of the identified hadrons, including strange particles, multi-strange baryons and antibaryons, characterizing entropy production and system temperature at freeze-out.
- Event-by-event fluctuations in multiplicity, charges, transverse momenta and $K/\pi$ ratios as a generic property of critical phenomena.
- Collective flow effects (directed, elliptic, and higher ones) for hadrons including strange particles.
- Hanbury Brown and Twiss (HBT) interferometry of identified particles and particle correlations (femtoscopy).

Femtoscopy, HBT correlations and event-by-event fluctuations reveal important information about the space-time history of the hot deconfined matter, and can help to understand the nature of the phase transition at high baryon density. Event-by-event fluctuations are also expected to provide a signature of the chiral critical point [17, 72].

In the second stage, electromagnetic probes (photons and dileptons) will be measured [17, 72]. Dileptons and photons will provide a powerful penetrating probe of dense QCD matter. They have been extensively and productively used at RHIC and SPS. Because of the small electromagnetic interaction cross sections, dileptons and photons do not suffer from final state interactions even in a dense medium and deliver unique information about the properties of QCD matter throughout its space-time history. The thermal electromagnetic radiation reveals the temperature of the hot QCD matter. Studies of low-mass dileptons enable access to the properties of

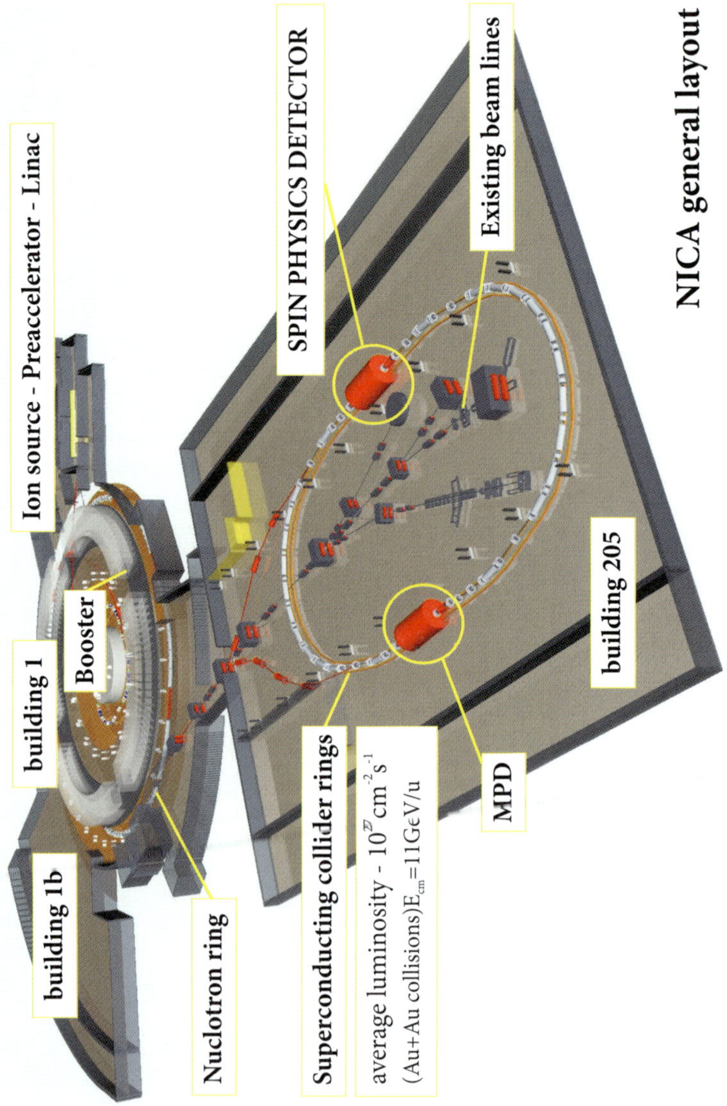


**Fig. 5.33** Location of the NICA collider in the JINR accelerator complex area [48, 17].

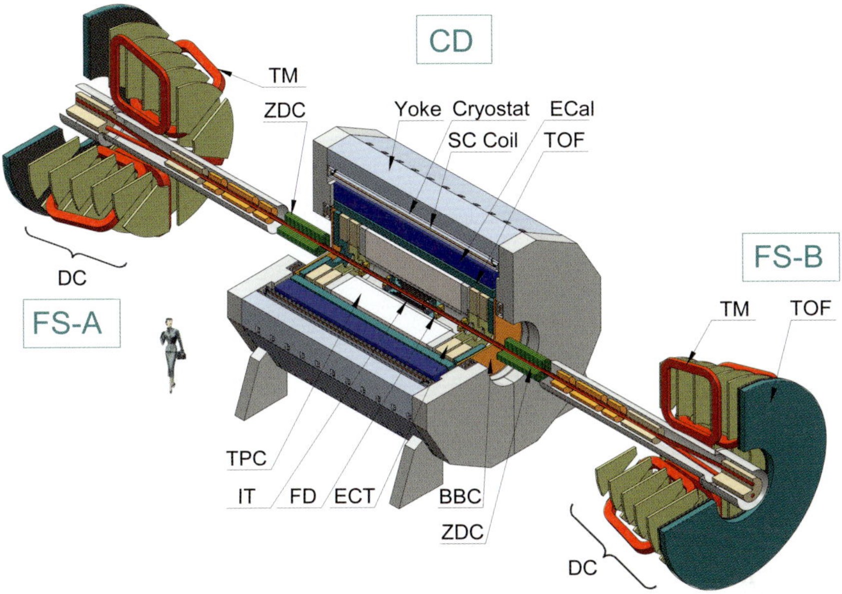


**Fig. 5.34** Schematic of the MPD: central detector (CD) and two forward spectrometers (FS-A and FS-B). BBC: beam–beam counter; DC: drift chambers; ECal: electromagnetic calorimeter; ECT: end-cap tracker; FD: fast-forward detector; IT: inner tracker; SC Coil: superconducting coil; TM: toroidal magnet: TOF: time-of-flight system; TPC: time projection chamber; ZDC: zero degree calorimeters. [72].

dense hadronic matter, and are sensitive to the restoration of chiral symmetry. The density of baryons is known to affect chiral symmetry restoration in a very significant way; therefore the study of electromagnetic probes at NICA energies, where a very high baryon density will be achieved, promises to be very valuable.

The beam energy of NICA is very much lower than that at the original RHIC (BNL) facility and that foreseen for ALICE at LHC (CERN). Yet the NICA energy range sits exactly on the top of the region where baryon density is expected to be highest (see Fig. 5.35). In this energy range the system occupies a maximal space-time volume in the mixed quark-hadron phase (the phase of coexistence of hadron and quark-qluon matter similar to the above-mentioned water-vapor coexistence-phase). The net baryon density at LHC energies is predicted to be lower. The energy region of NICA will allow analyzing the highest baryonic density under laboratory conditions.

Most likely, the chiral critical point (see Fig. 5.13) is located at a baryon density within the reach of the NICA collider, as well. A number of related phenomena can also be traced by experimental observations at NICA. In particular an interesting set of problems deals with the nonequilibrium and finite-size behavior in deconfined matter.

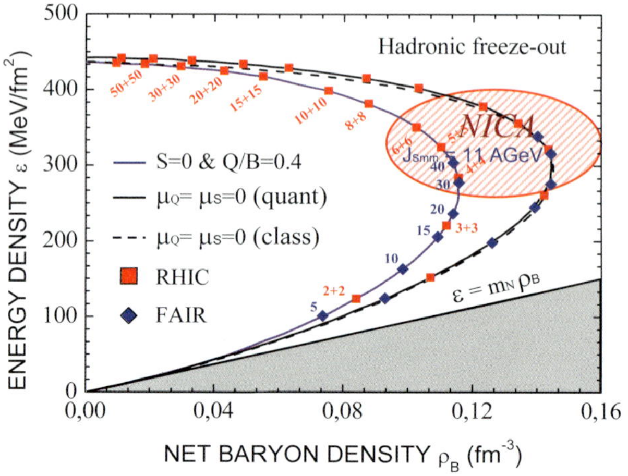


**Fig. 5.35** Freeze-out (ceasing of particle interactions in the system) estimated for different colliding energies [61]. The freeze-out baryon density is maximal at the collider energy $\sqrt{S_{NN}} = 4 + 4$ GeV. The blue numbers give the energy in the laboratory system, the red ones in the center-of-mass system.

The mechanism of multiparticle production by which the energy of the colliding hadrons or nuclei is converted into the final-state particles is a long-standing problem in physics. There has been significant progress in understanding such a mechanism at high energies, where the strong color fields, inside the colliding hadrons or nuclei, play an important role. The systematics of data on hadron production shows that such an approach formulated in terms of gluons and quarks can be applied starting from a center-of-mass energy of 15 GeV. Below this energy, the dynamics of multiparticle production should probably be described in terms of some effective low energy degrees of freedom. The deviations from the trends earlier established at higher energies at RHIC and SPS should provide an important clue to such dynamics.

An important part of the NICA program is also the physics of high multiplicity events, which are expected to provide access to QCD in the high parton density regime even at a moderate energy. It will be very interesting to explore the evolution of the produced system towards freeze-out through a possible mixed phase. Complementarity and duality in multiparticle production are appealing concepts that can be tested at NICA energies. A detailed study of multiparticle production in the energy range of the NICA collider will thus contribute to the understanding of QCD in the nonperturbative domain [17].

Cumulative processes are those that are kinematically forbidden for (quasi-)free nucleons. Therefore, they necessarily involve collective properties of the nuclear constituents and allow unique information to be extracted about the short-distance correlations of quarks and gluons in cold and hot hadronic matter.

Let us now turn to some other interesting aspects of high-energy-density physics that NICA will enable to be experimentally investigated.

The first concerns the existence of topological solutions, which is a distinctive and very important feature of QCD. Topological solutions arise due to the non-Abelian nature of the theory and lead to a number of important consequences for the properties of the vacuum state and for hadrons. In hot hadronic matter, topological fluctuations near the critical temperature can create space-time domains with locally broken P (parity) and/or CP (charge and parity) invariance.

In heavy-ion collisions, the local violation of parity can manifest itself through the separation of positive and negative hadrons with respect to the reaction plane; this charge separation would induce an electric dipole moment of the produced hot QGP. The charge separation stems from the interplay of a strong magnetic field (and/or angular momentum) in the early stage of the heavy-ion collision and the presence of topological configurations in hot matter ("the chiral magnetic effect"). Recent results from RHIC [84] provide experimental evidence for this phenomenon in heavy-ion c. This effect should be enhanced in the deconfined QGP phase, since it requires the separation of quarks over a large ("macroscopic" on the scale of hot matter size) distance. There are reasons to expect that the asymmetry will have a peak at some energy, possibly within the energy range of the NICA collider [17]. It should be emphasized that at high baryon density one can also expect the phenomenon of spontaneous parity violation with a number of unique signatures. The high-energy and low-energy scans at RHIC and the dedicated programs at NICA and FAIR would thus provide experimental information that is necessary and complementary.

The next theoretical aspect accessible to NICA concerns polarization effects and spin physics. Here there are two important issues:

- polarization effects in heavy ion collisions,
- the spin program with polarized beams.

Studies of polarization effects in heavy-ion collisions have just began. The conversion of the huge initial angular momentum available in a heavy-ion collision into the angular momentum carried by the spectators, by dense matter, and possibly by the spins of produced particles is an extremely interesting problem. Understanding it will undoubtedly allow us to penetrate deeper into collision dynamics. A number of observables have been recently proposed [17]. Most of them have never been studied before in heavy-ion experiments. Here, the physical issues include the dynamics of the collision, the chiral properties of the produced medium, and of that of the possible local violation of P and CP invariance discussed above.

The traditional spin program is also an important and integral part of the NICA project. Indeed, ever since the "spin crisis" of 1987, the composition of the nucleon spin in terms of fundamental constituents – quarks and gluons – remains at the focus of attention of physicists. Highlights of the NICA spin program include:

- measurements of Drell–Yan processes with longitudinally polarized proton and deuteron beams,

- spin effects in the inclusive and exclusive production of baryons,
- light and heavy mesons and direct photons,
- studies of helicity amplitudes and double spin asymmetries in elastic scattering.

It appears that the SPD (Fig 5.33) at NICA would allow significant contributions to the currently planned international programs in spin physics [17].

Many of these important physical issues will be addressed in the near future during the low-energy RHIC and CERN scans. The experimental results from these facilities will help the other two experimental projects (FAIR/CBM and NICA/MPD) to identify even more precisely the most promising directions for research. These would benefit, compared with low-energy RHIC, from the higher beam luminosity planned at NICA and from the dedicated detector design of MPD.

The NICA collider research program has substantial overlap with that of FAIR. However the approaches taken by FAIR and NICA are complementary to each other. For instance, the fixed target mode of FAIR makes it easier to access rare probes, whereas the collider setup of NICA provides a symmetrical, uniform, and energy-independent acceptance, which is beneficial for studies of various fluctuations and correlations (including P- and CP-odd ones). In conclusion, all the future low-energy scan experimental programs are complementary to each other and are interrelated. Their cooperation will be mutually beneficial.

In an apparently paradoxical manner, the contemporary energy downgrade of heavy-ion accelerators seems to be the outcome of the reverse running of H. G. Wells's time machine. Indeed, accelerators are being downgraded to, or planned for, energies that were accessible many years ago. Yet there is no paradox because the new low energy scans will be performed with highly improved high-tech modern detectors benefiting from new knowledge and frontier theoretical visions not available at earlier times. This implies the possibility of revealing critical phenomena that completely escaped detection in earlier scans. Moreover the new scans will certainly enable the correction of all mistakes and imprecise measurements of the past, leading, it is hoped, to a new-found appreciation of the real nature of strong interactions.

Let us finally remark that what is in general true for all experimental discoveries, namely that they should be confirmed in different laboratories at different times and by different research teams, is even more true in the case of such complex, subtle collective phenomena as those involved in the QCD phase diagram. Here, in order to obtained a well-established and firm understanding, complementary and overlapping information from different experiments is imperative. The recent developments in dense-matter physics have shown that we must be prepared for the unexpected!

## References

[1] Adams, J., Adler, C., Aggarwal, M.M., et al.: Azimuthal anisotropy at the relativistic heavy ion collider: the first and fourth harmonics. Phys. Rev. Lett. **92**(6), 062301 (2004). DOI 10.1103/PhysRevLett.92.062301

[2] Adams, J., Adler, C., Aggarwal, M.M., et al.: Particle-type dependence of azimuthal anisotropy and nuclear modification of particle production in Au+Au collisions at $\sqrt{s_{NN}}$ = 200 GeV. Phys. Rev. Lett. **92**(5), 052302 (2004). DOI 10.1103/PhysRevLett.92.052302

[3] Adler, C., Ahammed, Z., Allgower, C., et al.: Disappearance of back-to-back high-$p_T$ hadron correlations in central Au+Au collisions at $\sqrt{s_{NN}}$= 200 GeV. Phys. Rev. Lett. **90**(8), 082302 (2003). DOI 10.1103/PhysRevLett.90.082302

[4] Adler, S.S., Afanasiev, S., Aidala, C., et al.: Elliptic flow of identified hadrons in Au+Au collisions at $\sqrt{s_{NN}} = 200$ GeV. Phys. Rev. Lett. **91**(18), 182301 (2003). DOI 10.1103/PhysRevLett.91.182301

[5] Alt, C., Anticic, T., Baatar, B., et al.: Directed and elliptic flow of charged pions and protons in Pb+Pb collisions at 4*A* and 158*A* GeV. Phys. Rev. C **68**(3), 034903 (2003). DOI 10.1103/PhysRevC.68.034903. URL `http://link.aps.org/doi/10.1103/PhysRevC.68.034903`

[6] Anisimov, S.I., Prokhorov, A.M., Fortov, V.E.: Application of high-power lasers to study matter at ultrahigh pressures. Sov. Phys. – Usp. **27**(3), 181–205 (1984). DOI 10.1070/PU1984v027n03ABEH004036. URL `http://stacks.iop.org/0038-5670/27/181`

[7] Atzeni, S., Meyer-ter-Vehn, J.: The Physics of Inertial Fusion. Oxford University Press, Oxford (2004)

[8] Baldin, A.M., Malakhov, A.I., Sissakian, A.N.: Some problems of relativistic nuclear physics and multiple particle production [in Russian]. Phys. Elementary Part. At. Nucl. **32**(7), 6 (2001)

[9] Baumung, K., Bluhm, H.J., Goel, B., et al.: Shock-wave physics experiments with high-power proton beams. Laser Part. Beams **14**(2), 181 (1996)

[10] Baym, G.: Matter under extreme conditions. Presented at the ExtreMe Matter Institute EMMI Kick-Off Meeting & Symposium, July 16–17, 2008, GSI, Darmstadt, Germany

[11] Baym, G., Chin, S.A.: Can a neutron star be a giant MIT bag? Phys. Lett. B **62**(2), 241–244 (1976). DOI 10.1016/0370-2693(76)90517-7

[12] Chapline, G., Nauenberg, M.: Asymptotic freedom and the baryon–quark phase transition. Phys. Rev. D **16**(2), 450–456 (1977). DOI 10.1103/PhysRevD.16.450

[13] Collins, J.C., Perry, M.J.: Superdense matter: Neutrons or asymptotically free quarks? Phys. Rev. Lett. **34**(21), 1353–1356 (1975). DOI 10.1103/PhysRevLett.34.1353

[14] Csikor, F., Egri, G.I., Fodor, Z., et al.: The QCD equation of state at finite $T\mu$ on the lattice. Prog. Theor. Phys. Suppl. **153**, 93–105 (2004)

[15] Cuneo, M.E., Adams, R.G., Bailey, J.E., et al.: Generating high-brightness light ion beams for inertial fusion energy (IFP/14) (1998). URL `http://www.iaea.org/programmes/ripc/physics/html/iaeacn69.htm`

[16] Cuneo, M.E., Vesey, R.A., Bennett, G.R., et al.: Progress in symmetric ICF capsule implosions and wire-array Z-pinch source physics for double-pinch-

driven hohlraums. Plasma Phys. Control. Fusion **48**(2), R1–R35 (2006). DOI 10.1088/0741-3335/48/2/R01

[17] D. Blaschke et al. (Eds): Searching for a QCD mixed phase at the Nuclotron-based Ion Collider fAcility (NICA white paper) (2009). URL `http://theor.jinr.ru/twiki/pub/NICA/WebHome/Wh_Paper_dk6.pdf`

[18] Efremov, V.P., Pikuz Jr., S.A., Faenov, A.Y., et al.: Study of the energy release region of a heavy-ion flux in nanomaterials by X-ray spectroscopy of multicharged ions. JETP Lett. **81**(8), 378 (2005)

[19] Fodor, Z., Katz, S.D.: Critical point of QCD at finite $T$ and $\mu$, lattice results for physical quark masses. J. High Energy Phys. **2004**(04), 050 (2004). URL `http://stacks.iop.org/1126-6708/2004/i=04/a=050`

[20] Fortov, V., Iakubov, I., Khrapak, A.: Physics of Strongly Coupled Plasma. Oxford University Press, Oxford (2006)

[21] Fortov, V., Rudakov, L., Ni, A.: Application of intense relativistic electron beams in high dynamic pressure thermophysics. Sov. Thermal Phys. Rev. **371**, 589 (1992)

[22] Fortov, V.E.: Intense shock waves and extreme states of matter. Phys. Usp. **50**(4), 333 (2007). DOI 10.1070/PU2007v050n04ABEH006234. URL `http://ufn.ru/en/articles/2007/4/c/`

[23] Fortov, V.E., Hoffmann, D.H.H., Sharkov, B.Y.: Intense ion beams for generating extreme states of matter. Phys. Usp. **51**(2), 109 (2008). DOI 10.1070/PU2008v051n02ABEH006420. URL `http://ufn.ru/en/articles/2008/2/a/`

[24] Fortov, V.E., Ilkaev, R.I., Arinin, V.A., et al.: Phase transition in a strongly nonideal deuterium plasma generated by quasi-isentropical compression at megabar pressures. Phys. Rev. Lett. **99**(18), 185001 (2007). DOI 10.1103/PhysRevLett.99.185001. URL `http://link.aps.org/abstract/PRL/v99/e185001`

[25] Fortov, V.E., Ivlev, A.V., Khrapak, S.A., et al.: Complex (dusty) plasma: current status, open issues, perspectives. Phys. Rep. **421**(1), 1–103 (2005). DOI 10.1016/j.physrep.2005.08.007

[26] Fortov, V.E., Khrapak, A.G., Yakubov, I.T.: Fizika neideal'noi plazmy (Physics of Nonideal Plasma). Fizmatlit, Moscow (2004)

[27] Freedman, B.A., McLerran, L.D.: Fermions and gauge vector mesons at finite temperature and density. III. The ground-state energy of a relativistic quark gas. Phys. Rev. D **16**(4), 1169–1185 (1977). DOI 10.1103/PhysRevD.16.1169

[28] Gezerlis, A., Carlson, J.: Strongly paired fermions: Cold atoms and neutron matter. Phys. Rev. C **77**(3), 032801 (2008). DOI 10.1103/PhysRevC.77.032801. URL `http://link.aps.org/abstract/PRC/v77/e032801`

[29] Ginzburg, V.L.: The Physics of a Lifetime: Reflections on the Problems and Personalities of 20th Century Physics. Springer, Berlin, Heidelberg (2001)

[30] Glendenning, N.K.: Compact Stars: Nuclear Physics, Particle Physics, and General Relativity, 2nd edn. Springer, New York (2000)

[31] Gyulassy, M.: Quark gluon plasmas: Femto cosmology with A+A @ LHC. Presented at the ExtreMe Matter Institute EMMI Kick-Off Meeting & Symposium, July 16–17, 2008, GSI, Darmstadt, Germany
[32] Gyulassy, M., McLerran, L.: New forms of QCD matter discovered at RHIC. Nucl. Phys. A **750**(1), 30–63 (2005). DOI 10.1016/j.nuclphysa.2004.10.034
[33] Gyulassy, M., Plümer, M.: Jet quenching as a probe of dense matter. Nucl. Phys. A **527**, 641–644 (1991). DOI 10.1016/0375-9474(91)90173-4
[34] Gyulassy, M., Plümer, M., Thoma, M., Wang, X.N.: High $P_T$ probes of nuclear collisions. Nucl. Phys. A **538**, 37–49 (1992). DOI 10.1016/0375-9474(92)90756-A
[35] Hands, S.: The phase diagram of QCD. Contemp. Phys. **42**(4), 209–225 (2001). DOI 10.1080/00107510110063843. URL http://pdfserve.informaworld.com/104165_762903897_713806686.pdf
[36] Hoffmann, D.H.H., Fortov, V.E., Lomonosov, I.V., et al.: Unique capabilities of an intense heavy ion beam as a tool for equation-of-state studies. Phys. Plasmas **9**(9), 3651–3654 (2002). DOI 10.1063/1.1498260
[37] Jacobs, P., Klay, J.: Jets and high $p_T$ hadrons in dense matter: recent results from STAR (2003). URL http://arxiv.org/abs/nucl-ex/0308023
[38] Kalashnikov, O.K., Klimov, V.V.: Phase transition in the quark–gluon plasma. Phys. Lett. B **88**(3-4), 328–330 (1979). DOI 10.1016/0370-2693(79)90479-9
[39] Kanel, G.I., Rasorenov, S.V., Fortov, V.E.: Shock-Wave Phenomena and the Properties of Condensed Matter. Springer, New York (2004)
[40] Kapusta, J.I.: Quantum chromodynamics at high temperature. Nucl. Phys. B **88**(3-4), 461–498 (1979). DOI 10.1016/0550-3213(79)90146-9
[41] Karsch, F.: Lattice QCD at high temperature and the QGP (2006). URL http://arxiv.org/abs/hep-lat/0601013
[42] Knudson, M.D., Hanson, D.L., Bailey, J.E., et al.: Equation of state measurements in liquid deuterium to 70 GPa. Phys. Rev. Lett. **87**(22), 225501 (2001). DOI 10.1103/PhysRevLett.87.225501. URL http://link.aps.org/abstract/PRL/v87/e225501
[43] Krasnikov, N.V., Matveev, V.A.: The search for new physics at the Large Hadron Collider. Phys. Usp. **47**(7), 643 (2004). DOI 10.1070/PU2004v047n07ABEH001767. URL http://ufn.ru/en/articles/2004/7/a/
[44] Kruer, W.L.: The Physics of Laser Plasma Interactions. Addison-Wesley, Reading, MA (1988)
[45] Langanke, L.: A FAIR chance for nuclear astrophysics (2007). Kick-off event and symposium on the physics at FAIR
[46] Lindl, J.D.: Inertial Confinement Fusion. Springer, New York (1998)
[47] MacFarlane, J.J., Wang, P., Bailey, J., et al.: Analysis of K$\alpha$ line emission from aluminum plasmas created by intense proton beams. Phys. Rev. E **47**(4), 2748–2758 (1993). DOI 10.1103/PhysRevE.47.2748. URL http://link.aps.org/abstract/PRE/v47/p2748

[48] Meshkov, I., Sidorin, A. (Eds): Design and construction of Nuclotron-based Ion Collider fAcility (NICA), conceptual design report (2008). URL http://nica.jinr.ru/files/NICA_CDR.pdf
[49] Mesyats, G.A.: Impul'snaya energetika i elektronika (Pulse Power and Electronics). Nauka, Moscow (2004)
[50] Mintsev, V., Gryaznov, V., Kulish, M., et al.: Stopping power of proton beam in a weakly non-ideal xenon plasma. Contrib. Plasma Phys. **39**(1-2), 45–48 (1999). DOI 10.1002/ctpp.2150390111
[51] Mrowczynski, S., Thoma, M.H.: What do electromagnetic plasmas tell us about the quark–gluon plasma? Annu. Rev. Nucl. Part. Sci. **57**(1), 61–94 (2007). DOI 10.1146/annurev.nucl.57.090506.123124. URL http://arjournals.annualreviews.org/doi/abs/10.1146/annurev.nucl.57.090506.123124
[52] National Research Council: Frontiers in High Energy Density Physics. National Academies Press, Washington, DC (2003)
[53] NICA: URL http://theor.jinr.ru/twiki-cgi/view/NICA
[54] Novikov, I.D.: "Big Bang" echo (cosmic microwave background observations). Phys. Usp. **44**(8), 817 (2001). DOI 10.1070/PU2001v044n08ABEH000983. URL http://ufn.ru/en/articles/2001/8/h/
[55] O'Hara, K.M., Hemmer, S.L., Gehm, M.E., et al.: Observation of a strongly interacting degenerate Fermi gas of atoms. Science **298**(5601), 2179–2182 (2002). DOI 10.1126/science.1079107. URL http://www.sciencemag.org/cgi/content/abstract/298/5601/2179
[56] Okun', L.B.: Leptony i kvarki, 2nd edn. Nauka, Moscow (1990). [English Transl.: Leptons and Quarks. North-Holland, Amsterdam (1982)]
[57] Ollitrault, J.Y.: Anisotropy as a signature of transverse collective flow. Phys. Rev. D **46**(1), 229–245 (1992). DOI 10.1103/PhysRevD.46.229
[58] Pieranski, P.: Colloidal crystals. Contemp. Phys. **24**(1), 25–70 (1983). DOI 10.1080/00107518308227471
[59] Quigg, C.: The coming revolutions in particle physics. Sci. Am. **298**(2), 46 (2008)
[60] Quintenz, J., Sandia's Pulsed Power Team: Pulsed power team. In: Proc. 13th Int. Conf. on High Power Particle Beams. Nagaoka, Japan (2000)
[61] Randrup, J., Cleymans, J.: Exploring high-density baryonic matter: Maximum freeze-out density. In: Searching for a QCD mixed phase at the Nuclotron-based Ion Collider fAcility. (NICA White Paper), p. 16. JINR, Dubna, Russia (2009). URL http://theor.jinr.ru/twiki/pub/NICA/WebHome/Wh_Paper_dk6.pdf
[62] Riordan, M., Zajc, W.A.: The first few microseconds. Sci. Am. **294**(5), 34A–41 (2006)
[63] Rosmej, O.N., Blazevic, A., Korostiy, S., et al.: Charge state and stopping dynamics of fast heavy ions in dense matter. Phys. Rev. A **72**(5), 052901 (2005). DOI 10.1103/PhysRevA.72.052901. URL http://link.aps.org/abstract/PRA/v72/e052901

[64] Rubakov, V.A.: Large and infinite extra dimensions. Phys. Usp. **44**(9), 871 (2001). DOI 10.1070/PU2001v044n09ABEH001000. URL http://ufn.ru/en/articles/2001/9/a/
[65] Rubakov, V.A.: Introduction to cosmology. Proc. Sci. **RTN2005** (2005). URL http://pos.sissa.it/archive/conferences/019/003/RTN2005_003.pdf
[66] Rubakov, V.A.: Hierarchies of fundamental constants (to items Nos 16, 17, and 27 from Ginzburg's list). Phys. Usp. **50**(4), 390 (2007). DOI 10.1070/PU2007v050n04ABEH006240. URL http://ufn.ru/en/articles/2007/4/i/
[67] Russel, W.B., Saville, D.A., Schowalter, W.R.: Colloidal Dispersions. Cambridge University Press, Cambridge (1989)
[68] Sharkov, B.Y. (ed.): Yadernyi sintez s inertsionnym uderzhaniem (Inertial Confinement Nuclear Fusion). Fizmatlit, Moscow (2005)
[69] Shuryak, E.V.: Quark–gluon plasma and hadronic production of leptons, photons and psions. Phys. Lett. B **78**(1), 150–153 (1978). DOI 10.1016/0370-2693(78)90370-2
[70] Shuryak, E.V.: Quantum chromodynamics and the theory of superdense matter. Phys. Rep. **61**(2), 71–158 (1980). DOI 10.1016/0370-1573(80)90105-2
[71] Sissakian, A., Sorin, A.S.: The QCD Phase Diagram NICA, JINR Communication. JINR, Dubna, Russia (2009)
[72] Sissakian, A., et al.: The MultiPurpose Detector – MPD To Study Heavy Ion Collisions at NICA. Conceptual Design Report. JINR, Dubna, Russia (2009)
[73] Sissakian, A.N.: Frame projects breakthrough to future [in Russian]. Nauka v Rossii [Science in Russia] **6**, 4–11 (2008)
[74] Sissakian, A.N., Sorin, A.S.: The nuclotron-based ion collider facility (NICA) at JINR: new prospects for heavy ion collisions and spin physics. J. Phys. G: Nucl. Part. Phys. **36**(6), 064069 (2009). DOI 10.1088/0954-3899/36/6/064069
[75] Sissakian, A.N., Sorin, A.S., Suleymanov, M.K., et al.: Towards searching for a mixed phase of strongly interacting QCD matter at the JINR nuclotron (2006). URL http://arxiv.org/abs/nucl-ex/0601034
[76] Sissakian, A.N., Sorin, A.S., Suleymanov, M.K., et al.: Properties of strongly interacting matter and the search for a mixed phase at the JINR nuclotron. Phys. Part. Nucl. Lett. **5**(1), 8–17 (2008)
[77] Sissakian, A.N., Sorin, A.S., Toneev, V.D.: QCD Matter: A Search for a Mixed Quark–Hadron Phase (2006). URL http://arxiv.org/abs/nucl-th/0608032
[78] Sorensen, P.R.: Kaon and lambda production at intermediate $P_T$: Insights into the hadronization of the bulk partonic matter created in Au+Au collisions at RHIC. Ph.D. thesis, University of California, Los Angeles (2003)
[79] Spielman, R.B., Deeney, C., Chandler, G.A., et al.: Tungsten wire-array Z-pinch experiments at 200 TW and 2 MJ. Phys. Plasmas **5**(5), 2105–2111 (1998). DOI 10.1063/1.872881

[80] Stöcker, H., Hofmann, J., Maruhn, J.A., Greiner, W.: Shock waves in nuclear matter – proof by circumstantial evidence. Prog. Part. Nucl. Phys. **4**, 133–195 (1980). DOI 10.1016/0146-6410(80)90006-X

[81] Tahir, N.A., Deutsch, C., Fortov, V.E., et al.: Proposal for the study of thermophysical properties of high-energy-density matter using current and future heavy-ion accelerator facilities at GSI Darmstadt. Phys. Rev. Lett. **95**(3), 035001 (2005). DOI 10.1103/PhysRevLett.95.035001. URL `http://link.aps.org/abstract/PRL/v95/e035001`

[82] Tahir, N.A., Deutsch, C., Fortov, V.E., et al.: Studies of strongly coupled plasmas using intense heavy ion beams at the future FAIR facility: The HEDgeHOB collaboration. Contrib. Plasma Phys. **45**(3-4), 229–235 (2005). DOI 10.1002/ctpp.200510025

[83] Vitev, I., Gyulassy, M.: High-$P_T$ tomography of d+Au and Au+Au at SPS, RHIC, and LHC. Phys. Rev. Lett. **89**(25), 252301 (2002). DOI 10.1103/PhysRevLett.89.252301

[84] Voloshin, S.A., for the STAR Collaboration: Probe for the strong parity violation effects at RHIC with three particle correlations (2008). URL `http://arxiv.org/abs/0806.0029`

[85] Wang, X.N., Gyulassy, M.: Gluon shadowing and jet quenching in $A+A$ collisions at $\sqrt{s} = 200A$ GeV. Phys. Rev. Lett. **68**(10), 1480–1483 (1992). DOI 10.1103/PhysRevLett.68.1480

# Chapter 6
# Technical Applications of the Physics of High Energy Densities

The unique possibilities of the physics of extreme states of matter became quite obvious just after the successful experiments (Fig. 0.1) of David [1] conducted nearly three thousand years ago. Since that time the range of application in this field has continued to expand, steadily spreading higher up the scale of pressure and temperature, and has involved ever wider spheres of human activity. Defense, including chemical and thermonuclear charges; all the kinds of carbon, nuclear, and thermonuclear energy sources; the synthesis of superhard and heat-resistant materials (diamond, boron nitride, ceramics); explosive welding of metals; mining operations and applied geophysics; and lately even medicine – this is just an incomplete list of applications of modern high-energy-density physics.

Defense applications, in addition to energy applications, were and still are the first and main objects of application of extreme states of matter. The reason is that in order to strike an enemy it is necessary to apply pressures exceeding the yield strength of human body (ca. 0.1 atm) or its defense (armor, 50 kbar). That leads to the necessary speed at which a micro-projectile has to be launched: tens of meters or even kilometers per second. The process of combustion and detonation completely fulfills the required conditions, and the process of hydrodynamic (or electrodynamic) cumulation increases the shock velocity of ordinary ammunition.

The energy density of nuclear explosives is 6–7 orders higher than the energy capacity of chemical explosives. That is why the power of nuclear ammunition is many orders of magnitude greater [91] than that of chemical ammunition. It is important that a nuclear explosion not only generates ultraextreme densities in matter but also requires multi-megabar pressures in highly compressed nuclear fuel for its activation.

As for power generation, the necessity of increasing the pressure and temperature of the working medium in order to increase the energetic efficiency of processes of energy transformation became quite clear in the 19th century during the industrial revolution (the century of fire and steam). This transformation efficiency becomes clear by the energy flow divergence, which grows with increasing pressure and temperature of the working medium. For equilibrium thermal processes, the efficiency, according to Carnot's rule, also increases with the growth of the parameters of the

V.E. Fortov, *Extreme States of Matter*, The Frontiers Collection,
DOI 10.1007/978-3-642-16464-4_6, © Springer-Verlag Berlin Heidelberg 2011

working medium. So the transition to extreme, reachable pressures and temperatures was always the ideological basis of progress in power production. In Table 6.1 the main kinds of energy generators are listed together with an indication of the magnitude of flow of the transformed energy and characteristic parameters of the working medium.

**Table 6.1** The main kinds of energy generators with an indication of magnitude of flow of the transformed energy ($\pi$: Umov–Poynting vector, $\pi \sim \sqrt{T} \cdot P$) and characteristic parameters of the working medium.

| Generator | $\pi$ [au] | $T$ [K] | $P$ [bar] | Working medium |
|---|---|---|---|---|
| Renkine cycle: steam/gas turbines | $10^3$ | 800 | 250 | $H_2O$ |
| Renkine cycle: MHD (magnetohydrodynamic) generators | $10^4$ | 3 000 | 200 | gas, coals |
| Explosive MHD generators | $10^5$ | 30 000 | 1000 | Ar, Xe |
| Magnetic confinement fusion (ITER) | $10^6$ | $10^8$ | 100 | DT |
| Explosion-magnetic generators | $10^8$ | $10^4$ | $10^6$ | metal |
| Magnetic confinement fusion (Z-pinch) | $10^{10}$ | $10^8$ | $10^6$ | DT |
| Inertial confinement fusion | $10^{14}$ | $10^8$ | $10^{10}$ | DT |

Let us emphasize that for a nuclear ignition reaction in controlled thermonuclear fusion facilities with magnetic and, especially, inertial confinement, ultrahigh temperatures, pressures, and compression of the working medium are required. That is a most significant pragmatic motivation for research in high-energy-density physics.

## 6.1 Laser Inertial Confinement Fusion

### *6.1.1 Direct Drive*

Two directions in thermonuclear research are presently being developed: controlled thermonuclear fusion with either magnetic or inertial confinement (simply known as magnetic confinement fusion and inertial confinement fusion). With magnetic confinement, the thermonuclear plasma heated to $T \approx 10$ keV is retained under quasi-stationary conditions with the help of magnetic fields. The research status and prospects for this direction are represented in Fig. 6.1 and a drawing of the International Thermonuclear Experimental Reactor (ITER), which is now under construction and will use magnetic confinement, is shown in Fig. 6.2.

Controlled thermonuclear fusion with inertial confinement relies on the feasibility of obtaining positive thermonuclear energy release in the form of microexplosions initiated by laser, X-ray, or heavy-ion radiation with an energy of several megajoules and duration of about 1 ns. This line of research is the most signifi-

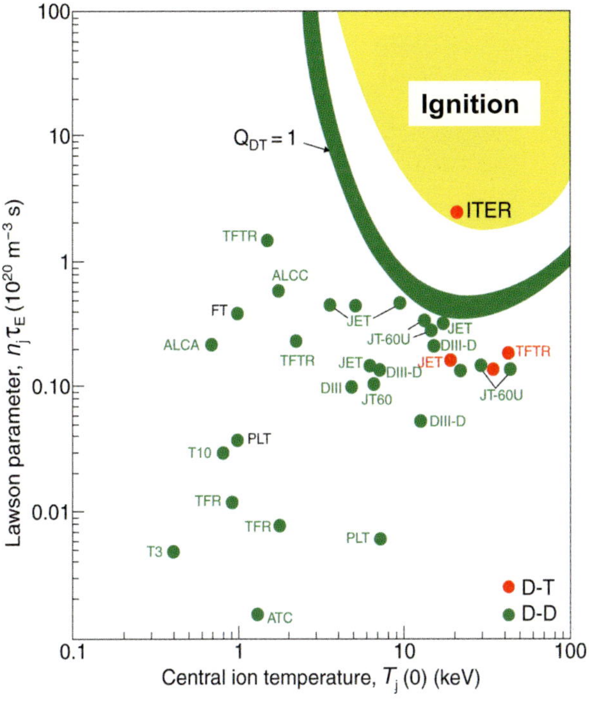


**Fig. 6.1** Parameters of thermonuclear fusion facilities with magnetic plasma confinement.

cant pragmatic motivation for research in high-energy-density physics [101, 84, 15, 90, 113, 28, 74]. Facilities constructed for this purpose and to be commissioned in the near future include the biggest laser facilities NIF (National Ignition FAcility, Livermore, CA, USA) [70, 99] and LMJ (Mégajoule Laser, Bordeaux, France) [27, 59], high-current Z-pinches [120, 33, 106], and later the new-generation relativistic heavy-ion accelerators, which will be employed for modeling the physical processes in thermonuclear weapons and studies in nuclear physics, along with the thermonuclear program.

The temperature dependence of the thermonuclear reaction rates (Fig. 6.3) makes deuterium–tritium fusion the reaction of choice; a temperature of $\approx$ 2–10 keV is required to make this reaction highly efficient. For a thermonuclear plasma expansion velocity of $10^8$ cm/s, this leads to a characteristic time of $\approx 10^{-9}$ s in the case of a spherical target $\approx 10^{-1}$ cm in diameter. The energy balance condition for this thermonuclear reaction (the Lawson criterion [84, 15, 90, 74]) is of the form

$$n\tau \approx \rho r/4c_s m_i \approx 2\times 10^{14}\ \text{s/cm}^3,$$

which corresponds to $\rho r \approx 0.1$–3 g/cm$^2$ and requires the thermonuclear fuel to be compressed to densities of hundreds of grams per cubic centimeter. At a temperature

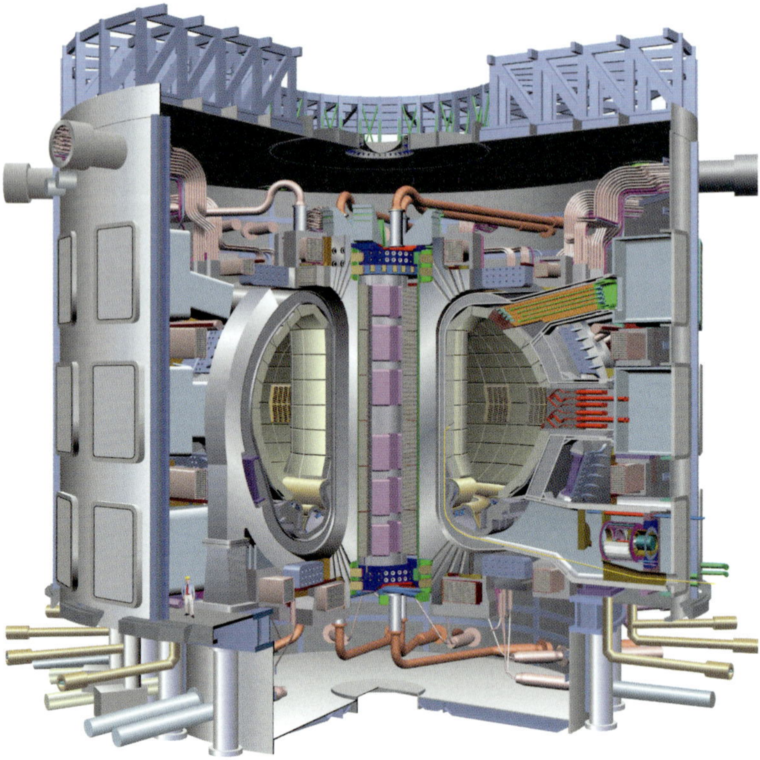

**Fig. 6.2** Experimental thermonuclear reactor ITER.

of 10 keV this corresponds to tremendous pressures of 100–200 Gbar. In particular, to release a thermonuclear energy equivalent to 0.1 tons of trinitrotoluene or 500 MJ requires three milligrams of the deuterium–tritium fuel, which corresponds to a microtarget $\approx 300\ \mu$m in diameter. As a result of laser irradiation, the fuel density at the target center amounts to 90 g/cm$^3$ for a compression energy of 30 kJ and a pressure of 13.5 Gbar, which may be achieved for an implosion rate of $1.4 \times 10^7$ cm/s. These parameters may vary significantly, depending on the specific target design and the "driver" selected.

The attainment of so high a compression is hindered by the high plasma temperature, which is lowered by way of "soft" quasi-adiabatic compression. To achieve this requires a special profiling of the laser radiation, which has to build up in time by the end of compression [84, 15, 90].

By and large, despite the relative simplicity of the initial idea and its practical realization in nuclear weapons 60 years ago, the technical problems encountered in the implementation of pulsed controlled thermonuclear fusion have turned out to be extremely complicated. The main problem consists in the efficient delivery of long-wavelength laser radiation energy to the dense thermonuclear plasma. In particular,

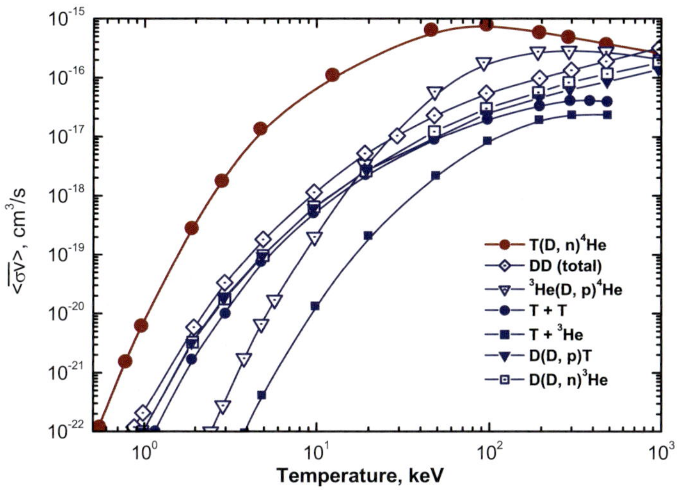


**Fig. 6.3** Thermonuclear reaction rate coefficient as a function of temperature [74]. One can see that the deuterium–tritium fusion reaction possesses the largest cross section and is therefore easiest to implement in practice.

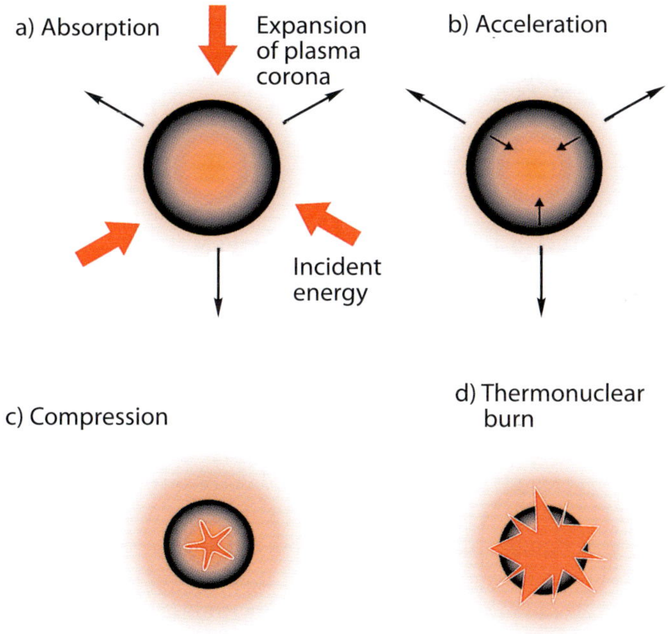


**Fig. 6.4** Thermonuclear target operating scheme [74].

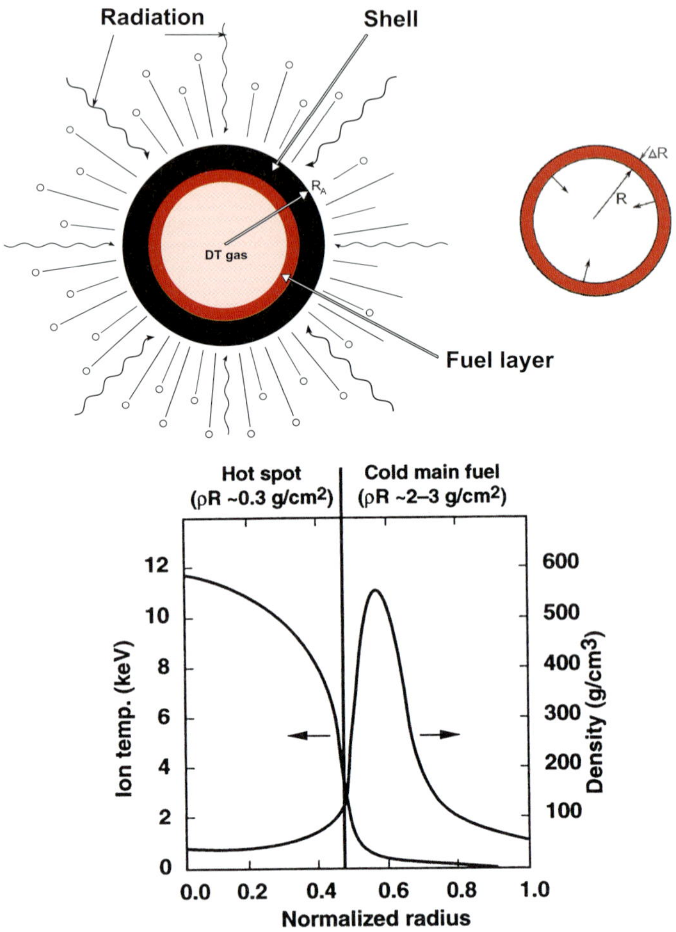


**Fig. 6.5** Variation of parameters in a thermonuclear target [74].

the 1-$\mu$m long-wavelength laser radiation is efficiently absorbed by the plasma with a critical density of $\approx 10^{21}$ cm$^{-3}$, which is more than one order of magnitude lower than the density of liquid hydrogen phase. This circumstance seriously hinders the direct laser heating of thermonuclear plasmas, making preferable the use of short-wavelength (X-ray) radiation. In addition, the thermonuclear reaction rate, which is proportional to $n^2$, will be low at low densities, requiring additional means for confinement.

Therefore, all schemes of pulsed thermonuclear fusion proposed to date involve compression of microtargets by irradiation of their external surface with intense directional energy fluxes (Figs. 6.4, 6.5), which leads to the heating and expansion of the outer part of a target and gives rise to ablation pressure responsible for the compression of the target towards its center (Fig. 6.6). This approach is favored by the circumstance that the requisite irradiation energy decreases ($\sim 1/n^2$) with increasing compression. At the same time, the coefficient of energy transfer to the target is small in the ablation scheme and does not exceed 10%, and the dynamic compression itself is hindered by the development of Rayleigh–Taylor instability, which disturbs the symmetry of compression and is responsible for the mixing of target layers. Furthermore, increasing the thermonuclear plasma density shortens the range of alpha particles, which "get stuck" in the dense target to form the thermonuclear burn wave traveling from the center of the microsphere to its periphery (Fig. 6.4).

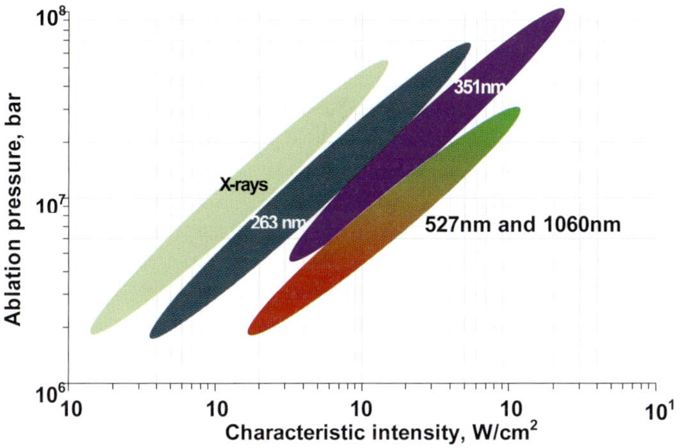


**Fig. 6.6** Dependence of the ablation pressure on the intensity and wavelength of the electromagnetic radiation of a driver [74].

Figure 6.4 shows the characteristic operating scheme of a simplified thermonuclear target. Under the action of the directional energy flux from the driver, the plasma corona heats up and expands towards the incident radiation. The recoil momentum emerging in this case forms compression waves in the target, which are focused at the target center.

At the instant of peak compression, at the target center there occurs a thermonuclear burn, which propagates from the target center to the periphery in the form of a wave (Fig. 6.4). The parameter distribution in such a target is depicted in Fig. 6.5. This scheme of controlled thermonuclear fusion with inertial confinement is referred to as direct-drive laser fusion, because the target is compressed and heated under ablation plasma pressure ($P \approx 100$ Mbar, $T \approx 10^6$ K) (Fig. 6.6) produced at its outer surface by focused laser radiation. A large number of direct-drive thermonuclear

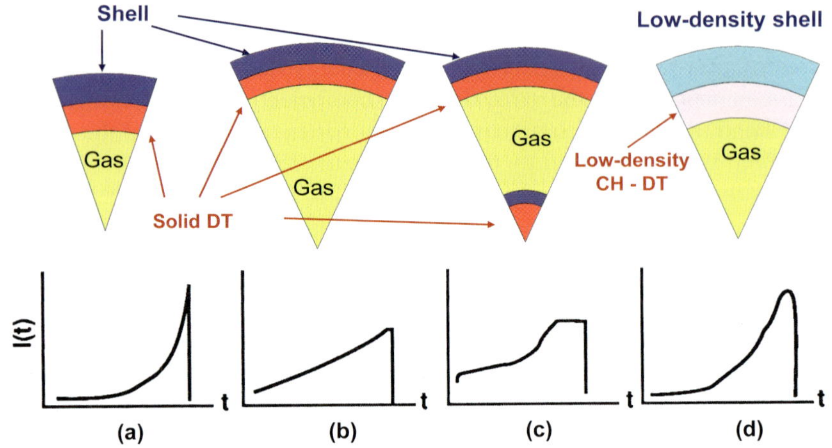


**Fig. 6.7** Structures of the thermonuclear targets for the CNF with inertial plasma confinement and the corresponding temporal profiles of laser irradiation [74].

target designs have been put forward (Figs 6.4, 6.7), where the energy density is as high as $10^{16}$–$10^{17}$ J/cm$^3$, which is comparable with astrophysical conditions. The dependence of the ablation pressure on the intensity of electromagnetic radiation of different wavelengths is depicted in Fig. 6.6. One can see that the short-wavelength radiation is the radiation of choice in the production of high energy densities in controlled nuclear fusion (CNF) plasmas [84, 15, 90].

Several specific thermonuclear targets with their corresponding temporal irradiation intensity profiles are given in Fig. 6.7.

## *6.1.2 Indirect Drive*

An important problem in the practical implementation of pulsed controlled thermonuclear fusion is, apart from the development of a high-power driver, the attainment of a high irradiation symmetry and accordingly the symmetry of dynamic plasma compression. To reduce the distortion effect of the Rayleigh–Taylor instability in the course of laser-driven compression, the so-called indirect-drive scheme [84, 15, 90, 28, 74] has been developed (Fig. 6.8), whereby the compression of the spherical target (Fig. 6.8b) is effected by the thermal soft X-rays from the side walls of a cylindrical capsule (Fig. 6.8) heated by laser radiation.

In the scheme (Fig. 6.8) and photograph (Fig. 6.9) of the thermonuclear NIF target, the spherical deuterium–tritium (DT) target (Fig. 6.8b) is accommodated in a gold cylinder irradiated from two sides by 192 laser beams. Upon incidence on the inner cylinder surface (Fig. 6.8a) (intensity of order $10^{15}$ W/cm$^2$) they vapor-

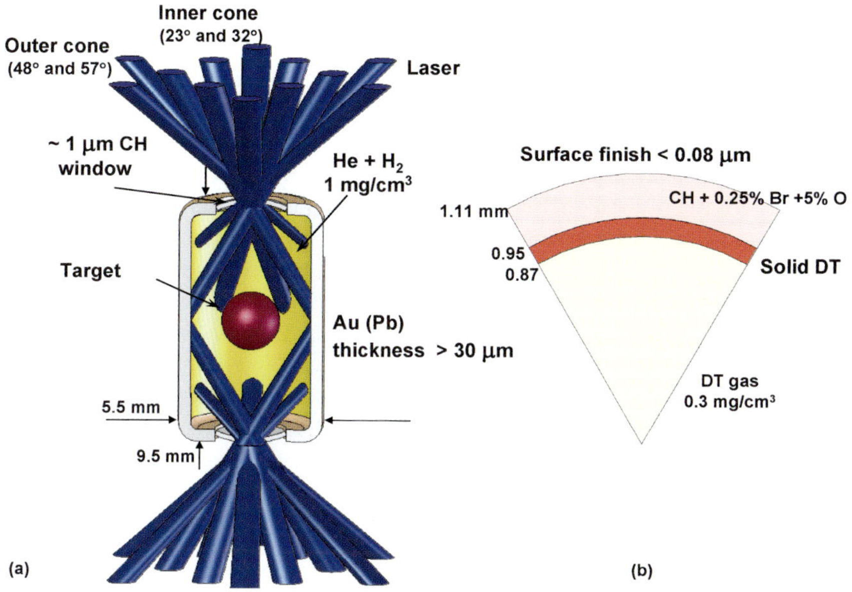


**Fig. 6.8** Scheme of an indirect-drive thermonuclear target (a) for the NIF laser system and cross section (b) of the thermonuclear sphere inside the target irradiated by X-rays.

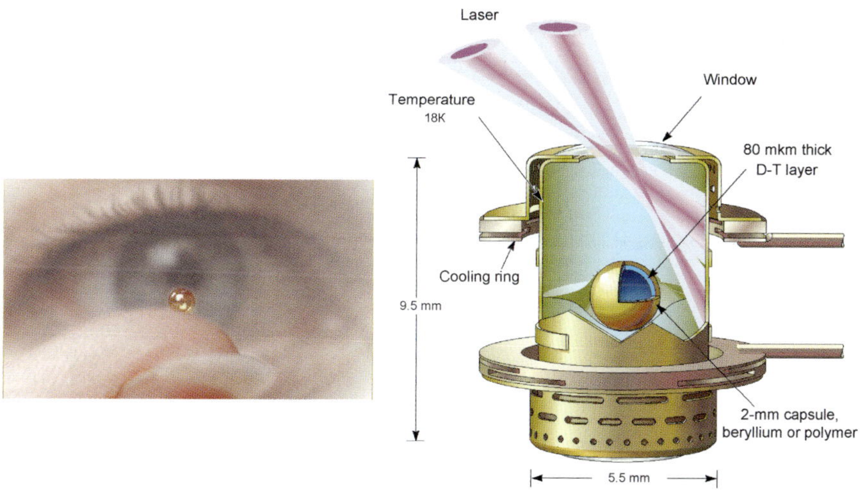


**Fig. 6.9** Photograph of an indirect-drive target and scheme of its irradiation by 192 laser beams of the NIF facility [101].

ize it to produce high-intensity soft X-rays in the cylindrical cavity. These X-rays in turn cause highly symmetric spherical compression of the thermonuclear target (Fig. 6.8b). To obviate the distortion effect of the edge of the plasma on the compression symmetry, the cylinder interior is filled with a low-density ($n \approx 10^{21}$ cm$^{-3}$) plasma produced by the ionization of aerogels. Similar schemes for the thermonuclear targets for heavy ions are described in [113, 53].

The operation of these thermonuclear targets is attended with complex nonlinear processes – beam filamentation and focusing, incoherent scattering, the generation of plasma waves (stimulated Raman and Brillouin scattering), nonthermal electrons and ions, two-plasmon decay, and many other phenomena, some of which have been adequately studied on smaller facilities and with the use of sophisticated computer codes.

The capabilities of the existing laser systems for inertial thermonuclear fusion and of those under construction are illustrated by Table 3.3 and Fig. 6.10, which show that the commissioning of the NIF and LMJ megajoule-level lasers will afford thermonuclear ignition conditions and energy-positive CNF conditions.

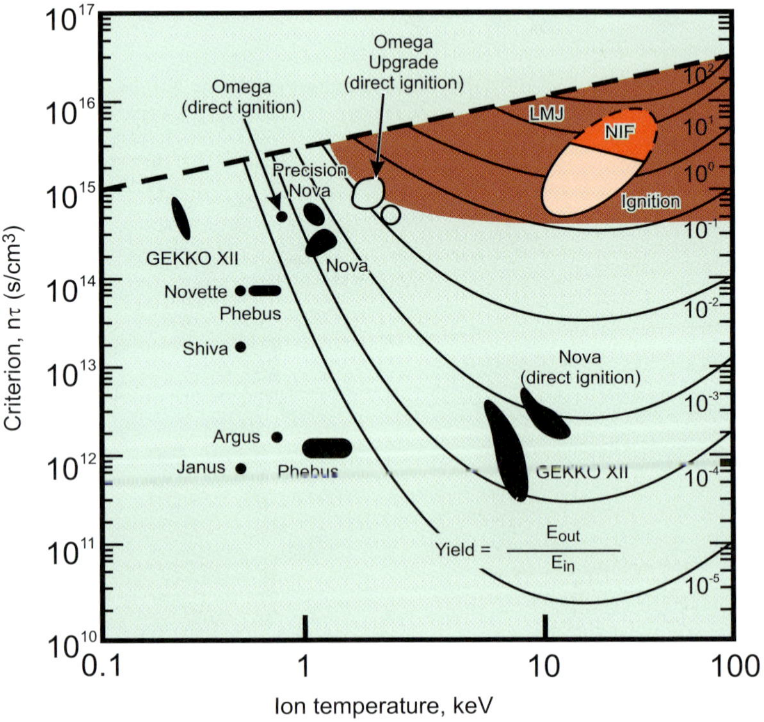


**Fig. 6.10** Capabilities of lasers for controlled thermonuclear fusion with inertial plasma confinement [74].

At the same time, the vigorous progress of laser technology permanently engenders new technical solutions and physical ideas, which strongly affect the development of the laser fusion program. In particular, the realization of crystal and glass laser pumping by semiconductor laser diodes raises the total system efficiency to 30–35%. As discussed in Chap. 4, the invention of the chirping technique has enabled relativistic plasmas to be generated, when the electrons in the electric field of the light wave are accelerated to velocities that approach the velocity of light. This lowers the critical frequency of plasma oscillations due to the relativistic increase in electron mass and therefore shifts the plasma transparency bound towards higher densities to improve the conditions for thermonuclear ignition. Nonlinear effects in the plasma, which are induced by the high intensities of laser radiation, are responsible for its conversion to high-order harmonics, which is also advantageous from the viewpoint of energy transport efficiency in dense plasmas. In any case, the transition to short-duration and high-intensity laser radiation is obviously expedient from the standpoint of laser fusion efficiency. Yet another idea for harnessing these short-pulse lasers in thermonuclear fusion is related to the notion of "fast ignition" [124].

### *6.1.3 Fast Ignition*

The advent of ultrahigh-power short-pulse lasers opens up new fascinating possibilities for laser fusion, making it possible to separate the processes of adiabatic compression and heating of the thermonuclear plasma [79, 15]. In this scheme, the thermonuclear target is compressed to high densities, but is heated to a temperature well below thermonuclear temperatures by a laser pulse of nanosecond duration. Then, the inner part of the compressed fuel is additionally heated by intense fluxes of megaelectronvolt electrons or ions arising from femtosecond laser irradiation to

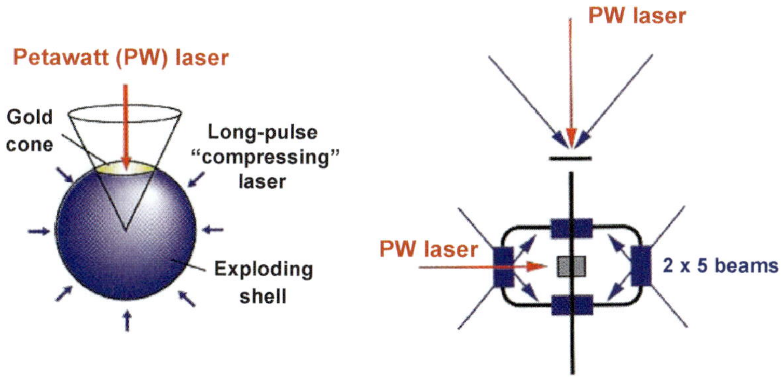


**Fig. 6.11** Fast ignition scheme. Red arrows indicate the petawatt laser radiation to heat the compressed target and the blue ones stand for the nanosecond laser radiation for target compression.

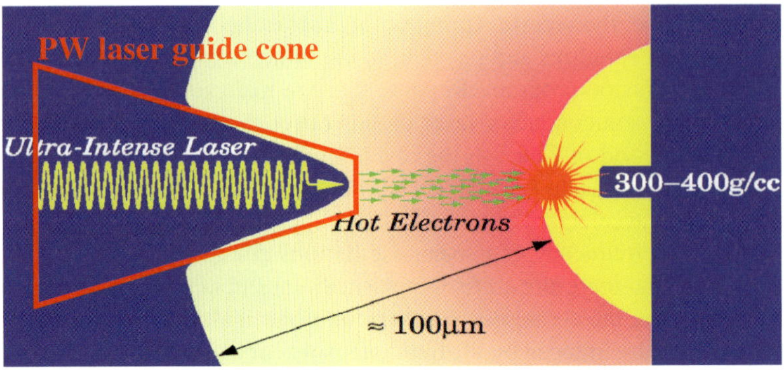


**Fig. 6.12** Scheme of "fast" thermonuclear target ignition. Electron current > 500 MA.

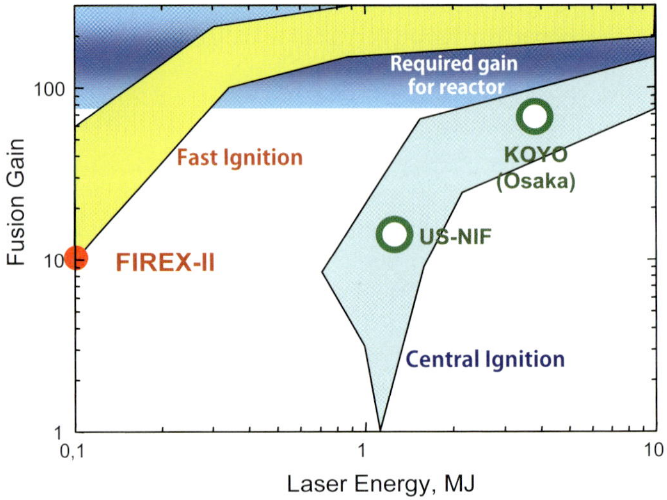


**Fig. 6.13** Thermonuclear gain coefficient versus laser energy [96].

initiate the thermonuclear burn of the entire target [124, 15, 90, 104] (Fig. 6.11). Furthermore, at extremely high intensities of laser irradiation one would expect effects promoting a deeper penetration of the light energy into the supercritical-density plasma. Owing to the action of the ponderomotive forces of the light wave on electrons, electrons and ions will be forced out of the laser beam region, resulting (owing to this self-focusing) in the formation of a low-density plasma channel. Via this channel the light energy will be delivered beyond the critical surface into the interior of the precompressed target and initiate its thermonuclear ignition (Fig. 6.12). This scheme, which is referred to as "fast ignition" [124], exhibits a higher thermonuclear yield and, what is highly important, a higher immunity to instabilities and mixing, because the stages of compression and heating are separated in time in this case. The requirement of optimal target operation leads to the conclusion that the duration of the laser pulse that compresses the thermonuclear fuel to 200–300 g/cm$^3$ should be equal to $\approx$ 10–20 ns.

The main problems of this scheme are associated with the efficient generation and transfer of ultrahigh-intensity energy fluxes into the interior of the compressed dense plasma. Estimates show [101] that it is necessary to heat a plasma volume of the order of the electron or $\alpha$-particle range ($\rho r \approx 0.5$ g/cm$^2$) to 10 keV. For a plasma density of $\approx$ 300 g/cm$^3$ this amounts to $\approx$ 10 $\mu$m in a time $\approx$ 10–20 ps [15, 90, 82]. The corresponding energy is equal to $\approx$ 3 kJ for a power of $4 \times 10^{14}$ W and an intensity of $\approx 10^{20}$ W/cm$^2$. In a different version of this scheme, use can be made of electrons or protons with an energy of 1–5 MeV, which are comparable to $\alpha$-particles in range.

Modern short-pulse lasers furnish the requisite intensities of $\approx 10^{15}$–$10^{20}$ W/cm$^2$ and generate a wide spectrum of effects useful for fast ignition in $10^{21}$–$10^{26}$ cm$^{-3}$ density plasmas – relativistic self-focusing and filamentation, quantum and sausage-type instabilities, as well as formation of vacuum channels and a set of new diverse particle acceleration mechanisms. Also possible is the generation of magnetic fields of superhigh intensity ($\approx 10^9$ G) and multimegaelectronvolt-energy ions [105].

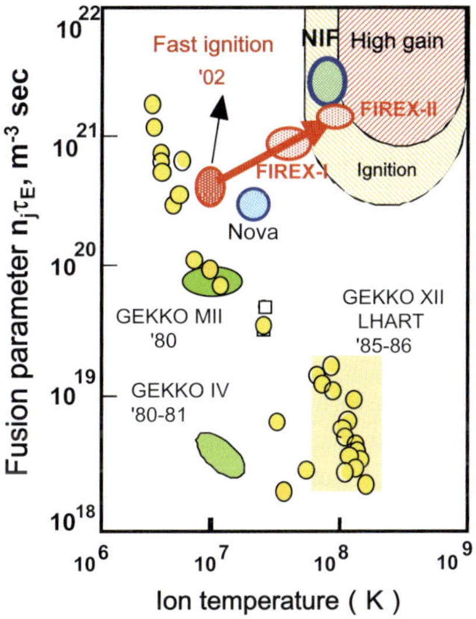


**Fig. 6.14** Comparison of "direct-drive" and "fast-ignition" thermonuclear schemes [96].

An important role in the study of these effects is played by computer simulations [104], which predict, in particular, a highly efficient (up to 30%) energy transfer from laser radiation to megavolt-range electrons (Fig. 6.13). In this case there emerge several new acceleration mechanisms – heating by oscillating ponderomotive forces, transverse-to-longitudinal electric field transformation in superdense plasma layers, acceleration of electrons in the betatron resonance in relativistic laser channels [104, 100]. The rich variety of these effects is currently the subject of careful studies. In particular, about 20% of the energy of a short-pulse laser was transferred to the compressed thermonuclear target in experiments conducted at Osaka University [81]. Kodama et al. [108, 23] came up with the idea of igniting compressed fuel not by electrons, but by the fast ions of the laser plasma. A comparison of "direct-drive" and "fast-ignition" thermonuclear schemes is shown in Fig. 6.14.

The action of high-power laser radiation on clusters has come under vigorous scrutiny [100, 37, 38, 40, 112, 139, 140, 85, 45, 102, 77, 39], because it permits the production of extremely high plasma temperatures, the investigation of strongly nonequilibrium effects, and the generation of neutron fluxes. The fact is that direct photoionization takes the place of thermal collisional ionization beginning with intensities of $10^{21}$–$10^{22}$ W/cm$^2$. In this case, the photoelectrons escape from a cluster to violate the fundamental condition of plasma electroneutrality [51], resulting in the generation of intense electric fields. A "Coulomb explosion" occurs, with the effect that the ions are accelerated to high ($\approx$ 1 MeV) energies [85, 45, 102] and fusion neutrons are generated [39]. This opens up interesting new possibilities for the development of compact laser-driven neutron generators for numerous technical and medical applications [80].

Another manifestation of "Coulomb explosion" was recorded in the experiments of Okihara et al. [103], which involved the laser irradiation of a plastic ($C_5H_{10}$) foam with a density of $\approx$ 10 mg/cm$^3$ at $q \approx 10^{18}$ W/cm$^2$. Ions with an energy of 200 keV were obtained, their parameters corresponding to the "Coulomb explosion" model.

## 6.2 Heavy-Ion Beam Fusion

The initiation of controlled thermonuclear reactions by heavy-ion ($A > 80$) beams [113] offers several potential advantages, for example the high efficiency and reliability of accelerators and that the ion accelerators developed for experiments in high-energy physics are ready for opertion. Considered for this purpose is the multigigaelectronvolt level of ion energies for a total beam energy ranging into the megajoules; these beams should be squeezed to 10 ns and focused onto millimeter-sized targets. To meet these conditions requires going beyond the limits defined by the space charge, which calls for the suppression of diverse instabilities and the study of collective effects in high-current beams as well as of several other complex processes. In particular, the rise in beam emittance (the rise of beam temperature) may be caused by collective processes (up to beam "crystallization" [109, 110]), the imperfection of magnetic field systems for the "reflection" from conducting surfaces, and inter-ion forces.

The beam propagation in the reactor chamber also invites careful analysis [101] of background plasma dynamics and reverse currents (stream and filamentation instabilities), "peeling" of the beam by the background plasma, its photoionization, etc.

Research into ion beam generation and high-energy-density physics involving ion beams is being vigorously pursued in several research centers: GSI, Germany (Fig. 5.7); Berkeley, USA; ITEF, Russia; IPKhF, ITES OIVT RAN, Russia [113, 52, 73, 53], where a wealth of interesting new data was obtained concerning the absorption of heavy ions by plasmas, shock-wave dynamics, isochoric target heating, plasma spectra, etc. [53, 107, 44, 97, 126, 125]. Of special interest is the project [125] to use the GSI heavy-ion accelerator (SIS-100) in combination with the petawatt PHELIX laser (Fig. 5.8). Work on the use of heavy ions in high-energy-density physics is described in greater detail in reviews ([73, 53]).

## 6.3 Laser-Plasma Acceleration of Charged Particles

Currently, many laboratories in the world have high-power short-pulse lasers (with a pulse duration $\tau_L$ shorter than a picosecond, $\tau_L < 10^{-12}$ s) (see Table 3.3), which are employed in studies of different processes and the development of numerous applications. One of these applications involves the elaboration of compact laser-plasma high-energy electron and ion accelerators on the basis of new-generation femtosecond lasers.

The dimensions of modern electron accelerators are determined by the intensity of the accelerating field, which is equal to $10^7$–$10^8$ V m$^{-1}$ and is limited by the breakdown in the acceleration system. That is why the problem of particle acceleration in plasmas rather than in vacuum has long been under discussion. On the one hand, in this case there is no breakdown-defined limitation and on the other the electric intensity in a relativistic charge-density wave (whose phase velocity is close to the velocity of light) may achieve huge magnitudes. Indeed, a simple estimate made with the aid of the Poisson equation permits this intensity to be related to the electron plasma density $n_e$:

$$E \cong \alpha(n_e\,[\mathrm{cm}^{-3}])^{1/2}\,[\mathrm{V\,cm}^{-1}],$$

where $\alpha = \delta n/n_e$ is the dimensionless amplitude of the plasma wave ($\delta n$ is the amplitude of electron density oscillations). For an electron density $n_e = 10^{17}$ cm$^{-3}$ and $\alpha = 0.3$, the accelerating field intensity in the plasma wave amounts to $10^8$ V cm$^{-1}$, which is two or three orders of magnitude higher than the acceleration rate in traditional radio-frequency accelerators. Figures 6.15 and 6.16 depict schematically one stage of a laser-plasma wakefield accelerator [4].

We consider several aspects of laser-driven acceleration following Belyaev et al. [20]. As shown in Chap. 4, the advent of new-generation solid-state lasers has led to unique conditions of laser target irradiation by 20–1000 fs long light pulses with radiation intensities in the $10^{17}$–$10^{21}$ W/cm$^2$ range. At these intensities previously unattainable superhigh electric fields are produced in a laer pulse, which far exceed the atomic electric field $E_{\mathrm{at}} \approx 5.14 \times 10^9$ V/cm. These conditions give rise to fascinating dynamics of laser radiation interaction with the plasma produced when the leading pulse edge or the prepulse interacts with a solid-state target. Laser radiation is rather efficiently transformed [20] to streams of fast charged particles – electrons

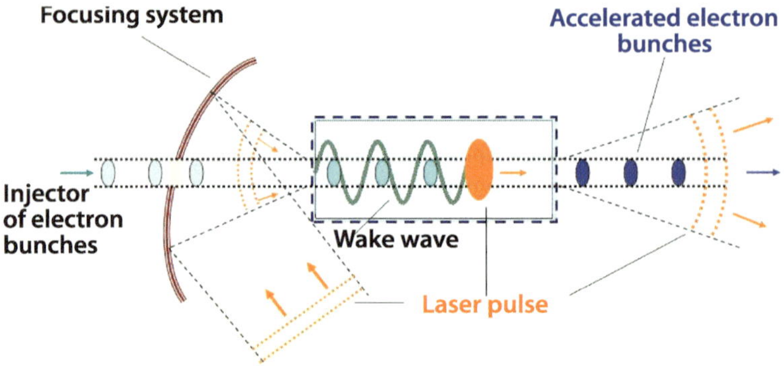


**Fig. 6.15** Schematic of one stage of a laser-plasma wakefield accelerator [4].

and atomic ions. Their interaction with the surrounding target material leads to hard X-ray generation in the ionization of inner atomic shells and to different nuclear and photonuclear reactions.

The relativistic nature of laser radiation intensity $I$ is realized when $\alpha > 1$, where $\alpha$ is the dimensionless momentum of an electron oscillating in the electric field of linearly polarized laser radiation expressed by the formula [20]

$$\alpha = \frac{eE}{mc\omega} = 0.85\lambda \left(\frac{I}{10^{18}}\right)^{1/2}$$

$$E = 27.7 I^{1/2}$$

Here, $e$ and $m$ are the electron charge and mass, $E$ is the electric amplitude of laser radiation field [V/cm], $\lambda$ is the radiation wavelength [$\mu$m] $\omega$ is the frequency of laser radiation, $c$ is the speed of light [cm/s], and $I$ is the radiation intensity [W/cm$^2$].

Terawatt-power laser systems of modest size enable intensities with $\alpha > 1$ to be attained, which corresponds to electric intensities above $10^{10}$ V/cm. Under these fields there occurs above-barrier ionization of atoms in atomic times on the order of $10^{-17}$ s, while the resultant electrons are accelerated to relativistic energies in the MeV range in the course of the laser pulse.

The acceleration of atomic ions in femto- and picosecond laser plasmas is a secondary process: it is caused by the emergence of high quasi-static electric fields arising from spatial charge separation (Fig. 6.17), which in turn is due to the motion of a fast electron bunch. At laser radiation intensities exceeding $I \geq 10^{18}$ W/cm$^2$ it is possible to obtain directional beams of high-energy ions with an energy $E_i > 1$ MeV.

At present, the generation of high-energy proton and ion beams in a laser plasma under ultrashort-pulse irradiation is a rapidly advancing area of research. This is due, in particular, to their practically important applications in the areas such as proton acceleration, structural material studies, proton radiography, short-lived ra-

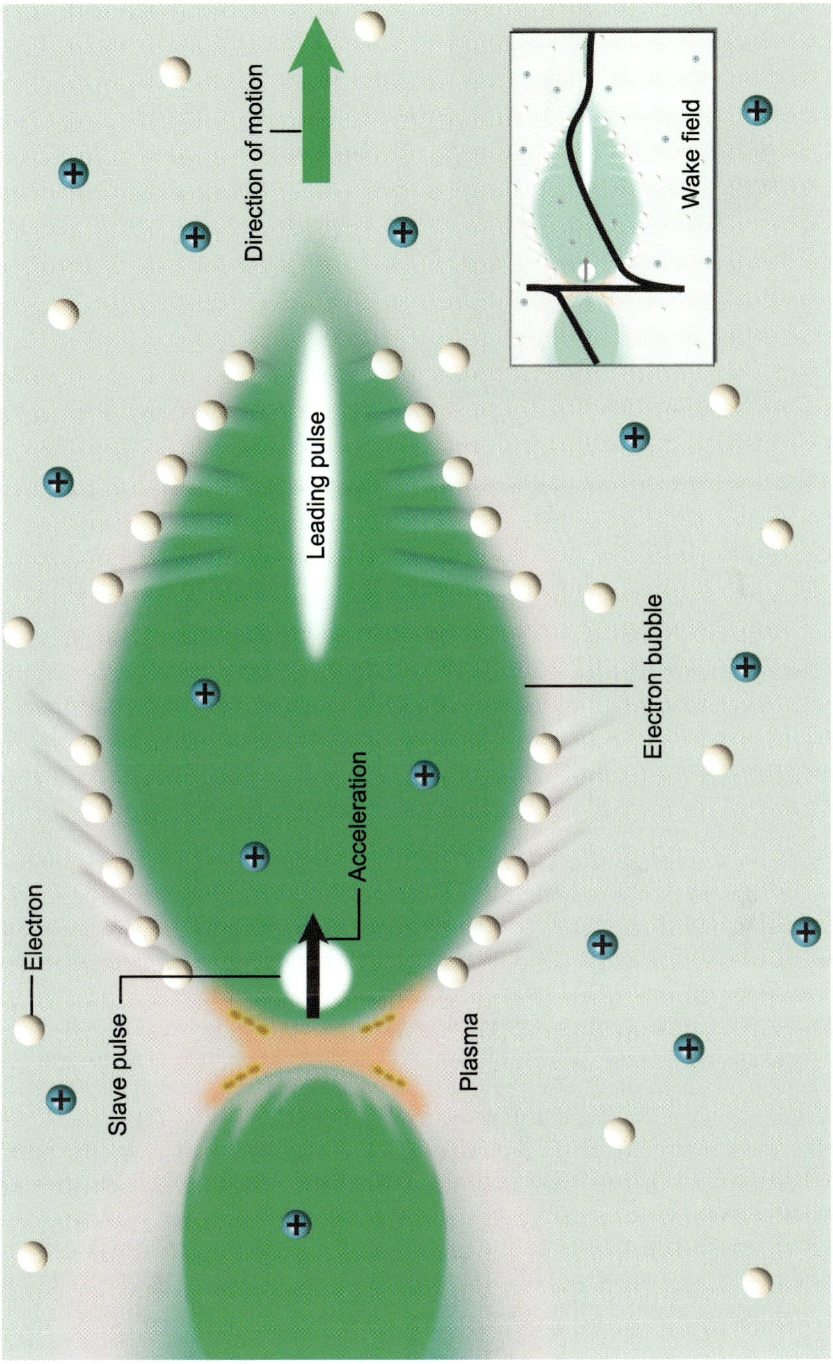


**Fig. 6.16** Wakefield acceleration scheme. The leading laser pulse expels electrons (shown in white) to produce a positively charged domain (green). The accelerating electric field propagates along the axis (bottom right). Reprinted, with permission, from [76]

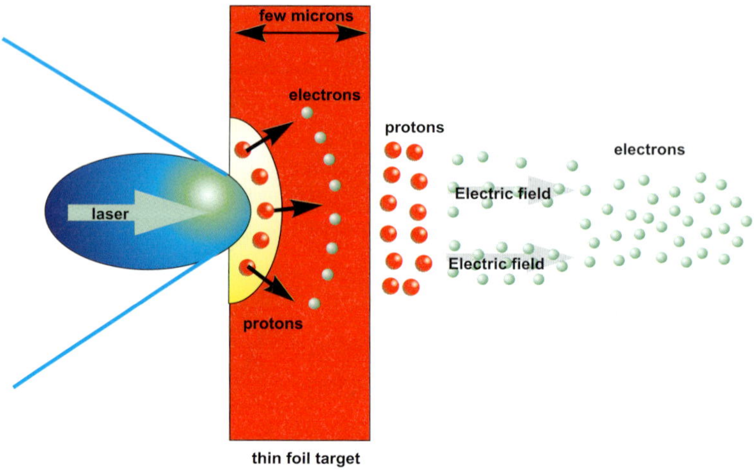


**Fig. 6.17** Scheme of electron and ion generation in a laser-produced plasma [78].

dioisotope production for medical applications, and controlled laser thermonuclear fusion. Laser radiation with intensities $I \geq 10^{18}$ W/cm$^2$ also makes possible the initiation of a number of nuclear reactions that were previously realizable only with the use of elementary particle accelerators.

Several mechanisms [20] of fast electron generation are presently under discussion that involve the action of a laser pulse on a plasma with a density much higher than the critical one. When a laser pulse does not have a prepulse (a high contrast ratio), the laser radiation interacts with a solid-state-density plasma with a sharp boundary. In this case the mechanism of "vacuum heating" is realized, as well as the so-called $v \times B$ mechanism ($B$ is the amplitude of the magnetic induction of the laser wave) caused by the longitudinal (along the direction of laser pulse propagation) ponderomotive force. The $v \times B$ mechanism becomes significant at relativistic intensities, when the electron oscillation energy is comparable to or higher than the electron rest energy $mc^2 = 511$ keV.

There also is a mechanism of fast electron generation at the plasma resonance on the critical plasma surface when the electric vector of laser radiation has a projection on the density gradient (normally at oblique incidence of laser radiation on a target), and the laser frequency coincides with the plasma frequency.

Unlike the ponderomotive $v \times B$ mechanism, the "vacuum heating" and the resonance absorption mechanism emerge for nonrelativistic (substantially lower, $\alpha < 1$) intensities as well.

Another mechanism of fast electron generation in the subcritical plasma region in front of the target is considered [20], which operates due to betatron resonance in the emerging magnetic field. In this regime, electrons are accelerated by the transverse ultrarelativistic electric field of the laser wave in the direction of wave polarization, while the azimuthal magnetic field induced by fast electron current produces the magnetic part of the Lorentz force. This force turns the electrons in such a way that they gradually reverse their direction of motion. For an exact betatron resonance, their reflection takes place at the moment the transverse electric field changes its direction, so that the electrons are permanently in the acceleration regime.

There also exist other electron acceleration mechanisms, which require specific experimental conditions; among them is acceleration in the wake wave, which we consider in greater detail below. In the case of resonance absorption, the field near the critical plasma surface is substantially higher than the field of incident laser radiation.

We do not set ourselves the task of describing numerous laser acceleration mechanisms at length (see the review [20]) but list the different mechanisms of electron heating in Table 6.2.

**Table 6.2** Comparison of different mechanisms of electron heating in dense media [20].

| Heating mechanism | When it applies |
|---|---|
| Stimulated inverse bremsstrahlung in the scattering of electrons from ions | Intensities below $10^{15}$ W/cm$^2$ |
| Longitudinal ponderomotive electron acceleration in the skin layer | Relativistic intensities above $10^{19}$ W/cm$^2$ |
| Vacuum heating | High contrast ratio, moderate intensities, short pulses |
| Resonance absorption of laser radiation | Low contrast ratio, long pulses |
| Electron acceleration by a laser wake wave | Gas targets, substantial subcritical plasma region, ultrashort pulses |
| Cyclotron mechanism | The presence of external constant magnetic field |
| Betatron mechanism | Vortical electric field produced by a varying magnetic flux penetrating the orbit of electrons |

The feasibility of harnessing lasers to excite relativistic charge-density waves in tenuous plasmas was first discussed by Tajima and Dawson [129], who, however, placed emphasis on the resonance method of excitation using two-frequency laser irradiation of moderate intensity. The possibility of employing high-intensity short pulses generated by modern lasers for the excitation of charge-density waves (so-called wake plasma waves) attracted considerable attention and provoked lively discussion after the publication of [68, 122] (see also [22, 21], the reviews [6, 46, 4], and references therein). The charge separation field, which occurs when the ponderomotive force expels electrons from the region occupied by the laser pulse, gives rise to a charge-density wave behind the laser pulse, the phase velocity of the wave being equal to the group velocity of the laser pulse. In a tenuous plasma, the latter is quite close to the velocity of light, and therefore the phase velocity of the wake plasma wave is also close to the velocity of light, just right to efficiently accelerate relativistic electrons (Figs. 6.18, 6.19). Even the first experiments [4] showed that such a wake wave is indeed excited by a short (subpicosecond) laser pulse quite ef-

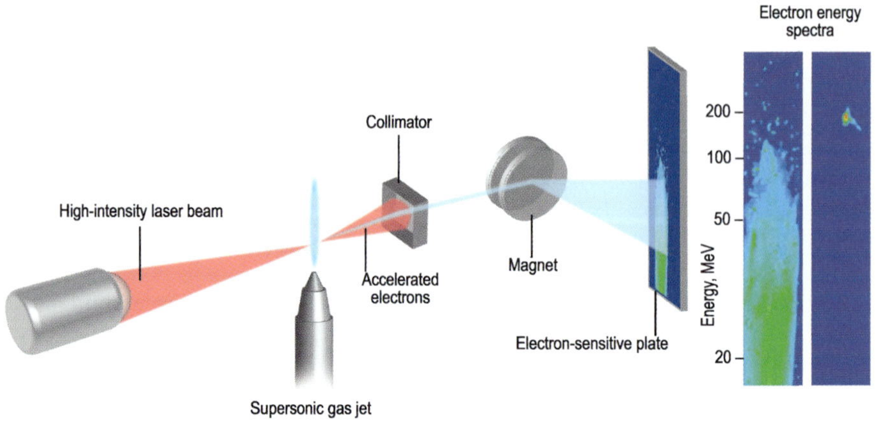


**Fig. 6.18** Laser wakefield accelerator scheme. Reprinted, with permission, from [76].

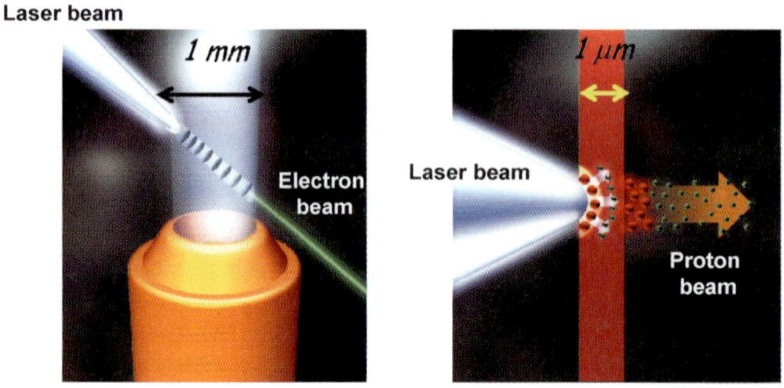


**Fig. 6.19** Direct (at the left) and wakefield laser acceleration.

ficiently and may accelerate electron bunches to high energies ranging into the tens and hundreds of MeV.

Recent successful experiments performed practically simultaneously in different laboratories in Europe and the USA had the effect that even skeptics turned their attention to the possibility of harnessing laser-plasma acceleration techniques in high-energy physics (Fig. 6.20). In particular, the year 2004 saw the demonstration of the possibility of producing quasi-monoenergetic electron bunches (Fig. 6.21) accelerated to an energy of about 100 MeV over a length of several millimeters and possessing a small emittance [58, 50], which is of significance for the use of such bunches in hard radiation sources and for multistage acceleration for high-energy physics.

The theoretical predictions about the importance of channeled laser-pulse propagation for the production of electron bunches ranging into the gigaelectronvolt

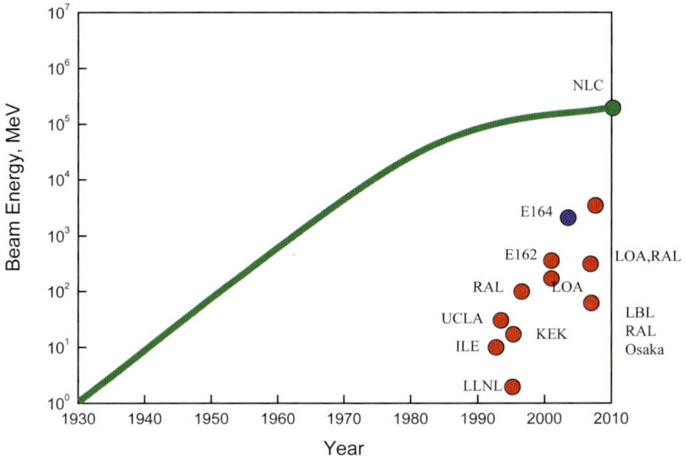


**Fig. 6.20** Progress in laser-driven electron acceleration.

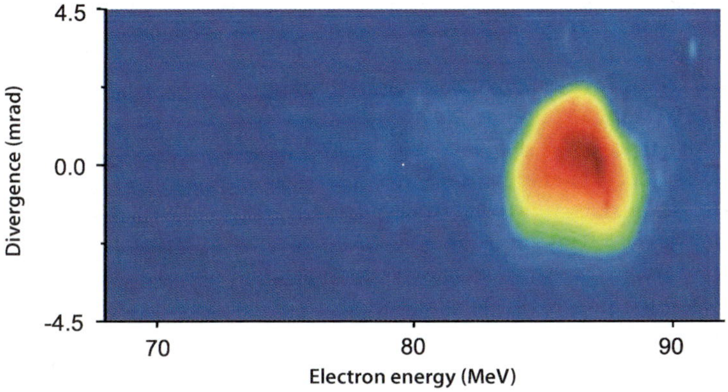


**Fig. 6.21** Energy spectrum and angular divergence of an accelerated bunch containing $2 \times 10^9$ electrons [58].

($10^9$ eV) energy range were amply borne out by the Lawrence Berkeley National Laboratory in 2006 [87] (Fig. 6.22).

The wake plasma wave is most efficiently generated by a laser pulse under resonance conditions when the pulse duration $\tau_e$ is close to a half-period of the wave, i.e., provided

$$c\tau_e \approx \lambda_p/2 \cong \pi c/\omega_p = \pi/k_p,$$

where $\omega_p = (4\pi e^2 n_e/m)^{1/2}$ is the electron plasma frequency, $k_p$ is the wave vector of the plasma wave, and $c$ is the velocity of light. These expressions show that the accelerating field intensity increases with plasma density and, accordingly, with

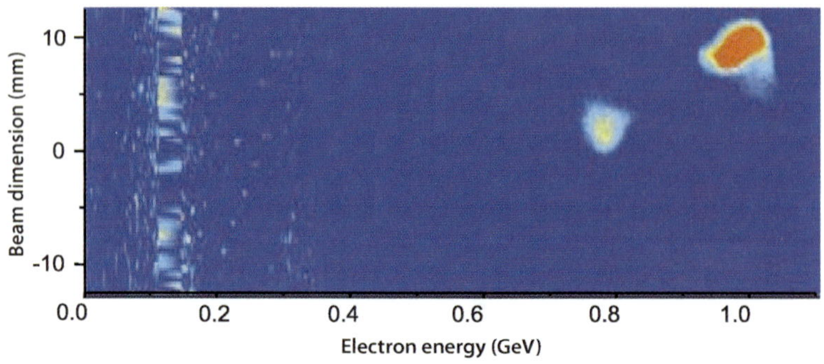


**Fig. 6.22** Energy spectrum and dimension of the accelerated electron bunch in the channeled propagation of a laser pulse through a capillary [87].

shortening of the laser pulse that excites the wake wave. In this case, however, it is well to bear in mind that increasing the plasma density lowers the group velocity of the pulse and, accordingly, the phase velocity of the wake plasma wave, which shortens the possible acceleration length (and decreases the maximum energy gain) of a relativistic electron due to its earlier escape from the accelerating phase of the wave. For instance, for laser radiation with a wavelength $\lambda_e \approx 1\ \mu$m and a plasma density $n_e \approx 10^{17}$ cm$^{-3}$, the maximum acceleration length is 1 m. For an acceleration rate $E \approx 10$ GV/m (for $\alpha \approx 0.3$) this corresponds to the maximum gain $\Delta\varepsilon_{max} \approx 6$ GeV in accelerated-electron energy. In this case, the duration of the laser pulse intended for the resonance excitation of the wake wave should be of the order of 100 fs.

The main difficulty encountered in attaining these parameters is the requirement that the high intensity of the laser pulse, which should be of the order of $10^{18}$ W/cm$^2$ in the example given above, be maintained throughout the entire propagation length ($\approx 1$ m). In the absence of any optical channeling mechanisms for the laser radiation, the propagation length for the high-intensity laser pulse (and, accordingly, the efficient acceleration length) is limited by the diffraction length. For realistic parameters of the lasers of the terawatt power range (bearing in mind the necessity of attaining the high radiation intensity), the diffraction length turns out to be much shorter than the maximum acceleration length. This is why attaining $\approx$ 1–10 GeV electron energies without channeling of the laser pulse requires its focal spot to be increased in order to increase its diffraction length, which calls for petawatt ($10^{15}$ W) laser power level to maintiain the high intensity [67, 89, 69]. When advantage is taken of channeled laser pulse propagation, the required power level may be lowered by more than an order of magnitude [87].

The optical channeling of a laser pulse as a way to overcome its diffraction spreading is possible owing to its self-focusing or when a plasma channel with an on-axis density minimum is produced to afford waveguided radiation propagation [43, 30]. A highly promising technique for propagating high-intensity laser pulses for tens and hundreds of Rayleigh lengths is the use of dielectric gas-filled capillaries, in which the plasma is produced either by a discharge [138, 75, 118], or by the optical ionization of the gas by the wake-wave-generating laser pulse itself [32, 31]. That is why much attention is presently being paid to studies of different ways of producing plasma channels and the propagation of short laser pulses through them, as well as to investigations into the generation mechanisms and structure of wake waves in gas-filled capillaries [117, 5, 54, 2, 11, 3].

The simple laws indicated above, which define the resonance excitation of the wake wave, are valid under the conditions of weak nonlinear dynamics of a laser pulse (under the conditions of weak relativity, when the oscillation energy of the electron in the laser field does not exceed its rest energy). The inclusion of the nonlinearity of laser pulse propagation has led to the discovery of new regimes of generation of high-intensity plasma fields when the linear resonance condition $c\tau_e \approx \lambda_p/2$ is not fulfilled. In particular, when the laser pulse length is far greater than the plasma wavelength, the mechanism of high-intensity wake-wave generation is due to the self-modulation of the laser pulse, with the result that the pulse intensity turns out to be longitudinally modulated with a scale length close to the plasma wavelength [6, 12, 13, 121, 83, 9, 98, 7, 34, 10, 8, 72]. The self-modulation stems from the development of a laser-pulse instability. Depending on the plasma and laser radiation parameters, this self-modulation may be related to either the self-focusing of the laser pulse or its forward stimulated Raman scattering, and is accompanied by the stimulated generation of a wake wave. As a consequence, the amplitude of the plasma wave may far exceed the magnitude generated by a short laser pulse under linear resonance conditions.

A strongly nonlinear regime of generation of the accelerating charge-separation field was recently discovered in the opposite limiting case of an ultrashort laser pulse ($c\tau_e < \lambda_p/2$) of ultrarelativistic intensity. Under these conditions, all plasma electrons immediately behind the laser pulse are expelled by the ponderomotive force (the so-called bubble regime, Fig. 4.9) [104, 29, 93], and the wake wave in the wave-breaking regime entrains the small fraction of background electrons and accelerates them with record acceleration rates.

When a relativistically intense laser pulse irradiates a thin foil, the laser energy may be efficiently converted to a well-collimated beam of accelerated ions [19, 16, 55, 47, 92, 48].

Among the applications of the laser-plasma ion acceleration that are being actively developed, mention should be made of the "fast ignition" in inertial thermonuclear fusion (Sect. 6.1.3) and the proton therapy of malignant tumors [128, 81, 16, 55, 56] (Fig. 6.23). The employment of protons in radiation therapy and oncology offers several significant advantages over other kinds of radiation. This is primarily due to the fact that the deceleration of prescribed-energy protons in a substance takes place in a localized region in the vicinity of the Bragg peak (for carbon, see Fig. 6.24 [113, 73, 53]), where the bulk of beam energy is deposited, thereby reducing the irradiation of healthy body tissue (Fig. 6.25).

We emphasize that the necessary condition for the successful use of accelerated ions for the above purposes is their high energy monochromaticity. In particular, the energy-spread in the proton beam intended for hadron therapy may not exceed

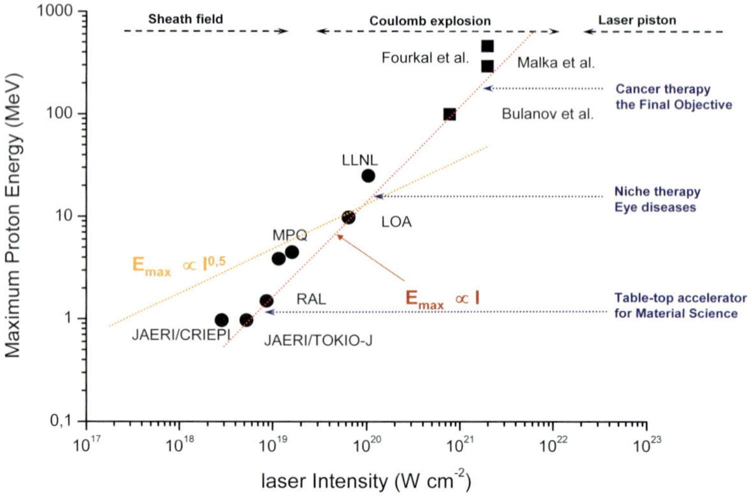


**Fig. 6.23** Parameters of laser ion accelerators.

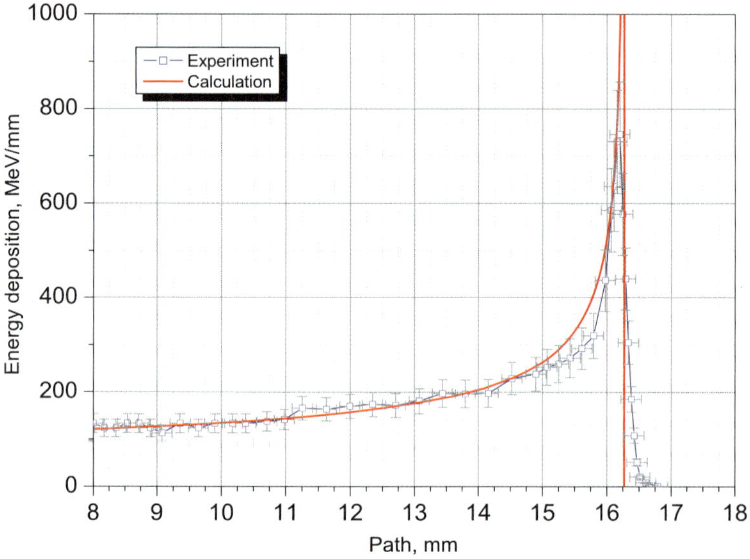


**Fig. 6.24** Energy deposition profile for a beam of 250 MeV/nucleon $C^{6+}$ nuclei in a copper target [53, 17].

2%. Esirkepov et al. [49] proposed two-layer targets for the production of quasi-monoenergetic beams of accelerated protons and analyzed them in a realistic three-

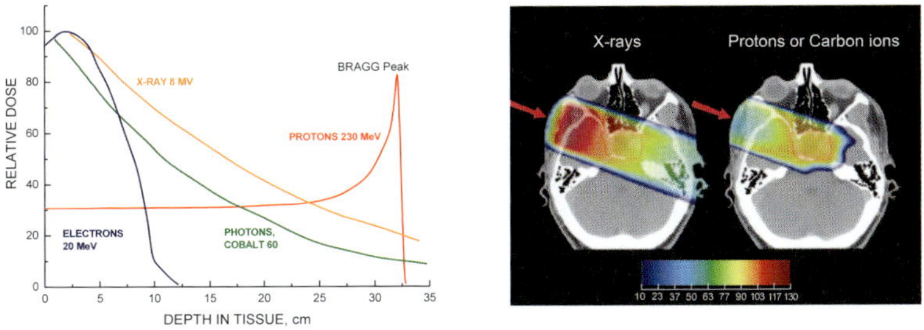


**Fig. 6.25** Proton and hadron therapy [78].

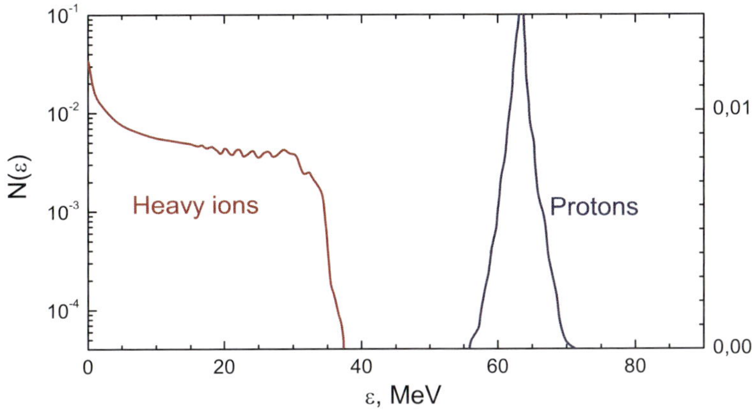


**Fig. 6.26** Spectra of accelerated protons and heavy ions [49].

dimensional geometry to demonstrate the feasibility of generating well-localized high-energy proton bunches with an energy-spread $\Delta E/E \approx 3\%$ (see Fig. 6.26).

## 6.4 Free-Electron Lasers and Ultrashort High-Intensity Radiation Sources

At present, free-electron lasers [42, 26, 134, 35] permit extremely intense radiation fluxes of femtosecond duration to be obtained in a broad (from the far-ultraviolet to soft X-ray) wavelength range (Fig. 6.27). These circumstances make free-electron lasers a promising tool for studying ultrafast chemical and biochemical processes with atomic-scale spatial resolution, for analyzing biological structures, for studying in real time the kinetics of physicochemical transformations, for investigating strongly nonideal and astrophysical plasmas, and for many other applications requiring atomic-scale spatial resolution and high temporal resolution.

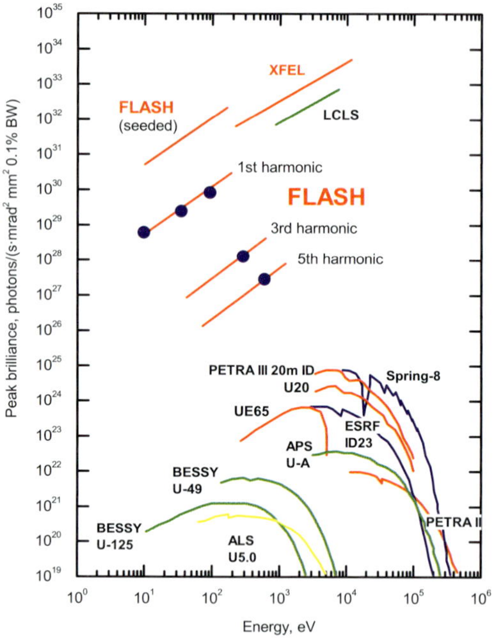


**Fig. 6.27** Peak brilliance (BW: bandwidth) of the experimental facilities FLASH and XFEL (Germany), LCLS, APS (USA), SPring-8 (Japan), ESRF (France), and SLS (Switzerland). The DESY facilities are indicated in red. Blue circles stand for the measured values. PETRA III began working in 2009 and the European XFEL will be commissioned in 2013 [134, 35, 36].

Figure 6.28 is a diagrammatic sketch of the FLASH free-electron laser at DESY, Germany [134, 35]. Today this facility is the champion in radiation brightness, and the brightness of this device projected for 2013 will exceed that of the majority of facilities of this kind by 6–9 orders of magnitude. In this case, the pulse duration will shorten from 50 to 10 fs for a total pulse energy of 10–50 mJ and a wavelength of 6.5–60 nm. Recall that in 10 fs the light traverses a distance of only $\approx 3\ \mu$m. The 260 m long FLASH facility with a power of 5 GW consists of a superconducting linear electron accelerator with a current of 1–2 kA, which imparts to the electrons a kinetic energy of 440–700 MeV and an energy-spread of $\approx 0.1\%$. This electron flux is applied to a 27 m long undulator section, in which there forms X-ray electromagnetic radiation in the form of sequences of 800 pulses with an energy up to 50 $\mu$J each, with a repetition rate of $\approx$ 5–10 Hz, an average radiation power of 100 mW, and a peak brightness of $10^{29}$–$10^{30}$ photons/(s mrad mm$^2$ 0.1%BW) [35].

With their unique radiation parameters, free-electron lasers permit a broad spectrum of investigations in different areas of science and technology that are quite

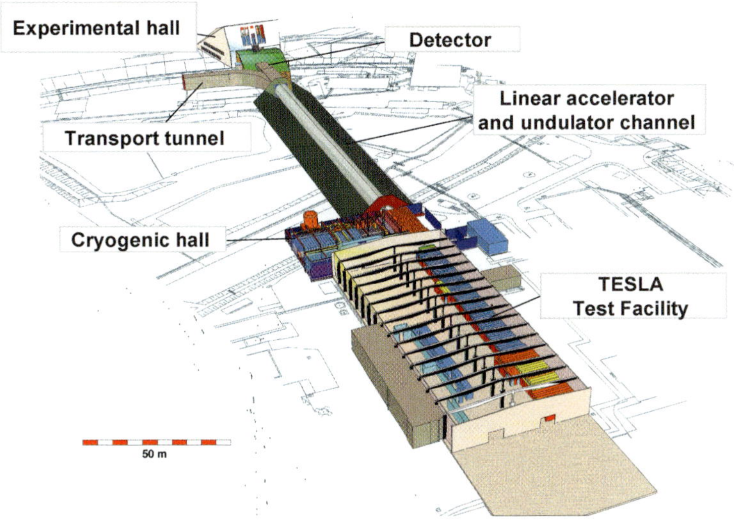


**Fig. 6.28** Diagrammatic sketch of the FLASH free-electron laser facility [134, 35].

frequently inaccessible by other means of generation and diagnostics. Owing to its short wavelength (comparable to atomic dimensions), this radiation is suitable for studying the structure of individual atoms and molecules, while its short duration makes it possible to follow the kinetics and mechanism of chemical and biological reactions by effecting the selective excitation of the selected degrees of freedom (Fig. 6.29). This will enable, in particular, studies of three-dimensional biological structures as well as determination of the location and role of its specific constituent structural elements, which is of great importance in developing new-generation medicines and polymers, as well as in constructing complex spatial molecular structures.

In the more distant future, the application of this technique is expected to enable researchers to trace the variation of the electronic spectrum, magnetic properties, reactivity, and other physicochemical properties of clusters under continuous number variation of their constituent atoms from several atoms to solid-state values, $\approx 10^{23}$ atoms.

The physics of shortwave electromagnetic radiation–plasma interactions exhibits significant special features and shows great promise from the standpoint of studies in high energy densities in laser plasma [35]. As we saw in Chap. 4 and Fig. 6.6, the passage to short-wavelength radiation is highly advantageous for this purpose, shifting the critical density, where $\omega_e \approx \omega_p$, to the higher-density domain – from $n_{ec} \approx 10^{21}$ cm$^{-3}$ for infrared to $n_{ec} \approx 10^{24}$ cm$^{-3}$ for ultraviolet radiation. This substantially improves the efficiency of laser energy delivery to the plasma and simplifies the laser interaction physics by shifting the instability thresholds, reducing the importance of nonthermal electrons, surface effects, etc. Owing to the short du-

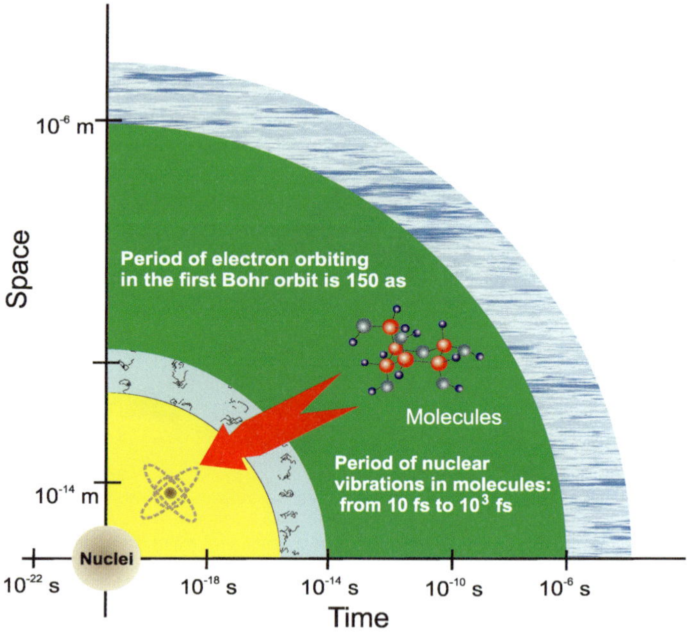


**Fig. 6.29** Spatiotemporal characteristics of atomic and nuclear objects [78].

ration of the laser pulse it heats only the electron component, leaving the ions cool and thereby avoiding hydrodynamic motion at the initial stage of the process and effecting a uniform spatial energy deposition in the substance.

Estimates show [35] that the use of free-electron lasers will enable (by way of bulk heating and the generation of shock and radiation waves) extremely high energy densities to be obtained and investigations of the equation of state, ionic composition, transient dynamics, and optical properties of the plasma under astrophysical conditions to be carried out.

Owing to the high photon energy of the X-ray laser radiation, it is especially well suited for studying multiphoton ionization and producing highly charged ions in laser plasmas. Furthermore, highly ionized states (such as $Fe^{+23}$) may by produced by means of electron ionization of ions in traps [35]. The electric fields in the vicinity of the nuclei of these ions are tremendous in magnitude, enabling the use of the corresponding spectra (including those occurring in the interaction of the ions with the high-intensity laser field) for a high-precision verification of quantum electrodynamics, which describes electromagnetic forces and is a part of the Standard Model of elementary particles.

The free-electron laser at the Jefferson Laboratory, USA, was used to advantage-for the simultaneous generation of high-power femtosecond X-ray radiation (by the interaction of Thomson scattering and the infrared radiation of the electron beam in the wiggler), infrared free-electron laser radiation, and terahertz radiation in the linear accelerator [42, 25, 26, 46].

Several schemes for producing high-intensity electromagnetic radiation with the use of lasers are presently under development. Considered in this connection are the above laser accelerators of femtosecond-long electron beams for the generation of electromagnetic radiation from the gigahertz range to X-rays. This radiation may emerge, for instance, as diffraction or transition radiation of a short particle beam. X-ray radiation may result from the interaction of a relativistic electron beam with a fraction of its parent laser radiation.

A particular lasing scheme is considered as applied to free-electron lasers. According to this scheme, the propagation of a high-intensity electron beam through a plasma with a low (in comparison with the beam density) density leads to the formation of a positively charged plasma cylinder owing to the ejection of plasma electrons in the radial direction by the leading edge of the laser beam. The relativistic electrons of the beam execute radial betatron oscillations in this ion channel to radiate electromagnetic energy in the X-ray range. This X-ray generation mechanism was proposed and realized in [133]. Similar betatron oscillations are also executed by the electrons captured by the wakefield of a short laser pulse (see Sect. 6.3). Hard X-ray radiation is also observed in this case.

Of considerable interest is electromagnetic radiation in the terahertz range, which may be employed to carry out diagnostics and spectroscopy of materials in which the transition energy between quantum states is quite low, about 0.01 eV, corresponding to photons with a frequency of the order of a teraherz and a radiation wavelength of 100 $\mu$m. These materials include, in particular, practically all organic materials, including biological and medical objects, many semiconductors, and high-temperature superconductors. Not only will the development of high-power terahertz radiation sources open up new vistas for medical diagnostics, but they may also become a reliable tool in the sphere of security (detection of drugs, explosives, etc.).

The terahertz-frequency range extends from far infrared radiation to microwave submillimeter radiation. The methods presently employed for the generation of both optical and microwave radiation are ineffective for the generation of this radiation. The output power of available semiconductor sources, which operate at low temperatures, is at a level of 0.1–0.01 W. With the aid of a free-electron laser it is possible to attain an output power of about 20 W. This power is insufficient for broad application of the terahertz radiation. This is precisely why attention is being paid to a different, new laser-plasma terahertz radiation source, which may exceed a megawatt in power.

The recording of low-frequency electromagnetic radiation of the terahertz range in the irradiation of gaseous and solid-state targets by high-intensity short laser pulses of femtosecond duration was first reported by Hamster et al. [71]. The authors attributed the observed effect to the conversion of terahertz-frequency plasma oscillations, which are excited in the plasma by a laser pulse, to electromagnetic radiation.

The work reported in [71] attracted the attention of researchers primarily from the standpoint of the physical mechanisms underlying the transformation of plasma oscillation energy to electromagnetic radiation and their efficiency, as well as from the standpoint of the properties and power of the electromagnetic radiation that may be obtained in the interaction of high-power short laser pulses with the plasma. The transformation of longitudinal plasma waves to transverse electromagnetic waves is possible, for instance, in the presence of plasma density nonuniformity or under imposition of an external magnetic field.

The generation of low-frequency (terahertz) electromagnetic radiation in the propagation of a short laser through a periodically nonuniform (stratified) plasma was studied in [61] (see also [123, 14, 62]). The effect is related to the transformation of wake plasma fields of the laser pulse to transverse electromagnetic waves by plasma density modulations. This problem, as applied to a plasma with random nonuniformities, when the laser pulse propagates through the plasma with density fluctuations, was considered in [62].

In a plasma with a regular nonuniformity, the low-frequency terahertz radiation of the laser pulse was investigated both numerically [116] and analytically [115, 114]. The excitation of terahertz radiation by the laser pulse in the plasma is also possible in the presence of an external magnetic field. In this case, the generation of terahertz wave fields is related to the Vavilov–Cherenkov effect, because the velocity of the laser pulse exceeds the phase velocity of the slow extraordinary wave excited by the pulse. This problem was investigated in the cases of transverse propagation of laser pulse relative to the external magnetic field [136, 119, 137, 41, 18] and longitudinal (along the magnetic field) propagation of the laser pulse [65].

The generation of terahertz-frequency electromagnetic waves at the intersection of a plasma boundary by an electric bunch accelerated and formed in the wake wave of the laser pulse was investigated in [86, 88, 131, 130, 111] and is due to the transition Ginzburg–Frank radiation. The low-frequency transition radiation of the laser pulse at its intersection with a tenuous plasma boundary is discussed in [64, 66]. The excitation of terahertz waves traveling from the boundary into vacuum and into the plasma interior is related in this case to the ponderomotive action of the laser pulse and the emergence of an eddy current at the plasma boundary. Along with the generation of bulk electromagnetic waves there also occurs the transition radiation of terahertz surface waves, which travel along the plasma boundary [57].

Terahertz radiation may be generated in a tenuous plasma in the interaction of two counterpropagating short laser pulses [63]. In this case a small-scale standing plasma wave is excited in the interaction region, which produces radiation at double the plasma frequency. The process under discussion is based on the elementary process of two-plasmon coalescence with the production of a photon. The generation of terahertz radiation also occurs in the optical breakdown of a gas when the laser pulse is focused with an axicon. When an external permanent electric field [60] or a microwave field [24] is applied, the ionization wave excites a plasma wave traveling at a supraluminal speed, which is radiated due to the Vavilov–Cherenkov effect.

The phenomenon of terahertz and subterahertz radiation generation in the propagation of a low-intensity laser pulse through the atmosphere has been observed in many experiments (see, for instance, [132]). The authors of [123, 14] attribute the occurrence of this radiation to density modulations in the plasma filament produced in the ionization of the air.

It is pertinent to note that other laser-assisted methods of terahertz radiation production are also discussed in the literature, which rely on nonlinear (four-wave) wave interactions [135].

## 6.5 Plasma in Accelerators

Charged-particle beams in modern high-energy accelerators are yet another typical example of high-energy density physics [101]. Although the density of these accelerated bunches is not high, their acceleration and the collective interaction between themselves and the accelerating fields corresponds to substantial energy densities. Moreover, slight variation of relative particle velocities after laser or electron cooling results in high values of the nonideality parameter of such beam-produced plasma.

A demonstration example is "crystalline" beams [109, 110], which emerge under the laser cooling of ions accelerated in a synchrotron. When the relative ion velocity corresponds to a temperature on the order of millikelvins, the Coulomb interaction effects become significant and give rise to the "crystallization" of this plasma (Fig. 6.30).

Other mechanisms of plasma generation in accelerators are also possible [101]. The plasma may emerge under the action of the cyclotron radiation of accelerated particles, in the ionization of the residual gas, or when the beam finds itself at the wall. In the last-mentioned case, for the GSI and CERN LHC accelerators the energy releases involved stretch into the megajoule range [53, 127]. In this case, the emergent electrons with a density of $10^5$–$10^7$ cm$^{-3}$ arrive at the center of the acceleration channel to give rise to a two-stream "beam–plasma" instability, cause a loss of the particles under acceleration, and impair the beam emittance. In many respects these processes are similar to the effects occurring in laser plasmas and may be analyzed by the methods of high-energy-density physics.

A broad spectrum of plasma states with high energy densities emerges in the operation of pulsed high-current accelerators and high-power microwave generators [94, 95]. The application of a high (megaelectronvolt) pulsed voltage to the vacuum gap in these devices results in the electric microexplosion of cathodic micropoints and the explosive emission of high-current electron beams. The energy densities attainable in this case amount to $\approx$ 1–10 MJ/cm$^3$ for pressures ranging into megabars. The accelerators and microwave generators designed on this principle find wide industrial application and are a demonstration of the practical use of basic research in high-energy-density physics.

The next generation of high-power lasers will make it possible to achieve power densities $> 10^{23}$ W/cm$^2$, whereby the longitudinal electric field will be $>$1 TV/cm. Particles in these fields may be accelerated to relativistic energies over short distances, which lengthens their lifetime.

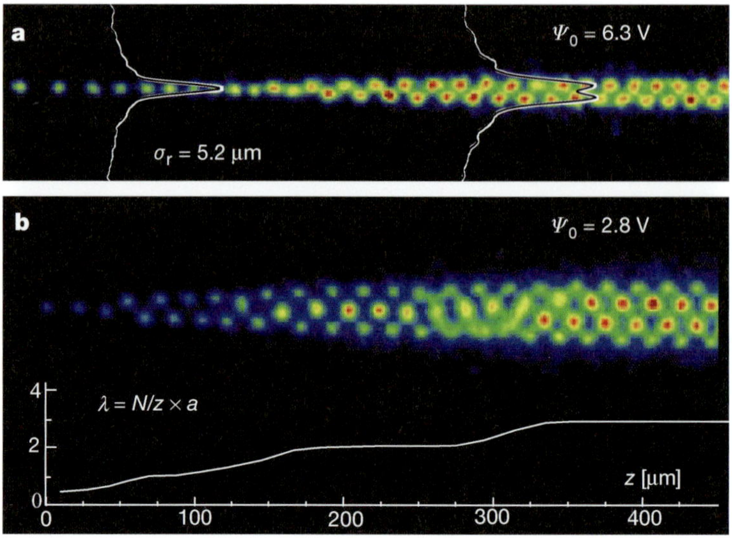


**Fig. 6.30** Crystalline beams in accelerators [109, 110].

Calculations [104] show that the irradiation of a metallic target at an intensity of $10^{23}$ W/cm$^2$ may produce 5 GeV protons, which in turn may give rise to pions with a lifetime of $\approx$20 ns, the laser irradiation lasting for $\approx$10 fs. By employing the laser acceleration mechanism it is possible to rapidly (in several picoseconds) raise the energy and lifetime of the pions by a factor of 100 – to 15 GeV and 2 $\mu$s. This lengthened lifetime basically makes it possible to take advantage of conventional acceleration techniques to accelerate pions to the gigaelectronvolt range, where pions will decay into muon and neutrino beams.

A pion source would be of interest for the development of accelerators of pions as particles without quark structure and producing a weaker synchrotron radiation. Devices for studying gamma–gamma collisions might become another elegant direction of combined use of accelerators [101] and lasers.

## References

[1] Bible. Old Testament, 1 Samuel, 17: 34, 40, 43, 48–51

[2] Andreev, N.E., Chizhonkov, E.V., Frolov, A.A., Gorbunov, L.M.: On laser wakefield acceleration in plasma channels. Nucl. Instrum. Methods Phys. Res. A **410**(3), 469–476 (2002). DOI 10.1016/S0168-9002(98)00181-8

[3] Andreev, N.E., Cros, B., Gorbunov, L.M., et al.: Laser wakefield structure in a plasma column created in capillary tubes. Phys. Plasmas **9**(9), 3999–4009

(2002). DOI 10.1063/1.1497165
[4] Andreev, N.E., Gorbunov, L.M.: Laser-plasma acceleration of electrons. Phys. Usp. **42**(1), 49 (1999). DOI 10.1070/PU1999v042n01ABEH000447. URL http://ufn.ru/en/articles/1999/1/f/
[5] Andreev, N.E., Gorbunov, L.M., Kirsanov, V.I., et al.: Structure of the wake field in plasma channels. Phys. Plasmas **4**(4), 1145–1153 (1997). DOI 10.1063/1.872186
[6] Andreev, N.E., Gorbunov, L.M., Kirsanov, V.I., et al.: Resonant excitation of wakefields by a laser-pulse in a plasma. JETP Lett. **55**(10), 571–576 (1992)
[7] Andreev, N.E., Gorbunov, L.M., Kirsanov, V.I., et al.: The theory of laser self-resonant wake field excitation. Phys. Scr. **49**(1), 101–109 (1994). URL http://stacks.iop.org/1402-4896/49/101
[8] Andreev, N.E., Gorbunov, L.M., Kirsanov, V.I., Sakharov, A.S.: Self-modulation of high-intensity laser pulses in underlense plasmas and plasma channels. AIP Conf. Proc. **396**(1), 61–74 (1997). DOI 10.1063/1.52974. URL http://link.aip.org/link/?APC/396/61/1
[9] Andreev, N.E., Kirsanov, V.I., Gorbunov, L.M.: Stimulated processes and self-modulation of a short intense laser pulse in the laser wake-field accelerator. Phys. Plasmas **2**(6), 2573–2582 (1995). DOI 10.1063/1.871219
[10] Andreev, N.E., Kirsanov, V.I., Sakharov, A.S., et al.: On the phase velocity of plasma waves in a self-modulated laser wake-field accelerator. Phys. Plasmas **3**(8), 3121–3128 (1996). DOI 10.1063/1.871659
[11] Andreev, N.E., Nishida, Y., Yugami, N.: Propagation of short intense laser pulses in gas-filled capillaries. Phys. Rev. E **65**(5), 056407 (2002). DOI 10.1103/PhysRevE.65.056407. URL http://link.aps.org/abstract/PRE/v65/e056407
[12] Antonsen Jr., T.M., Mora, P.: Self-focusing and Raman scattering of laser pulses in tenuous plasmas. Phys. Rev. Lett. **69**(15), 2204–2207 (1992). URL http://link.aps.org/abstract/PRL/v69/p2204
[13] Antonsen Jr., T.M., Mora, P.: Self-focusing and Raman scattering of laser pulses in tenuous plasmas. Phys. Fluids B **5**(5), 1440–1452 (1993). DOI 10.1063/1.860884. URL http://dx.doi.org/10.1063/1.860884
[14] Antonsen Jr., T.M., Palastro, J., Milchberg, H.M.: Excitation of terahertz radiation by laser pulses in nonuniform plasma channels. Phys. Plasmas **14**(3), 033107 (2007). DOI 10.1063/1.2715864
[15] Atzeni, S., Meyer-ter-Vehn, J.: The Physics of Inertial Fusion. Oxford University Press, Oxford (2004)
[16] Atzeni, S., Temporal, M., Honrubia, J.J.: A first analysis of fast ignition of precompressed ICF fuel by laser-accelerated protons. Nucl. Fusion **42**(3), L1–L4 (2002). URL http://stacks.iop.org/0029-5515/42/L1
[17] Bakhmetjev, I.E., Fertman, A.D., Golubev, A.A., et al.: Research into the advanced experimental methods for precision ion stopping range measurements in matter. Laser Part. Beams **21**(1), 1–6 (2003). DOI 10.1017/S0263034602211015

[18] Bakunov, M.I., Bodrov, S.B., Maslov, A.V., Sergeev, A.M.: Two-dimensional theory of Cherenkov radiation from short laser pulses in a magnetized plasma. Phys. Rev. E **70**(1), 016401 (2004). DOI 10.1103/PhysRevE.70.016401. URL http://link.aps.org/abstract/PRE/v70/e016401

[19] Bamber, C., Boege, S.J., Koffas, T., et al.: Studies of nonlinear QED in collisions of 46.6 GeV electrons with intense laser pulses. Phys. Rev. D **60**(9), 092004 (1999). DOI 10.1103/PhysRevD.60.092004. URL http://link.aps.org/abstract/PRD/v60/e092004

[20] Belyaev, V.S., Krainov, V.P., Lisitsa, V.S., Matafonov, A.P.: Generation of fast charged particles and superstrong magnetic fields in the interaction of ultrashort high-intensity laser pulses with solid targets. Phys. Usp. **51**(8), 793 (2008). DOI 10.1070/PU2008v051n08ABEH006541. URL http://ufn.ru/en/articles/2008/8/b/

[21] Berezhiani, V.I., Murusidze, I.G.: $e^+e^-$ – Pair production by a focused laser pulse in vacuum. Phys. Lett. A **148**(6-7), 338–340 (1990). DOI 10.1016/0375-9601(90)90813-4

[22] Bulanov, S.V., Kirsanov, V.I., Sakharov, A.S.: Excitation of ultrarelativistic plasma waves by pulse of electromagnetic radiation. JETP Lett. **50**(4), 198 (1991)

[23] Bychenkov, V.Y., Rozmus, W., Maksimchuk, A., et al.: Fast ignitor concept with light ions. Plasma Phys. Rep. **27**(12), 1017–1020 (2001). DOI 10.1134/1.1426135

[24] Bystrov, A.M., Vvedenskii, N.V., Gildenburg, V.B.: Generation of terahertz radiation upon the optical breakdown of a gas. JETP Lett. **82**(12), 753–757 (2005)

[25] Carr, G.L., Martin, M.C., McKinney, W.R., et al.: High-power terahertz radiation from relativistic electrons. Nature **420**(6912), 153–156 (2002). DOI 10.1038/nature01175

[26] Carr, G.L., Martin, M.C., McKinney, W.R., et al.: Very high power THz radiation at Jefferson Lab. Phys. Med. Biol. **47**(21), 3761–3764 (2002). DOI 10.1088/0031-9155/47/21/313

[27] Cavailler, C.: Inertial fusion with the LMJ. Plasma Phys. Control. Fusion **47**(12B), B389–B403 (2005). DOI 10.1088/0741-3335/47/12B/S28

[28] Chen, F.F.: Introduction to Plasma Physics and Controlled Fusion, Vol. 1, 2nd edn. Springer, New York (1984)

[29] Clark, E.L., Krushelnick, K., Davies, J.R., et al.: Measurements of energetic proton transport through magnetized plasma from intense laser interactions with solids. Phys. Rev. Lett. **84**(4), 670–673 (2000). DOI 10.1103/PhysRevLett.84.670. URL http://link.aps.org/abstract/PRL/v84/p670

[30] Clark, T.R., Milchberg, H.M.: Optical mode structure of the plasma waveguide. Phys. Rev. E **61**(2), 1954–1965 (2000). DOI 10.1103/PhysRevE.61.1954. URL http://link.aps.org/abstract/PRE/v61/p1954

[31] Courtois, C., Couairon, A., Cros, B., et al.: Propagation of intense ultrashort laser pulses in a plasma filled capillary tube: Simulations and experiments. Phys. Plasmas **8**(7), 3445–3456 (2001). DOI 10.1063/1.1378327

[32] Cros, B., Courtois, C., Malka, G., et al.: Excitation of accelerating wakefields in inhomogeneous plasmas. IEEE Trans. Plasma Sci. **28**(4), 1071–1077 (2000). DOI 10.1109/27.893291

[33] Cuneo, M.E., Vesey, R.A., Bennett, G.R., et al.: Progress in symmetric ICF capsule implosions and wire-array Z-pinch source physics for double-pinch-driven hohlraums. Plasma Phys. Control. Fusion **48**(2), R1–R35 (2006). DOI 10.1088/0741-3335/48/2/R01

[34] Decker, C.D., Mori, W.B., Tzeng, K.C., Katsouleas, T.C.: Modeling single-frequency laser-plasma acceleration using particle-in-cell simulations: the physics of beam breakup. IEEE Trans. Plasma Sci. **24**(2), 379–392 (1996). DOI 10.1109/27.510002

[35] DESY: FLASH. URL `http://flash.desy.de/`

[36] DESY: PETRA III. URL `http://petra3.desy.de/`

[37] Ditmire, T., Springate, E., Tisch, J.W., et al.: Explosion of atomic clusters heated by high-intensity femtosecond laser pulses. Phys. Rev. A **57**(1), 369–382 (1998). DOI 10.1103/PhysRevA.57.369. URL `http://link.aps.org/abstract/PRA/v57/p369`

[38] Ditmire, T., Tisch, J.W.G., Springate, E., et al.: High-energy ions produced in explosions of superheated atomic clusters. Nature **386**(6620), 54–56 (1997). DOI 10.1038/386054a0

[39] Ditmire, T., Zweiback, J., Yanovsky, V.P., et al.: Nuclear fusion from explosions of femtosecond laser-heated deuterium clusters. Nature **398**(6727), 489–492 (1999). DOI 10.1038/19037

[40] Ditmire, T., Zweiback, J., Yanovsky, V.P., et al.: Nuclear fusion in gases of deuterium clusters heated with a femtosecond laser. Phys. Plasmas **7**(5), 1993–1998 (2000). DOI 10.1063/1.874020

[41] Dorranian, D., Starodubtsev, M., Kawakami, H., et al.: Radiation from high-intensity ultrashort-laser-pulse and gas-jet magnetized plasma interaction. Phys. Rev. E **68**(2), 026409 (2003). DOI 10.1103/PhysRevE.68.026409. URL `http://link.aps.org/abstract/PRE/v68/e026409`

[42] Douglas, D.R., Jordan, K.C., Merminga, L., et al.: Experimental investigation of multibunch, multipass beam breakup in the Jefferson Laboratory free electron laser upgrade driver. Phys. Rev. ST Accel. Beams **9**(6), 064403 (2006). DOI 10.1103/PhysRevSTAB.9.064403. URL `http://link.aps.org/abstract/PRSTAB/v9/e064403`

[43] Durfee III, C.G., Milchberg, H.M.: Light pipe for high intensity laser pulses. Phys. Rev. Lett. **71**(15), 2409–2412 (1993). URL `http://link.aps.org/abstract/PRL/v71/p2409`

[44] Efremov, V.P., Pikuz Jr., S.A., Faenov, A.Y., et al.: Study of the energy release region of a heavy-ion flux in nanomaterials by X-ray spectroscopy of multicharged ions. JETP Lett. **81**(8), 378 (2005)

[45] Eloy, M., Azambuja, R., Mendonca, J.T., Bingham, R.: Interaction of ultrashort high-intensity laser pulses with atomic clusters. Phys. Plasmas **8**(3), 1084–1086 (2001). DOI 10.1063/1.1345709

[46] Esarey, E., Sprangle, P., Krall, J., Ting, A.: Overview of plasma-based accelerator concepts. IEEE Trans. Plasma Sci. **24**(2), 252–288 (1996). DOI 10.1109/27.509991

[47] Esirkepov, T., Borghesi, M., Bulanov, S.V., et al.: Highly efficient relativistic-ion generation in the laser-piston regime. Phys. Rev. Lett. **92**(17), 175003 (2004). DOI 10.1103/PhysRevLett.92.175003. URL http://link.aps.org/abstract/PRL/v92/e175003

[48] Esirkepov, T., Yamagiwa, M., Tajima, T.: Laser ion-acceleration scaling laws seen in multiparametric particle-in-cell simulations. Phys. Rev. Lett. **96**(10), 105001 (2006). DOI 10.1103/PhysRevLett.96.105001. URL http://link.aps.org/abstract/PRL/v96/e105001

[49] Esirkepov, T.Z., Bulanov, S.V., Nishihara, K., et al.: Proposed double-layer target for the generation of high-quality laser-accelerated ion beams. Phys. Rev. Lett. **89**(17), 175003 (2002). DOI 10.1103/PhysRevLett.89.175003. URL http://link.aps.org/abstract/PRL/v89/e175003

[50] Faure, J., Glinec, Y., Pukhov, A., et al.: A laser-plasma accelerator producing monoenergetic electron beams. Nature **431**(7008), 541–544 (2004). DOI 10.1038/nature02963

[51] Fortov, V.E. (ed.): Entsiklopediya nizkotemperaturnoi plazmy (Encyclopedia of Low-Temperature Plasma). Nauka, Moscow (2000)

[52] Fortov, V.E.: Intense Shock Waves and Extreme States of Matter. Bukos, Moscow (2005)

[53] Fortov, V.E., Hoffmann, D.H.H., Sharkov, B.Y.: Intense ion beams for generating extreme states of matter. Phys. Usp. **51**(2), 109 (2008). DOI 10.1070/PU2008v051n02ABEH006420. URL http://ufn.ru/en/articles/2008/2/a/

[54] Fortov, V.E., Ivlev, A.V., Khrapak, S.A., et al.: Complex (dusty) plasma: Current status, open issues, perspectives. Phys. Rep. **421**(1), 1–103 (2005). DOI 10.1016/j.physrep.2005.08.007

[55] Fourkal, E., Li, J.S., Xiong, W., et al.: Intensity modulated radiation therapy using laser-accelerated protons: a Monte Carlo dosimetric study. Phys. Med. Biol. **48**(24), 3977–4000 (2003). DOI 10.1088/0031-9155/48/24/001

[56] Fourkal, E., Shahine, B., Ding, M., et al.: Particle in cell simulation of laser-accelerated proton beams for radiation therapy. Med. Phys. **29**(12), 2788–2798 (2002). DOI 10.1118/1.1521122. URL http://dx.doi.org/10.1118/1.1521122

[57] Frolov, A.A.: Excitation of surface waves at a plasma boundary by a short laser pulse. Plasma Phys. Rep. **33**(3), 179–188 (2007). DOI 10.1134/S1063780X07030026

[58] Geddes, C.G.R., Tóth, C., van Tilborg, J., et al.: High-quality electron beams from a laser wakefield accelerator using plasma-channel guiding. Nature **431**(7008), 538–541 (2004). DOI 10.1038/nature02900

[59] Giorla, J., Bastian, J., Bayer, C., et al.: Target design for ignition experiments on the laser Mégajoule facility. Plasma Phys. Control. Fusion **48**(12B), B75–B82 (2006). DOI 10.1088/0741-3335/48/12B/S0

[60] Golubev, S.V., Suvorov, E.V., Shalashov, A.G.: On the possibility of terahertz wave generation upon dense gas optical breakdown. JETP Lett. **79**(8), 361–364 (2004)

[61] Gorbunov, L.M., Frolov, A.A.: Emission of low-frequency electromagnetic waves by a short laser pulse in stratified rarefied plasma. JETP **83**(5), 967–973 (1996)

[62] Gorbunov, L.M., Frolov, A.A.: Emission of low-frequency electromagnetic waves by a short laser pulse propagating in a plasma with density fluctuations. Plasma Phys. Rep. **26**(8), 646–656 (2000). DOI 10.1134/1.1306994

[63] Gorbunov, L.M., Frolov, A.A.: Electromagnetic radiation at twice the plasma frequency emitted from the region of interaction of two short laser pulses in a rarefied plasma. JETP **98**(3), 527–537 (2004). DOI 10.1134/1.1705705

[64] Gorbunov, L.M., Frolov, A.A.: Low-frequency transition radiation from a short laser pulse at the plasma boundary. JETP **102**(6), 894–901 (2006). DOI 10.1134/S1063776106060033

[65] Gorbunov, L.M., Frolov, A.A.: On the theory of Cherenkov emission from a short laser pulse in a magnetized plasma. Plasma Phys. Rep. **32**(6), 500–513 (2006). DOI 10.1134/S1063780X06060079

[66] Gorbunov, L.M., Frolov, A.A.: Transition radiation generated by a short laser pulse at a plasma–vacuum interface. Plasma Phys. Rep. **32**(10), 850–865 (2006). DOI 10.1134/S1063780X06100059

[67] Gorbunov, L.M., Kalmykov, S.Y., Mora, P.: Laser wakefield acceleration by petawatt ultrashort laser pulses. Phys. Plasmas **12**(3), 033101 (2005). DOI 10.1063/1.1852469

[68] Gorbunov, L.M., Kirsanov, V.I.: The excitation of plasma waves by an electromagnetic wave packet. JETP **93**, 509 (1987). (In Russian)

[69] Gordienko, S., Pukhov, A.: Scalings for ultrarelativistic laser plasmas and quasimonoenergetic electrons. Phys. Plasmas **12**(4), 043109 (2005). DOI 10.1063/1.1884126

[70] Hammel, B.A., National Ignition Campaign Team: The NIF ignition program: progress and planning. Plasma Phys. Control. Fusion **48**(12B), B497–B506 (2006). DOI 10.1088/0741-3335/48/12B/S47

[71] Hamster, H., Sullivan, A., Gordon, S., et al.: Subpicosecond, electromagnetic pulses from intense laser-plasma interaction. Phys. Rev. Lett. **71**(17), 2725–2728 (1993). URL `http://link.aps.org/abstract/PRL/v71/p2725`

[72] Hegelich, B.M., Albright, B.J., Cobble, J., et al.: Laser acceleration of quasi-monoenergetic MeV ion beams. Nature **439**, 441–444 (2006). DOI 10.1038/nature04400

[73] Hoffmann, D.H.H., Fortov, V.E., Lomonosov, I.V., et al.: Unique capabilities of an intense heavy ion beam as a tool for equation-of-state studies. Phys. Plasmas **9**(9), 3651–3654 (2002). DOI 10.1063/1.1498260

[74] Hogan, W.J. (ed.): Energy from Inertial Fusion. IAEA, Vienna, Austria (1995)

[75] Hooker, S.M., Spence, D.J., Smith, R.A.: Guiding of high-intensity picosecond laser pulses in a discharge-ablated capillary waveguide. J. Opt. Soc. Am. B **17**(1), 90–98 (2000). DOI 10.1364/JOSAB.17.000090. URL `http://josab.osa.org/abstract.cfm?URI=josab-17-1-90`

[76] Joshi, C.: Plasma accelerators. Sci. Am. **294**(2), 40–47 (2006)

[77] Kaplan, A.E., Dubetsky, B.Y., Shkolnikov, P.L.: Shock shells in Coulomb explosions of nanoclusters. Phys. Rev. Lett. **91**(14), 143401 (2003). DOI 10.1103/PhysRevLett.91.143401. URL `http://link.aps.org/abstract/PRL/v91/e143401`

[78] Khazanov, E.A., Sergeev, A.M.: Petawatt laser based on optical parametric amplifiers: Their state and prospects. Phys. Usp. **51**(9), 969 (2008). DOI 10.1070/PU2008v051n09ABEH006612. URL `http://ufn.ru/en/articles/2008/9/h/`

[79] Kirzhnits, D.A.: Extremal states of matter (ultrahigh pressures and temperatures). Sov. Phys. – Usp. **14**(4), 512–523 (1972). DOI 10.1070/PU1972v014n04ABEH004734. URL `http://stacks.iop.org/0038-5670/14/512`

[80] Kishimoto, Y., Masaki, T., Tajima, T.: High energy ions and nuclear fusion in laser–cluster interaction. Phys. Plasmas **9**(2), 589–601 (2002). DOI 10.1063/1.1418433

[81] Kodama, R., Norreys, P.A., Mima, K., et al.: Fast heating of ultrahigh-density plasma as a step towards laser fusion ignition. Nature **412**(6849), 798–802 (2001). DOI 10.1038/35090525

[82] Konyukhov, A.V., Likhachev, A.P., Oparin, A.M., et al.: Numerical modeling of shock-wave instability in thermodynamically nonideal media. JETP **98**(4), 811–819 (2004). DOI 10.1134/1.1757680

[83] Krall, J., Ting, A., Esarey, E., Sprangle, P.: Enhanced acceleration in a self-modulated-laser wake-field accelerator. Phys. Rev. E **48**(3), 2157–2161 (1993). DOI 10.1103/PhysRevE.48.2157. URL `http://link.aps.org/abstract/PRE/v48/p2157`

[84] Kruer, W.L.: The Physics of Laser Plasma Interactions. Addison-Wesley, Reading, MA (1988)

[85] Last, I., Schek, I., Jortner, J.: Energetics and dynamics of Coulomb explosion of highly charged clusters. J. Chem. Phys. **107**(17), 6685–6692 (1997). DOI 10.1063/1.474911. URL `http://link.aip.org/link/?JCP/107/6685/1`

[86] Leemans, W.P., Geddes, C.G.R., Faure, J., et al.: Observation of terahertz emission from a laser-plasma accelerated electron bunch crossing a plasma-vacuum boundary. Phys. Rev. Lett. **91**(7), 074802 (2003). DOI 10.1103/PhysRevLett.91.074802. URL `http://link.aps.org/abstract/PRL/v91/e074802`

[87] Leemans, W.P., Nagler, B., Gonsalves, A.J., et al.: GeV electron beams from a centimetre-scale accelerator. Nat. Phys. **2**(10), 696–699 (2006). DOI 10.1038/nphys418

[88] Leemans, W.P., van Tilborg, J., Faure, J., et al.: Terahertz radiation from laser accelerated electron bunches. Phys. Plasmas **11**(5), 2899–2906 (2004). DOI 10.1063/1.1652834

[89] Lifschitz, A.F., Faure, J., Malka, V., Mora, P.: GeV wakefield acceleration of low energy electron bunches using petawatt lasers. Phys. Plasmas **12**(9), 093104 (2005). DOI 10.1063/1.2010347

[90] Lindl, J.D.: Inertial Confinement Fusion. Springer, New York (1998)

[91] Loborev, V.M., Pertsev, V.V., Sudakov, V.E., et al. (eds.): Fizika jadernogo vzryva (The Physics of Nuclear Explosions], vol. 1. FizMatLit, Moscow (2009)

[92] Maksimchuk, A., Flippo, K., Krause, H., et al.: Plasma phase transition in dense hydrogen and electron—hole plasmas. Plasma Phys. Rep. **30**(6), 473–495 (2004). DOI 10.1134/1.1768582

[93] Maksimchuk, A., Gu, S., Flippo, K., et al.: Forward ion acceleration in thin films driven by a high-intensity laser. Phys. Rev. Lett. **84**(18), 4108–4111 (2000). DOI 10.1103/PhysRevLett.84.4108. URL `http://link.aps.org/abstract/PRL/v84/p4108`

[94] Mesyats, G.A.: Impul'snaya energetika i elektronika (Pulse Power and Electronics). Nauka, Moscow (2004)

[95] Mesyats, G.A., Yalandin, M.I.: High-power picosecond electronics. Phys. Usp. **48**(3), 211 (2005). DOI 10.1070/PU2005v048n03ABEH002113. URL `http://ufn.ru/en/articles/2005/3/a/`

[96] Mima, K., Fast Ignition Research Group: Present status and future prospects of laser fusion and related high energy density plasma research. AIP Conf. Proc. **740**(1), 387–397 (2004). DOI 10.1063/1.1843522. URL `http://link.aip.org/link/?APC/740/387/1`

[97] Mintsev, V., Gryaznov, V., Kulish, M., et al.: Stopping power of proton beam in a weakly non-ideal xenon plasma. Contrib. Plasma Phys. **39**(1-2), 45–48 (1999). DOI 10.1002/ctpp.2150390111

[98] Mori, W.B., Decker, C.D., Hinkel, D.E., Katsouleas, T.: Raman forward scattering of short-pulse high-intensity lasers. Phys. Rev. Lett. **72**(10), 1482–1485 (1994). URL `http://link.aps.org/abstract/PRL/v72/p1482`

[99] Moses, E.I., Bonanno, R.E., Haynam, C.A., et al.: The National Ignition Facility: path to ignition in the laboratory. Eur. Phys. J. D **44**(2), 215–218 (2006). DOI 10.1140/epjd/e2006-00106-3

[100] Mourou, G.A., Tajima, T., Bulanov, S.V.: Optics in the relativistic regime. Rev. Mod. Phys. **78**(2), 1804–1816 (2006). DOI 10.1103/RevModPhys.78.309. URL `http://link.aps.org/abstract/RMP/v78/p309`

[101] National Research Council: Frontiers in High Energy Density Physics. National Academies Press, Washington, DC (2003)

[102] Nishihara, K., Amitani, H., Murakami, M., et al.: High energy ions generated by laser driven Coulomb explosion of cluster. Nucl. Instrum. Meth. Phys. Res. A **464**(1-3), 98–102 (2001). DOI 10.1016/S0168-9002(01)00014-6

[103] Okihara, S., Esirkepov, T.Z., Nagai, K., et al.: Ion generation in a low-density plastic foam by interaction with intense femtosecond laser pulses. Phys. Rev. E **69**(2), 026401 (2004). DOI 10.1103/PhysRevE.69.026401. URL `http://link.aps.org/abstract/PRE/v69/e026401`

[104] Pukhov, A.: Strong field interaction of laser radiation. Rep. Prog. Phys. **66**(1), 47–101 (2003). DOI 10.1088/0034-4885/66/1/202

[105] Pukhov, A., Meyer-ter-Vehn, J.: Laser wake field acceleration: the highly non-linear broken-wave regime. Appl. Phys. B **74**(4-5), 355–361 (2002). DOI 10.1007/s003400200795

[106] Quintenz, J., Sandia's Pulsed Power Team: Pulsed power team. In: Proc. 13th Int. Conf. on High Power Particle Beams. Nagaoka, Japan (2000)

[107] Rosmej, O.N., Blazevic, A., Korostiy, S., et al.: Charge state and stopping dynamics of fast heavy ions in dense matter. Phys. Rev. A **72**(5), 052901 (2005). DOI 10.1103/PhysRevA.72.052901. URL `http://link.aps.org/abstract/PRA/v72/e052901`

[108] Roth, M., Cowan, T.E., Key, M.H., et al.: Fast ignition by intense laser-accelerated proton beams. Phys. Rev. Lett. **86**(3), 436–439 (2001). DOI 10.1103/PhysRevLett.86.436

[109] Schatz, T., Schramm, U., Habs, D.: Crystalline ion beams. Nature **412**(6848), 717–720 (2001). DOI 10.1038/35089045

[110] Schramm, U., Schatz, T., Bussmann, M., Habs, D.: Cooling and heating of crystalline ion beams. J. Phys. B **36**(3), 561–571 (2003). DOI 10.1088/0953-4075/36/3/314

[111] Schroeder, C.B., Esarey, E., van Tilborg, J., Leemans, W.P.: Theory of coherent transition radiation generated at a plasma-vacuum interface. Phys. Rev. E **69**(1), 016501 (2004). DOI 10.1103/PhysRevE.69.016501. URL `http://link.aps.org/abstract/PRE/v69/e016501`

[112] Shao, Y.L., Ditmire, T., Tisch, J.W.G., et al.: Multi-keV electron generation in the interaction of intense laser pulses with Xe clusters. Phys. Rev. Lett. **77**(16), 3343–3346 (1996). DOI 10.1103/PhysRevLett.77.3343. URL `http://link.aps.org/abstract/PRL/v77/p3343`

[113] Sharkov, B.Y. (ed.): Yadernyi sintez s inertsionnym uderzhaniem (Inertial Confinement Nuclear Fusion). Fizmatlit, Moscow (2005)

[114] Sheng, Z.M., Mima, K., Zhang, J.: Powerful terahertz emission from laser wake fields excited in inhomogeneous plasmas. Phys. Plasmas **12**(12), 123103 (2005). DOI 10.1063/1.2136107

[115] Sheng, Z.M., Mima, K., Zhang, J., Sanuki, H.: Emission of electromagnetic pulses from laser wakefields through linear mode conversion. Phys. Rev. Lett. **94**(9), 095003 (2005). DOI 10.1103/PhysRevLett.94.095003. URL `http://link.aps.org/abstract/PRL/v94/e095003`

[116] Sheng, Z.M., Wu, H.C., Li, K., Zhang, J.: Terahertz radiation from the vacuum-plasma interface driven by ultrashort intense laser pulses. Phys.

Rev. E **69**(2), 025401 (2004). DOI 10.1103/PhysRevE.69.025401. URL http://link.aps.org/abstract/PRE/v69/e025401

[117] Shvets, G., Wurtele, J.S., Chiou, T.C., Katsouleas, T.C.: Excitation of accelerating wakefields in inhomogeneous plasmas. IEEE Trans. Plasma Sci. **24**(2), 351–362 (1996). DOI 10.1109/27.509999

[118] Spence, D.J., Butler, A., Hooker, S.M.: Gas-filled capillary discharge waveguides. J. Opt. Soc. Am. B **20**(1), 138–151 (2003). URL http://josab.osa.org/abstract.cfm?URI=josab-20-1-138

[119] Spence, N., Katsouleas, T., Muggli, P., et al.: Simulations of Cerenkov wake radiation sources. Phys. Plasmas **8**(11), 4995–5005 (2001). DOI 10.1063/1.1408625

[120] Spielman, R.B., Deeney, C., Chandler, G.A., et al.: Tungsten wire-array Z-pinch experiments at 200 TW and 2 MJ. Phys. Plasmas **5**(5), 2105–2111 (1998). DOI 10.1063/1.872881

[121] Sprangle, P., Esarey, E., Krall, J., Joyce, G.: Propagation and guiding of intense laser pulses in plasmas. Phys. Rev. Lett. **69**(15), 2200–2203 (1992). URL http://link.aps.org/abstract/PRL/v69/p2200

[122] Sprangle, P., Esarey, E., Ting, A., Joyce, G.: Laser wakefield acceleration and relativistic optical guiding. Appl. Phys. Lett. **53**(22), 2146–2148 (1988). DOI 10.1063/1.100300. URL http://link.aip.org/link/?APPLAB/53/2146/1

[123] Sprangle, P., Penano, J.R., Hafizi, B., Kapetanakos, C.A.: Ultrashort laser pulses and electromagnetic pulse generation in air and on dielectric surfaces. Phys. Rev. E **69**(6), 066415 (2004). DOI 10.1103/PhysRevE.69.066415. URL http://link.aps.org/abstract/PRE/v69/e066415

[124] Tabak, M., Hammer, J., Glinsky, M.E., et al.: Ignition and high gain with ultrapowerful lasers. Phys. Plasmas **1**(5), 1626–1634 (1994). DOI 10.1063/1.870664

[125] Tahir, N.A., Deutsch, C., Fortov, V.E., et al.: Proposal for the study of thermophysical properties of high-energy-density matter using current and future heavy-ion accelerator facilities at GSI Darmstadt. Phys. Rev. Lett. **95**(3), 035001 (2005). DOI 10.1103/PhysRevLett.95.035001. URL http://link.aps.org/abstract/PRL/v95/e035001

[126] Tahir, N.A., Deutsch, C., Fortov, V.E., et al.: Studies of strongly coupled plasmas using intense heavy ion beams at the future FAIR facility: The HEDgeHOB collaboration. Contrib. Plasma Phys. **45**(3-4), 229–235 (2005). DOI 10.1002/ctpp.200510025

[127] Tahir, N.A., Kain, V., Schmidt, R., et al.: The CERN Large Hadron Collider as a tool to study high-energy density matter. Phys. Rev. Lett. **94**(13), 135004 (2005). DOI 10.1103/PhysRevLett.94.135004. URL http://link.aps.org/abstract/PRL/v94/e135004

[128] Tajima, T.: Summary of Working Group 7 on "Exotic acceleration schemes". AIP Conf. Proc. **569**(1), 77–81 (2001). DOI 10.1063/1.1384337. URL http://link.aip.org/link/?APC/569/77/1

[129] Tajima, T., Dawson, J.M.: Laser electron accelerator. Phys. Rev. Lett. **43**(4), 267–270 (1979). URL http://link.aps.org/abstract/PRL/v43/p267

[130] van Tilborg, J., Schroeder, C.B., Esarey, E., Leemans, W.P.: Pulse shape and spectrum of coherent diffraction-limited transition radiation from electron beams. Laser Part. Beams **22**, 415–422 (2004). DOI 10.1017/S0263034604040078

[131] van Tilborg, J., Schroeder, C.B., Filip, C.V., et al.: Temporal characterization of femtosecond laser-plasma-accelerated electron bunches using terahertz radiation. Phys. Rev. Lett. **96**(1), 014801 (2006). DOI 10.1103/PhysRevLett.96.014801. URL http://link.aps.org/abstract/PRL/v96/e014801

[132] Tzortzakis, S., Mechain, G., Patalano, G., et al.: Coherent subterahertz radiation from femtosecond infrared filaments in air. Opt. Lett. **27**(21), 1944–1946 (2002). DOI 10.1364/OL.27.001944. URL http://www.opticsinfobase.org/abstract.cfm?URI=ol-27-21-1944

[133] Wang, S., Clayton, C.E., Blue, B.E., et al.: X-Ray emission from betatron motion in a plasma wiggler. Phys. Rev. Lett. **88**(13), 135004 (2002). DOI 10.1103/PhysRevLett.88.135004. URL http://link.aps.org/abstract/PRL/v88/e135004

[134] XFEL Project Group at DESY: The European X-ray laser project XFEL. URL http://xfel.desy.de/

[135] Yampolsky, N.A., Fraiman, G.M.: Conversion of laser radiation to terahertz frequency waves in plasma. Phys. Plasmas **13**(11), 113108 (2006). DOI 10.1063/1.2372462

[136] Yoshii, J., Lai, C.H., Katsouleas, T., et al.: Radiation from Cerenkov wakes in a magnetized plasma. Phys. Rev. Lett. **79**(21), 4194–4197 (1997). DOI 10.1103/PhysRevLett.79.4194. URL http://link.aps.org/abstract/PRL/v79/p4194

[137] Yugami, N., Higashiguchi, T., Gao, H., et al.: Experimental observation of radiation from Cherenkov wakes in a magnetized plasma. Phys. Rev. Lett. **89**(6), 065003 (2002). DOI 10.1103/PhysRevLett.89.065003. URL http://link.aps.org/abstract/PRL/v89/e065003

[138] Zigler, A., Ehrlich, Y., Cohen, C., et al.: Optical guiding of high-intensity laser pulses in a long plasma channel formed by a slow capillary discharge. J. Opt. Soc. Am. B **13**(1), 68–71 (1996). DOI 10.1364/JOSAB.13.000068. URL http://josab.osa.org/abstract.cfm?URI=josab-13-1-68

[139] Zweiback, J., Cowan, T.E., Smith, R.A., et al.: Characterization of fusion burn time in exploding deuterium cluster plasmas. Phys. Rev. Lett. **85**(17), 3640–3643 (2000). DOI 10.1103/PhysRevLett.85.3640. URL http://link.aps.org/abstract/PRL/v85/p3640

[140] Zweiback, J., Ditmire, T.: Femtosecond laser energy deposition in strongly absorbing cluster gases diagnosed by blast wave trajectory analysis. Phys. Plasmas **8**(10), 4545–4550 (2001). DOI 10.1063/1.1394778

# Chapter 7
# Astrophysical Aspects of High Energy Densities

High-energy-density physics underlies the contemporary understanding of the structure of astrophysical objects and their evolution, which takes place under the action of gravitational forces and thermonuclear energy release [105].

The scale and diversity of these phenomena are astounding, and comprehending them calls for the latest ideas and data from the cutting edge of modern physics. In this case, the observations of astrophysical objects are carried out over a broad range of the electromagnetic spectrum, from hard $\gamma$-rays to waves with meter-long wavelength (Figs. 7.1, 7.2), providing unique information about the physical processes at ultrahigh energies unattainable by modern accelerators. According to Academician Ya.B. Zel'dovich's figurative statement, the universe is "an accelerator for the poor", which does not demand from them multibillion expenses for the construction of experimental facilities.

The range of matter parameter variations in the universe is extremely broad [75, 214]: from empty space[1] and the tenuous intergalactic gas with a density of $10^{-30}$ g/cm$^3$ to extremely high neutron star densities of $10^{14}$–$10^{17}$ g/cm$^3$ (Table 7.1). The temperature of the intergalactic gas with a density $n \approx 10^{-4}$–$10^{-3}$ cm$^{-3}$ amounts to $10^7$–$10^8$ K, but may range up to a billion degrees when heated by shock waves (arising from the shedding of stellar shells, stellar collisions and explosions, the collisions of gas clouds, etc.). Inside neutron stars it is equal to $10^8$–$10^{11}$ K [105]. 99% of visible matter is heated to temperatures exceeding $10^5$ K.

While the magnetic field amounts to $\approx 10^{-9}$ Gs in intergalactic space and $10^{-6}$ Gs near the galactic plane, at the surface of neutron stars it is 22 orders of magnitude higher. The record is held by the recently discovered magnetars – neutron stars produced in supernovae bursts. The magnetars exhibit a tremendous magnetic field – up to $10^{15}$ Gs, which corresponds to densities of $\approx 10^8$ g/cm$^3$, approaching the density of nuclear matter [124, 117]. Gigantic black holes entirely absorb stellar systems and hot galactic nuclei. It is not unlikely that the magnetic tunnels ("wormholes") [102], which have recently come under discussion, connect our and

[1] This figure follows from the measurements of gravitational effects in empty space and is consistent with the notion of the upper bound, which follows from the lower limit of the curvature of space [40].

V.E. Fortov, *Extreme States of Matter*, The Frontiers Collection,
DOI 10.1007/978-3-642-16464-4_7, 

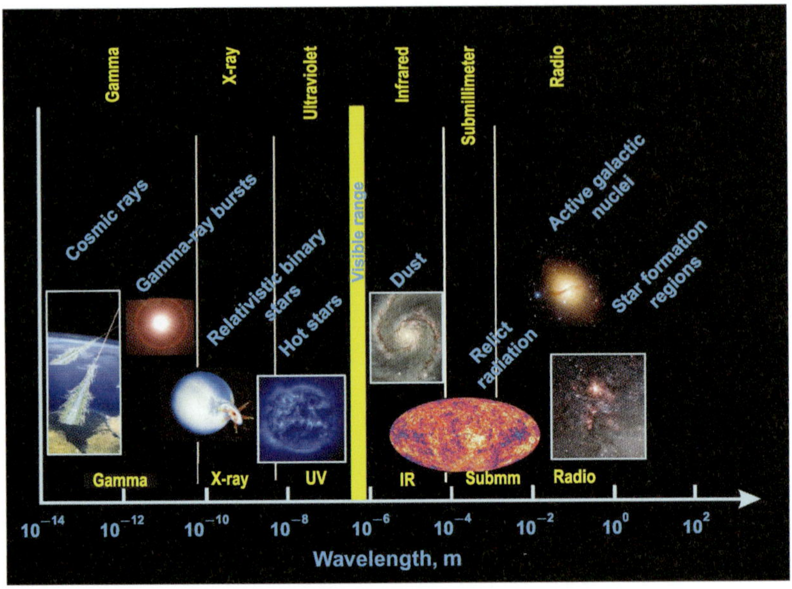


Fig. 7.1 Spectral regions of observations of astrophysical objects [233].

Fig. 7.2 Cartwheel Galaxy images in different spectral regions [233].

**Table 7.1** Characteristic parameters of matter in nature and in the laboratory.

| Object | $T$ [K] | $\rho$ [g/cm$^3$] | $P$ [bar] |
|---|---|---|---|
| Intergalactic gas | $10^7$–$10^8$ | $10^{-30}$–$10^{-3}$ | $10^{-17}$–$10^{-7}$ |
| Earth, center | $5\times10^3$ | 10–20 | $3.6\times10^6$ |
| Jupiter, center | $(1.5–3)\times10^4$ | 5–30 | $(3–6)\times10^7$ |
| Exoplanets | $10^3$–$10^5$ | 1–30 | $10^7$–$10^8$ |
| Diamond anvils | $4\times10^3$ | 5–20 | $5\times10^6$ |
| Shock waves | $10^7$ | 13–50 | $5\times10^9$ |
| Controlled thermonuclear fusion, magnetic confinement | $10^8$ | $3\times10^{-9}$ | 50 |
| Controlled thermonuclear fusion, inertial confinement | $10^8$ | 150–200 | $2\times10^{11}$ |
| Sun | $1.5\times10^7$ | 150 | $10^{11}$ |
| Red giant | $(2–3)\times10^7$ | $10^3$–$10^4$ | $5\times10^{12}$ |
| White dwarf | $10^7$ | $10^6$–$10^9$ | $10^{16}$–$10^{22}$ |
| Relativistic collision of gold nuclei, 100 GeV/nucleon, Brookhaven | $2\times10^7$–$7\times10^{13}$ | $10^{15}$ | $10^{30}$ |
| Neutron star, black hole, gamma bursts | $10^8$–$10^{11}$ | $10^{14}$–$5\times10^{15}$ | $10^{25}$–$10^{27}$ |
| Early universe (Planck conditions) | $10^{32}$ | $10^{94}$ | $10^{106}$ |

other universes. The gravitational accretion of matter gives rise to highly collimated jets, beams of charged particles accelerated to ultrahigh energies. Supernovae explosions generate shock waves, plasma ejections, turbulent plasma, and dust clouds, producing the material for the formation of stars [145, 75, 227, 214]. Neutron stars measuring several kilometers rotate at kilohertz frequencies and act on the plasma with their gigantic magnetic fields to generate high-intensity X-ray radiation.

Detailed astronomical observations testify to the predominance of new forms of matter such as dark matter and dark energy. As this proceeds, we recognize that the structure of even the closest planets of the Solar System is as yet imperfectly understood.

It is desired to reproduce to some extent these exotic states and the transformations of matter in laboratory conditions by way of relativistic nuclear collisions, at the focus of ultrahigh-power lasers, in imploding plasma pinches, or in supercomputer simulations [145]. In this case, the difference in laboratory and astrophysical scales amounts to many (up to 25) orders of magnitude, calling for the selection of adequate dimensionless variables and a careful analysis of similarity criteria [184, 47]. The existing and fundamentally possible technical capabilities along with the prospects of high energy density cumulation were described in the previous chapters and constitute the technical basis for laboratory astrophysics [47, 184, 207], a rapidly developing area of high-energy-density physics. Laboratory astrophysics has to do with the experimental study of the equation of state, intense shock, radiative, and magnetohydrodynamic waves, the hydrodynamics of mixing, supersonic

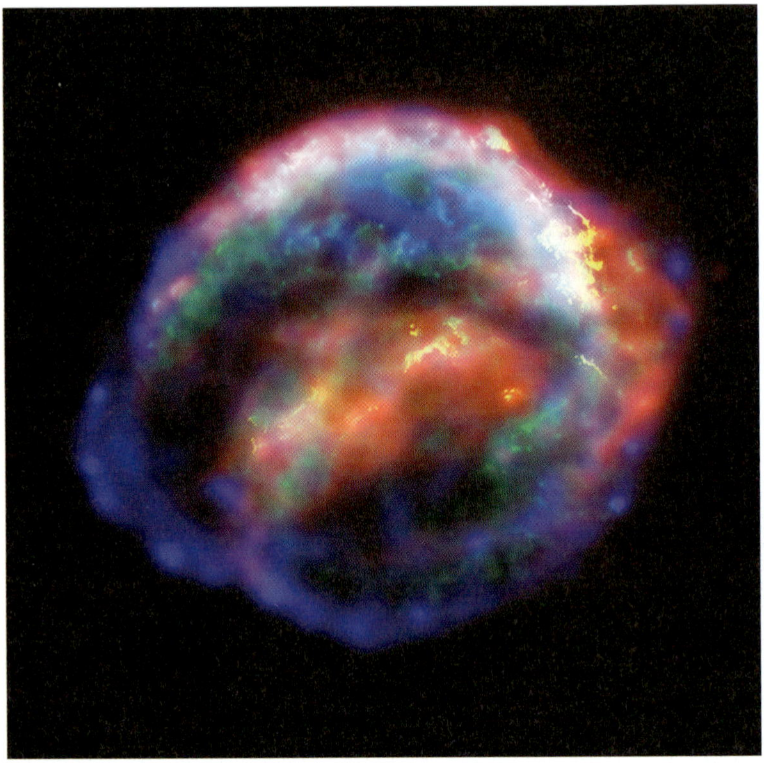

**Fig. 7.3** Kepler supernova remnant (in X-ray and infrared radiation) [143].

and relativistic jets, radiating streams, the optical properties of photoionized plasmas, the transport and kinetic properties of strongly compressed hot matter, and relativistic plasmas. These processes are intimately related to a wide class of astrophysical objects and phenomena, such as supernovae (Figs. 7.3, 7.4), neutron stars, astrophysical jets (Fig. 7.5) accelerated by radiation, molecular and dust clouds, black holes, "wormholes", $\gamma$-ray bursts, planets, and exoplanets.

The central element for constructing the structure and evolution of astrophysical objects is the physical properties of strongly compressed and heated matter [105]. These are the equations of state, the composition of plasma, its optical properties, and its transfer coefficients – viscosity, thermal conductivity, diffusion, electrical conductivity, stopping power for particles, etc. These data are required in a wide range of the parameters of state, with only narrow bands of these parameters attainable by laboratory measurements today. At the same time, theoretical models developed for hot plasmas or cold dense matter operate in rather broad ranges [105, 66, 62, 71, 145].

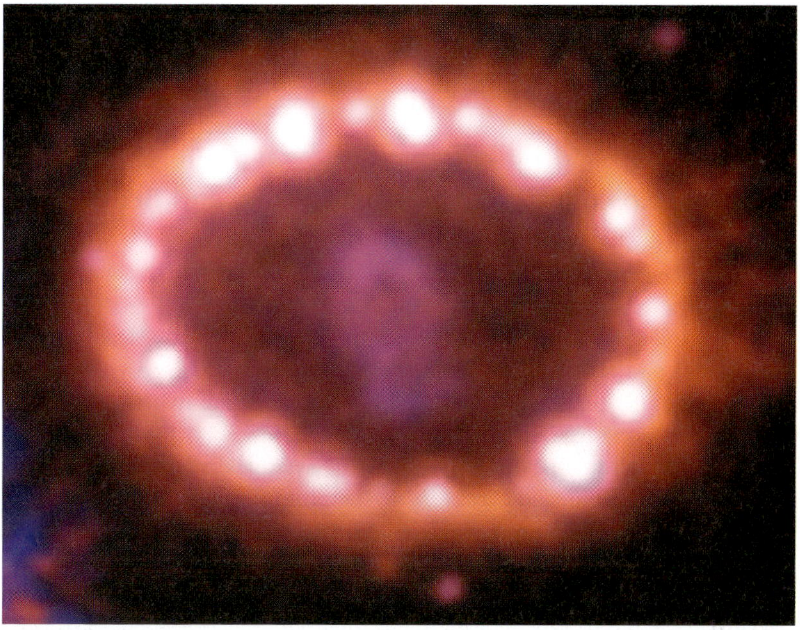

**Fig. 7.4** Gas ring heated by an intense shock wave around the supernova 1987A (SN1987A) [143].

**Fig. 7.5** Sketch of the binary object SS433 [214]. At the left is a massive star, whose matter flows to a compact object at the right (a neutron star or a black hole), which give rise to two relativistic jets with velocities of $\approx 1/4$ the speed of light.

Let us briefly consider examples of astrophysical objects; in doing this we shall primarily follow [145] and view them only from the standpoint of high-energy-density physics.

## 7.1 Planets, Exoplanets, Substars, and White and Brown Dwarfs

The comprehensive information about the giant planets of the Solar System obtained by unmanned stations as well as the discovery of hundreds of planets outside of the Solar System (exoplanets) has lent strong impetus to planetary research [31, 170]. The evolutional and structural models elaborated in this area rely on qualitative information about the physical properties of compressed hot matter in the megabar and ultramegabar pressure ranges (Figs. 2.2, 3.32, Table 7.1).

Figure 7.6 [175] shows the dimensions and the average distance to the Sun for the planets of the solar system, and Figs. 7.7, 7.8 their masses and composition. The data points in Fig. 7.7 are located near the average density of $\approx 3$ g/cm$^3$, although they fall within the range between 0.5 g/cm$^3$ for cometary nuclei and 7.7 g/cm$^3$ for metal asteroids and meteorites. The biggest planet – Jupiter – is ten times smaller in size than the Sun, but it has a density close to the solar one (1.33 and 1.41 g/cm$^3$, respectively). Saturn is close to Jupiter in size, but its density is about half – 0.70 g/cm$^3$. The densities of Uranus and Neptune are 1.27 and 1.64 g/cm$^3$; together with Jupiter and Saturn they make up the group of giant planets of the Solar System. The Earth (5.52 g/cm$^3$), Venus (5.24 g/cm$^3$), Mars (3.94 g/cm$^3$), and Mercury (5.43 g/cm$^3$) constitute the terrestrial group of planets with a high average matter density. This circumsolar matter in its turn accounts for 0.134% of the mass of the Solar System, the overwhelming amount of which (99.866%) is accounted for by the Sun – a star, a yellow dwarf.

Figures 7.9 and 7.10 show the structure and characteristic parameters of several giant planets of the Solar System, which account for 99.5% of the mass of circumsolar matter. One can see that complex structures and diverse physical processes are involved, which researchers should learn to reproduce in laboratories and describe by theoretical models of dense plasmas at megabar pressures.

Figure 7.11 shows the location of planetary matter on the phase diagram and the conditions accessible to dynamic experiments. Figure 7.12 presents the phase diagram of iron, where the terrestrial states and Fe Hugoniot curves are shown. The internal structure of the Earth is shown in Fig. 7.13.

Much less observational information has been gained about the planets that are outside of the Solar System. To date, more than 230 such objects have been discovered since 1992. Astronomers detect these objects from the lowering of stellar brightness at the instant the planet crosses in front of the stellar disc, between the star and the terrestrial observer ("shadowing" method, Fig. 7.14 [186]). Figure 7.15 shows a rare picture of the exoplanet-satellite of the star Gliese 229 made by the Hubble space telescope. The bright halo at the right is the optical exposure of the photodetector. The mass of this exoplanet is 20–60 times the mass of Jupiter.

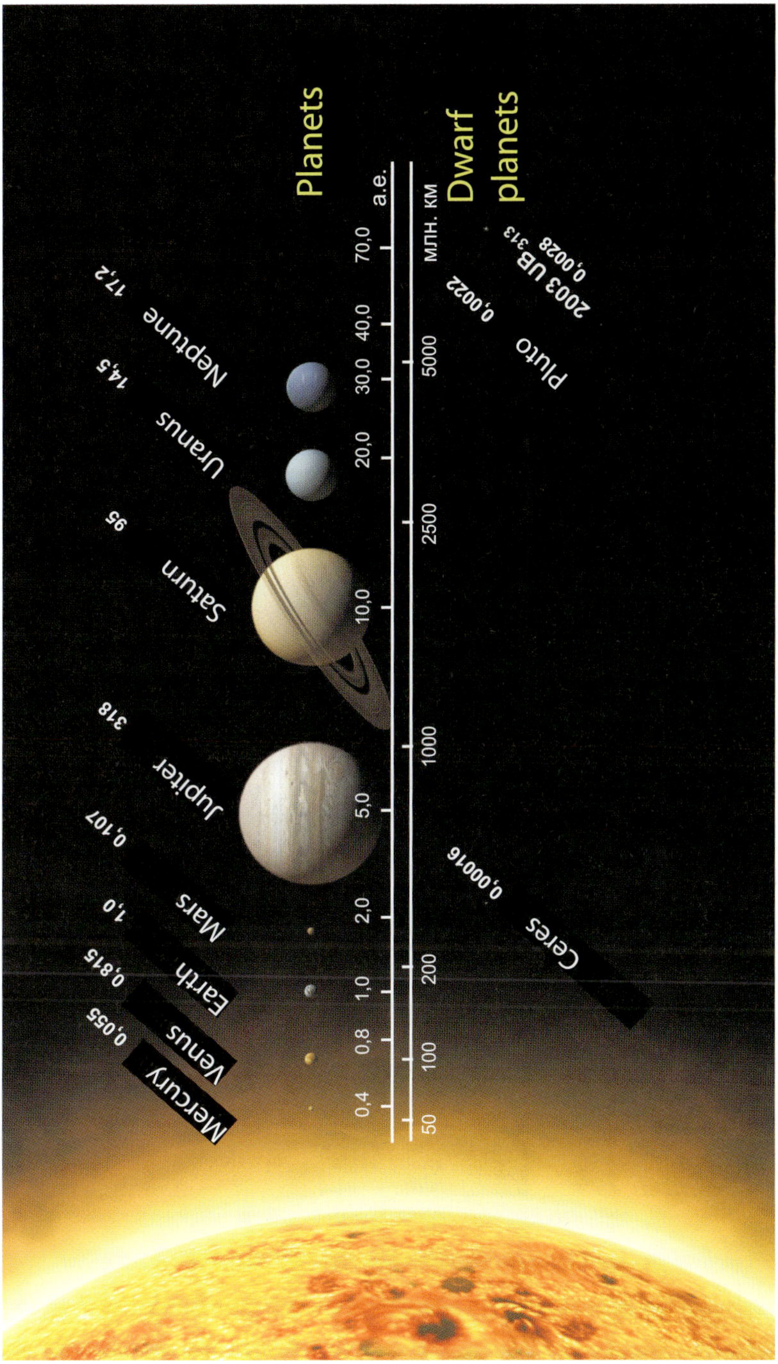


**Fig. 7.6** Masses of planets (in units of terrestrial mass) and their average distances to the Sun. Reprinted, with permission, from [175].

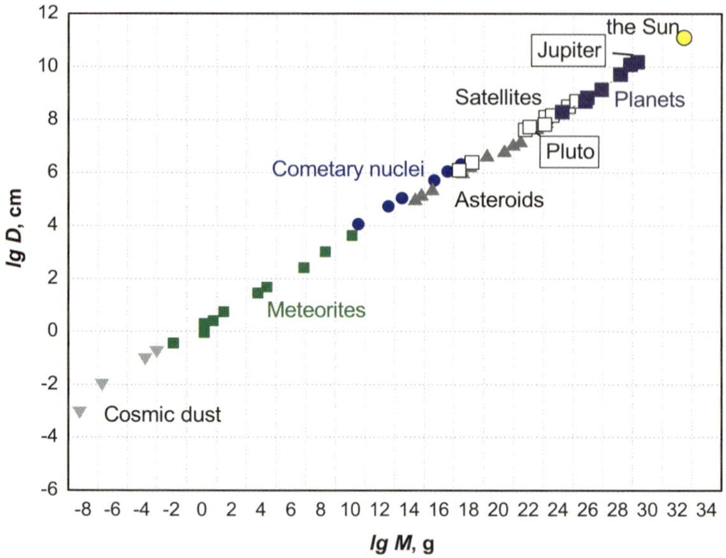


**Fig. 7.7** Masses and dimensions of the objects of the Solar System. All of them are located almost in one line, corresponding to an average density of about 3 g/cm$^3$. Reprinted, with permission, from [198].

Only one of the planets discovered – Gliese 581, which is at a distance of 20.5 light years from the Earth – only vaguely resembles our planet (temperature

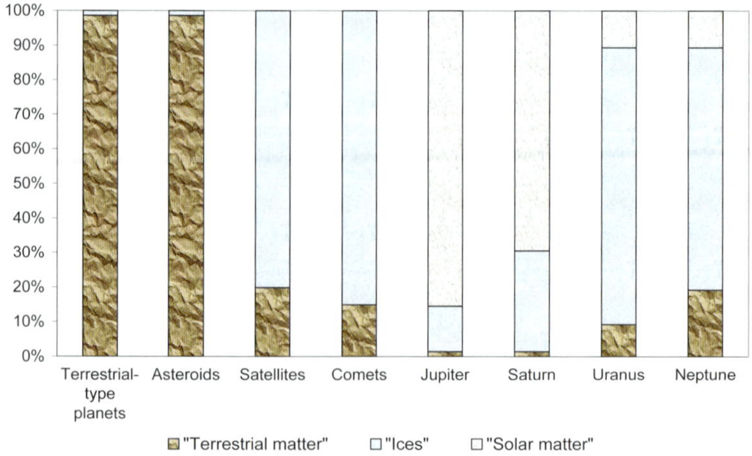


**Fig. 7.8** Relative content of different types of matter in the objects of the Solar System. Reprinted, with permission, from [198].

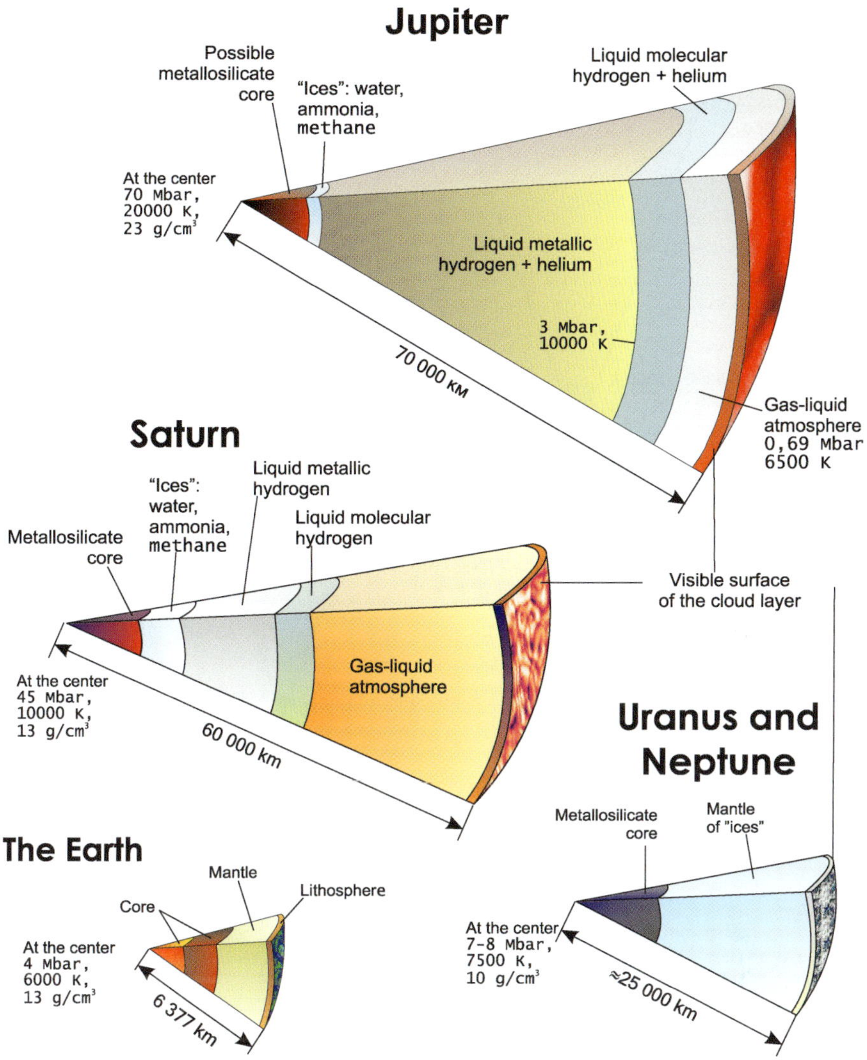


**Fig. 7.9** Diagram of the internal structure of the giant planets as compared with terrestrial structure [175].

of 15–55 °C). The NASA's Kepler space mission, launched in March 2009, is engaged in the search for Earth-like exoplanets. The exoplanets discovered to date are primarily classed as quite massive objects (see Fig. 7.16 [31, 170, 74]) with a broad spectrum of orbit dimensions, which is responsible for the wide diversity of the physical conditions of their matter.

In particular, the exoplanet DH2094586, which was discovered in 1998, resembles Jupiter in structure and by parameters (somewhat exceeding it in temperature),

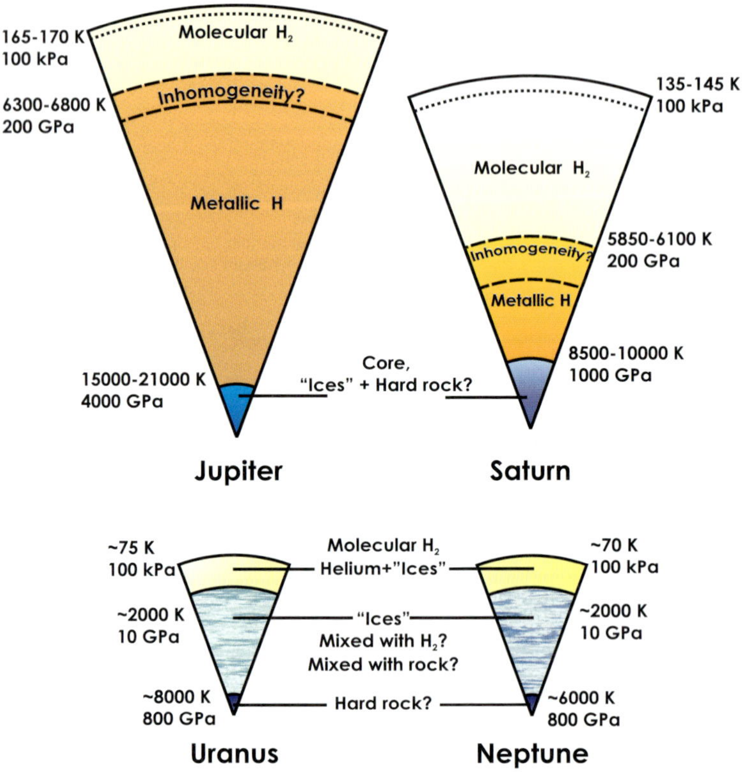


**Fig. 7.10** Structure and physical conditions of the giant planets of the Solar System [74].

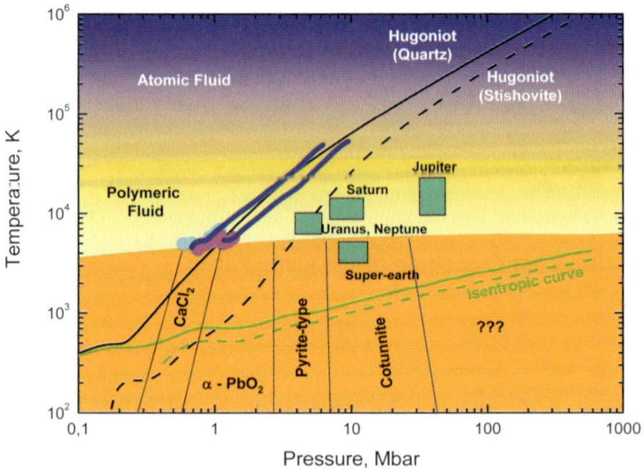


**Fig. 7.11** Location of substances in the phase diagram.

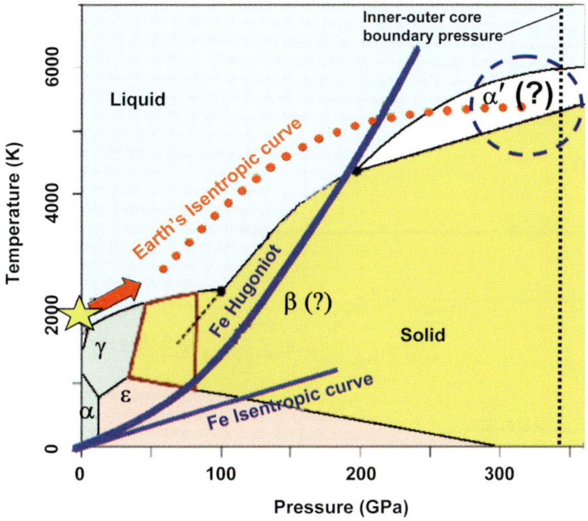


**Fig. 7.12** Phase diagram of iron with indication of terrestrial states and the Hugoniot adiabatic curve.

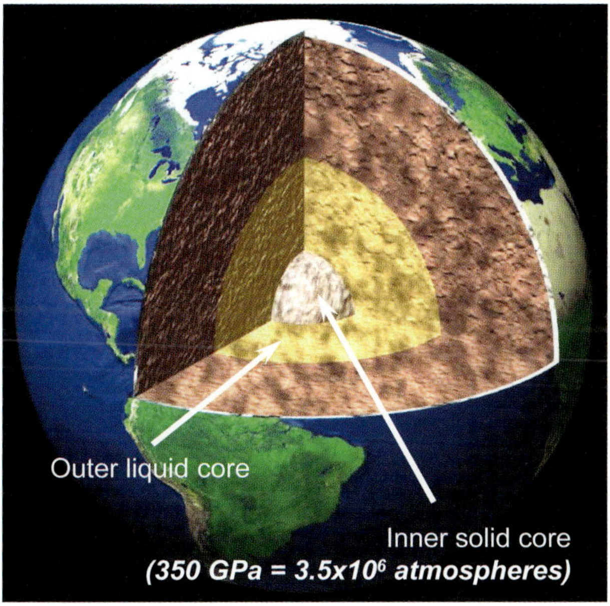


**Fig. 7.13** Internal structure of the Earth.

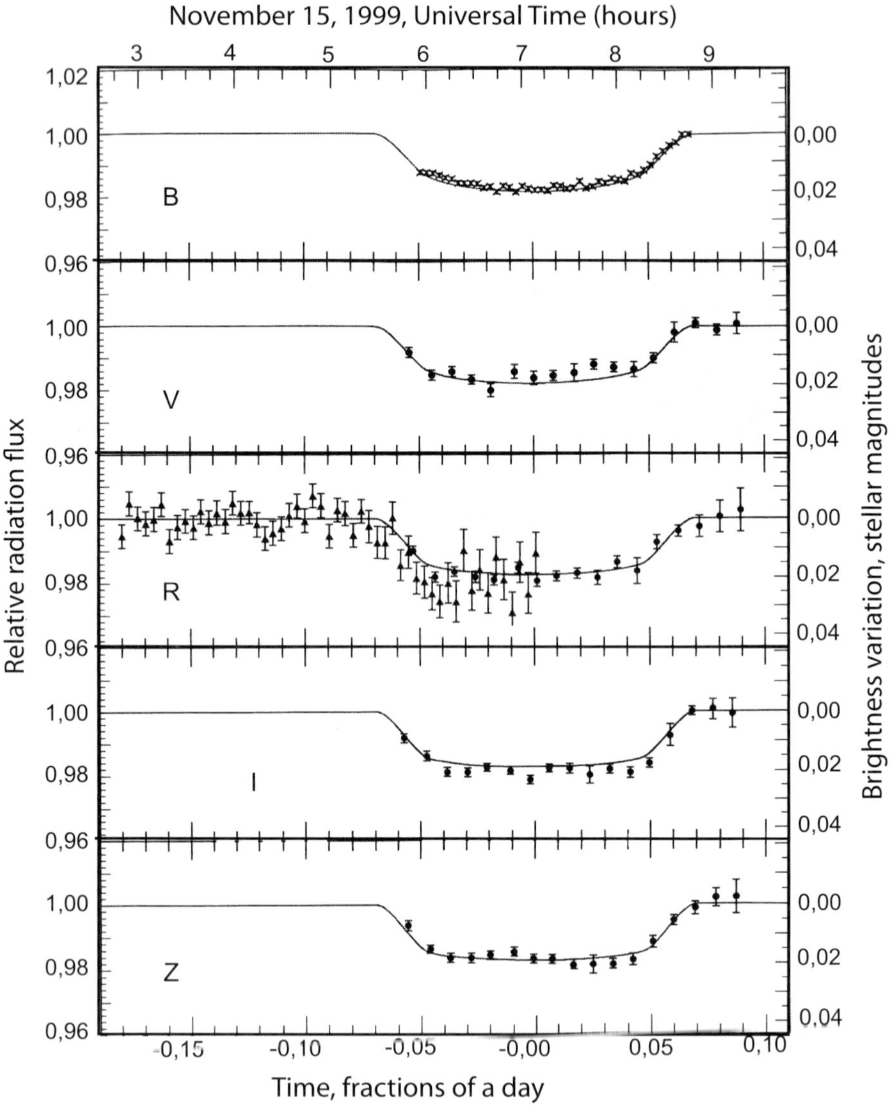


**Fig. 7.14** V376 Pegasi was the first variable star in our galaxy that was found to vary in brightness owing to a transit of a planet across its disk. The star is dimmed in the transit of the planet by several percent in all spectral ranges (B, V, R, I, Z). Reprinted, with permission, from [186]

with a domain of a plasma phase transition and the phase layering of hydrogen and helium plasmas. Researchers predict [74], in the majority of instances, the occurrence of a massive core containing heavy elements (with a mass up to 100 terrestrial masses). Figures 7.16 and 7.17 show the dimensions and masses of a number of

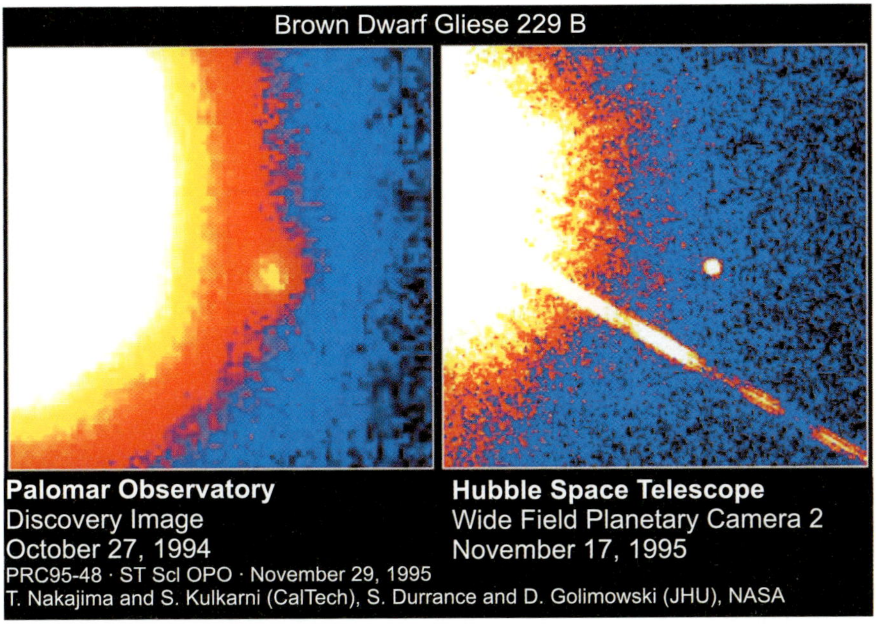


**Fig. 7.15** Photograph of a satellite of the star Gliese 229. The image was obtained by the Hubble Space Telescope [197].

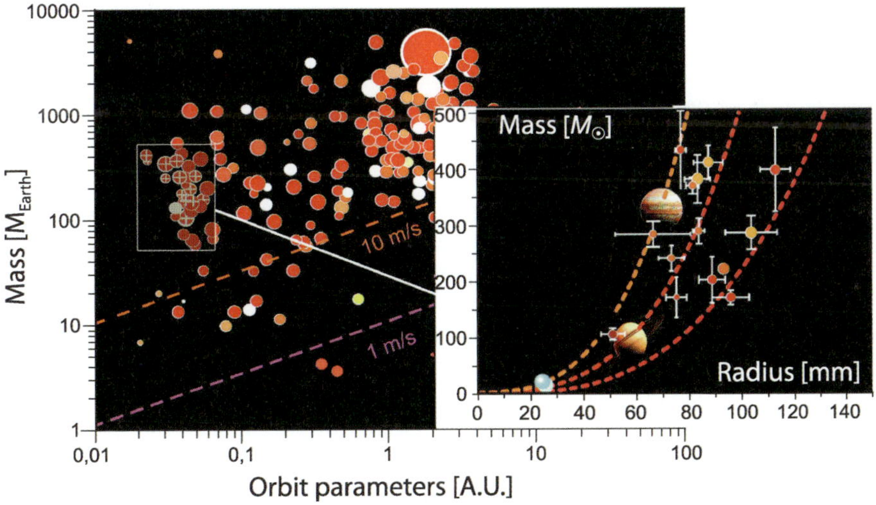


**Fig. 7.16** Characteristics of exoplanets [74].

giant planets and exoplanets as well as their chemical composition in comparison with several planets of the Solar System.

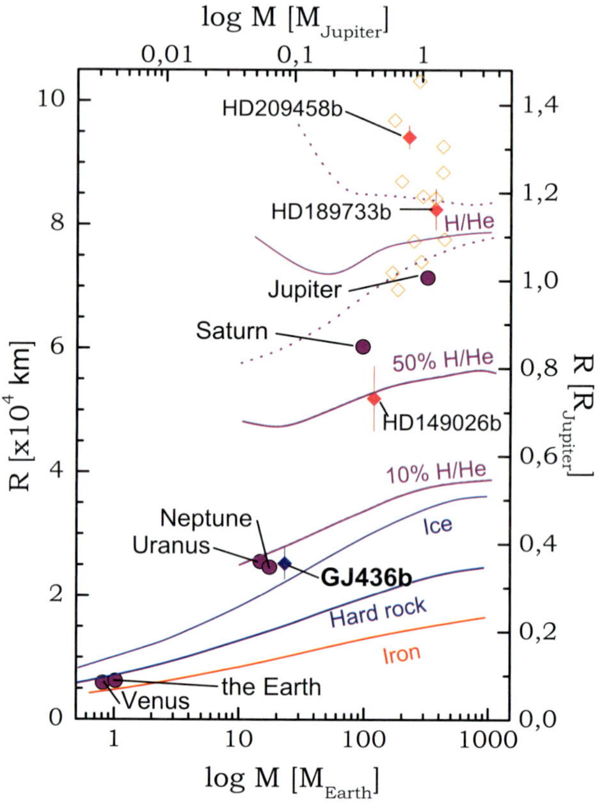


**Fig. 7.17** Characteristics and chemical composition of exoplanets compared with several planets of the Solar System [74].

A continuation of active investigations of giant planets and exoplanets with spacecraft is planned for the near future. The Cassini space probe will measure the gravitational field of Saturn with high precision; the CoRoT probe and Kepler (launched in December 2006 and March 2009, respectively) should discover tens of new exoplanets; the Juno spacecraft (to be launched in 2011) will measure the gravitational and magnetic fields of Jupiter with high precision; and Kronos (launch after 2015) is intended to measure the parameters of Saturn.

To interpret the data of terrestrial and space measurements as well as to construct, on their basis, models of evolution, structure, and energetics of these objects requires reliable data about the physical properties of nonideal plasmas in the megabar pressure range (Figs. 7.9, 7.10). At issue is a dense multicomponent plasma with a strong collective interparticle interaction, where pressure-induced ionization – so-called "cold" ionization – is of paramount importance along with thermal ionization effects. The pressures at the center of Jupiter (Figs. 7.9, 7.17, 7.21) amount to ≈40–60 Mbar for a temperature of $(15\text{–}20)\times 10^3$ K, while the pressure at the center of the brown dwarf GL 229 is equal to $\approx 10^5$ Mbar [227].

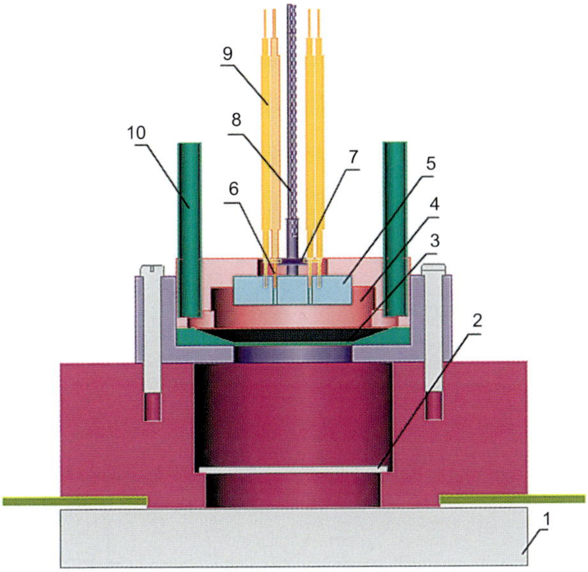


**Fig. 7.18** Explosion generator for the shock-wave compression of plasmas [72]. 1: explosive charge, 2: steel plate, 3: bottom of experimental plate, 4: substance under investigation, 5: leucosapphire window, 6: indium electrodes, 7: shunt resistance, 8: quartz-quartz optical fiber, 9: coaxial electric cables, 10: gas supply lines.

Experiments on the multiple (quasi-adiabatic) shock compression of hydrogen, helium, and their mixtures performed with pneumatic guns [147] and explosive loading devices (Fig. 7.18) [62, 71, 72, 68] yield the required information about pressure-induced ionization (Fig. 7.19) [62, 71, 72, 64] and about the phase transition (Fig. 7.20) [68] in this plasma. The measurements suggest that the plasma phase transition takes place on the deuterium isentropic curve at $P \approx 1.2$ Mbar and the pressure-induced ionization occurs in a similar pressure range for a plasma density of $\approx 0.5\text{–}1.0$ g/cm$^3$. This permitted the radius in Jupiter at which metallization occurs to be measured, shifting it towards greater radii (Fig. 7.21) [146, 66].

The magnitude of "cold"-ionization pressure is of consequence in estimating the convective effects and the generation of Jupiter's high (10–15 Gs) magnetic field. The occurrence of a plasma phase transition is of interest in estimating the energy (including gravitational energy) liberation in the phase separation of helium and heavy elements (Fig. 7.22), as well as in estimating thermal fluxes. Therefore, the phase diagrams of hydrogen, helium, and hydrogen–helium mixture plasmas,

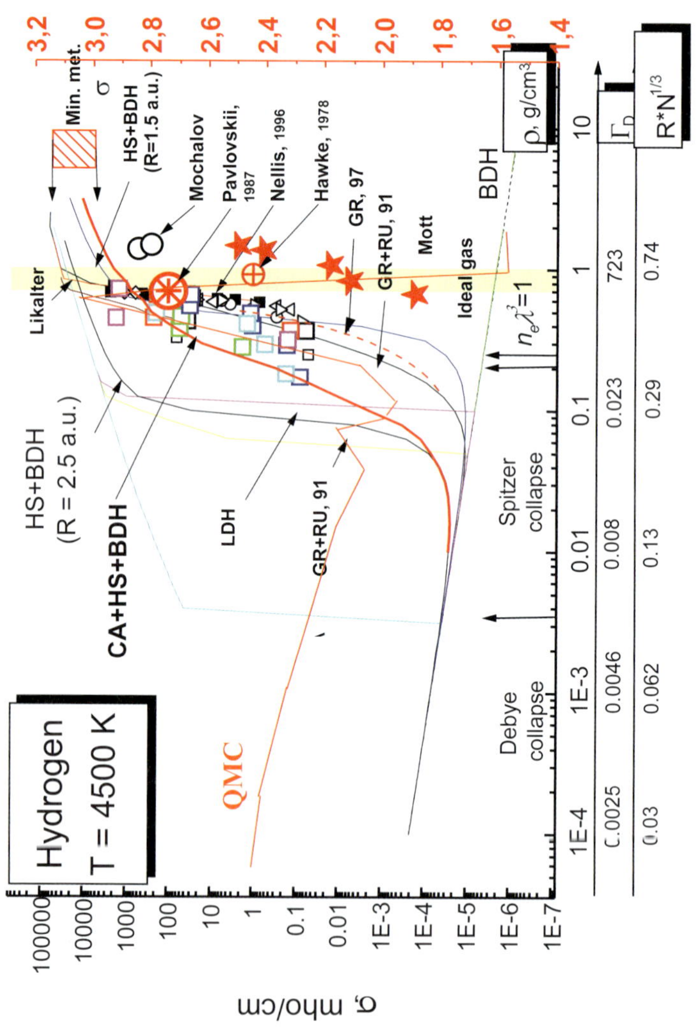


**Fig. 7.19** Pressure-induced ionization of nonideal hydrogen plasmas [72, 147, 87, 160]. The area of a thermodynamic phase transition [68] is marked in yellow; asterisks indicate the data of density measurements by pulsed X-ray radiography [68]. QMC: Calculations by a quantum Monte Carlo technique [61, 60, 24].

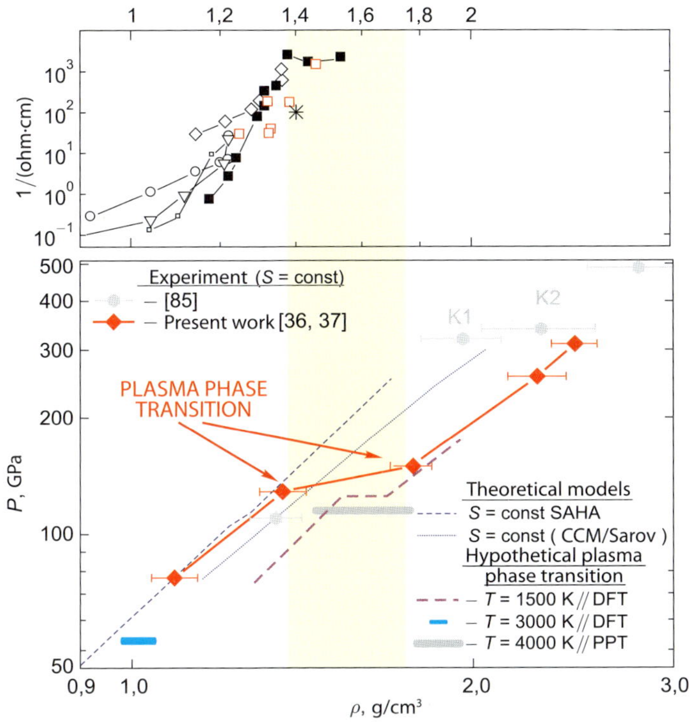


**Fig. 7.20** Recording of the adiabatic compressibility of a deuterium plasma [68]. The domain of the plasma phase transition is marked in yellow. Top: $D_2$-density-scaled electric conductivity data (Fig. 7.19).

plasma phase transitions, metallization boundaries, the mutual solubility of chemical elements and nonideal plasmas of different composition, as well as the possibility of "helium rain" occurring at high $P$ and $T$ invite further investigation [145].

Close to planetary objects in dimensions are brown dwarfs, low-mass stars, and substars, in the interior of which nuclear reactions have died out [16] owing to the insufficiency of mass (commonly this is 0.07–0.09 of the solar mass and a size comparable to that of Jupiter). These "unsuccessful stars" range from Jupiter to the Sun in mass and consist of a hydrogen–helium degenerate or partly degenerate nonideal plasma with a pressure of about $10^5$ atmospheres at the center, while the plasma in white dwarfs is completely degenerate [5, 33, 112, 170]. As in the case of planets, studying and constructing atmospheric models of brown dwarfs and other substellar objects is an extremely intricate task, which calls for comprehensive thermodynamic and spectral calculations of molecular multicomponent plasmas (millions of spectral lines and bands). Account should also be taken of the occurrence of a condensate, the shift and broadening of spectral lines, as well as the presence of metals and their compounds. All of this is required, in particular, for calculating the luminosity of

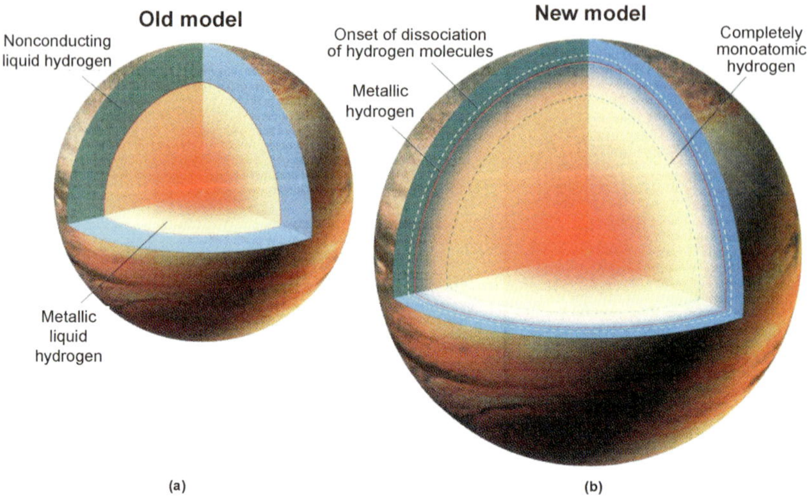


**Fig. 7.21** Schematic of Jupiter's structure before (a) and after (b) measurements of electric conductivity in shock-compressed hydrogen [146]. The zone of the metallic core has shifted from 0.75 to $\approx 0.9$ of Jupiter's radius (b).

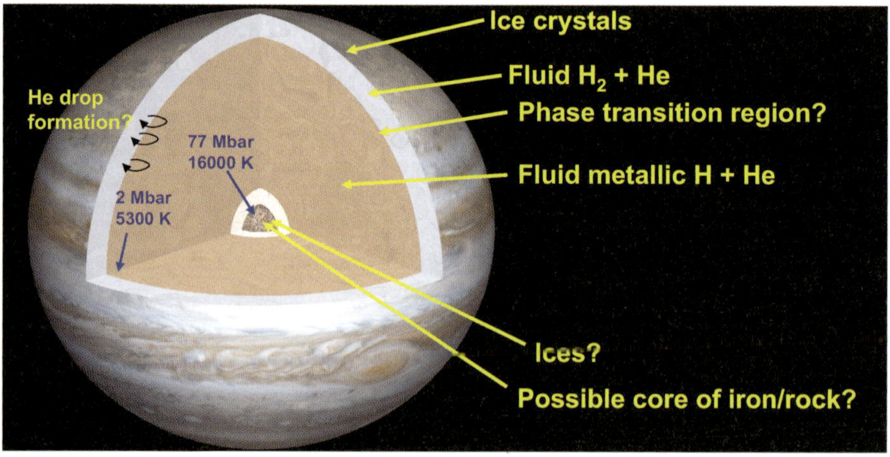


**Fig. 7.22** Detailed structure of Jupiter.

these scarcely observable objects and determining their contribution to the hidden galactic mass.

By and large, the optical properties of stellar plasmas are one of the central areas of application of high-energy-density physics to astrophysics [173], because radiation determines the energy transfer inside the stars, their evolution, and observable luminosity, yielding the bulk of observational information about these objects.

Sophisticated computer programs, spectroscopic databases, and high-level radiation-transfer programs combined with hydrodynamic motion codes have been developed for this purpose. In particular, [137] contains a list of 6 million lines of $H_2O$ and [159] 300 million lines. But even they do not provide a complete description of water vapor bands for late-type stars and brown dwarfs. An adequate inclusion of nonideality effects in compressed plasmas and condensate formation in the form of clusters and dust constitute a special problem in this case [5].

As an example of calculations of this kind, we give an analysis of optical phenomena on the impact of the comet Shoemaker–Levy 9 on Jupiter [107, 67], which enabled the composition and structure of Jupiter's atmosphere to be determined more precisely, revealing the effect of shock acceleration in an exponential atmosphere, and explaining the optical signals (Fig. 7.23) recorded on the Earth and by space probes (Figs. 7.24, 7.25).

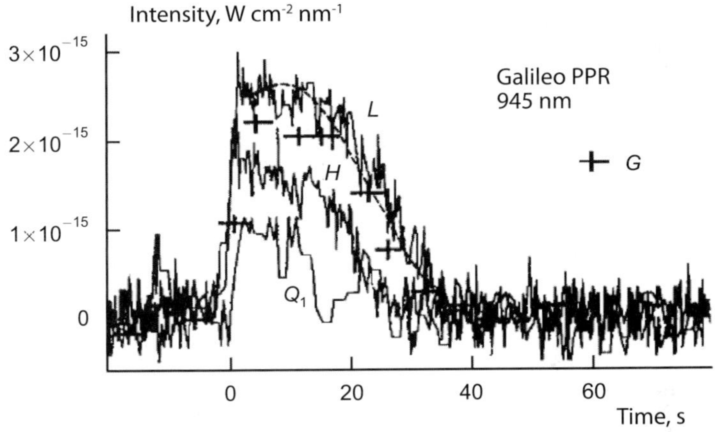


**Fig. 7.23** Brightness curves recorded by the Galileo spacecraft for the impacts of comet Shoemaker–Levy 9. The observations were carried out at a wavelength $\lambda = 945$ nm. The dashed curve is an analytical solution [107, 67].

Not only do the optical plasma properties determine the brightness, spectra, and structure of stellar objects, but they may also lead to large-scale pulsations [227] with a loss of stellar stability, like the long-period (11 month) oscillations of the star Mira (Omicron Ceti) discovered in 1596 (Fig. 7.26). In such stars (Cepheids) there develop oscillations of the radius and radiation temperature due to an increase in optical opacity coefficient $\kappa(T,\rho,Z)$ of the plasma, which is a complex function of the plasma composition, temperature, and density [63]. An increase in $\kappa(T,\rho,Z)$ blocks the thermal flux from the center to the outer layers of the star to give rise to a pressure growth, which in turn generates a compression wave, leading to the

**Fig. 7.24** Result of the impacts of fragments of comet Shoemaker–Levy 9 on Jupiter. Below the red spot in the lower part one can see white vortices, the impact marks [107, 67].

expansion of the shell, its cooling, and a decrease in density. If $\kappa(T,\rho,Z)$ decreases on expansion, this increases the energy flux, which gives rise to radial stellar oscillations and pulsations of the observable temperature. Clearly the dynamics of this process is determined by the plasma composition and the ionization of helium and hydrogen for $4\times10^3$ K and $1.5\times10^4$ K.

Radiative processes are of fundamental importance in describing the origin, evolution, and structure of interstellar objects. For definiteness we consider [204] the mass of an interstellar cloud of one solar mass ($1\ M_\odot$) that transforms into a star owing to gravitational instability. For a temperature of 5–10 K and a density $n\approx10^5$ cm$^{-3}$ its radius is $R\approx3\times10^6$ AU $\approx10^9R_{\rm sun}$. Initially on the collapse of the cloud the gravitational force increases the kinetic energy of the particles moving towards the center, while the temperature remains constant, because collisions play only a minor role and the thermal energy is removed by infrared radiation, for which the cloud is still transparent. During this isothermal compression, the gravity is $\sim R^{-2}$ and the matter-expelling Archimedean force is $\sim R^{-1}$. So, while early in the collapse these forces are of the same order of magnitude, the role of the latter drops sharply with decreasing radius. The isothermal compression, which corresponds to free fall, lasts for $\approx10^5$ years, following which the protostar loses transmittance in the IR range, its radiative cooling is hindered, and the compression passes into the adiabatic regime accompanied by a rise in temperature. The increase of temperature and pressure in this regime slows down the compression of

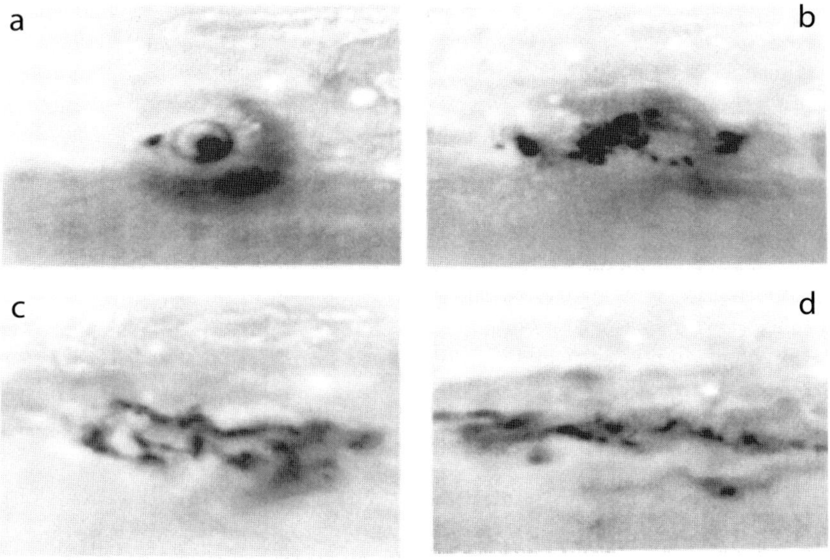


**Fig. 7.25** Photographs showing the evolution of vortices at the point of impact of the fragments of comet Shoemaker–Levy 9 on Jupiter: a) July 18, 1994 (approximately 1 h after the impact); b) July 23, 1994; c) July 30, 1994; d) August 24, 1994.

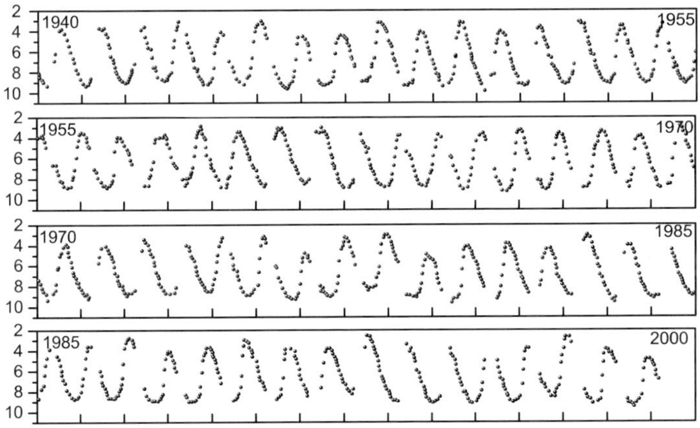


**Fig. 7.26** Mira (Omicron Ceti) brightness variations over 60 years, according to the information of the American Association of Variable Star Observers. Indicated is the visual stellar magnitude averaged over 10 days. Reprinted, with permission, from [186].

the cloud, in which there occur dissociation and ionization reactions at temperatures of $\approx 10^3$ K and $10^4$ K to additionally moderate the increase of matter temperature. However, on completion of the dissociation and ionization, the process becomes adiabatic once again. The protostar goes over to a quasi-equilibrium state, whereby the force of gravity is balanced by the plasma pressure and the radiative losses are compensated by a slow compression, which leads to the release of gravitational energy. In a characteristic time of $\approx 5 \times 10^7$ years, the gravitational compression heats the core of the protostar to thermonuclear temperatures, so that the star becomes a main-sequence star.

Nowadays, this qualitative model has been replaced by sophisticated two- and three-dimensional models of the evolution of protostars with the inclusion of rotation, convection, thermal and density-induced ionization, and a comprehensive description of the optical properties of multicomponent plasmas [176].

In the transformation of a gravitationally unstable cloud into a star there occur titanic changes in the physical conditions of the matter: the density rises by 20 and the temperature by 6 orders of magnitude. The matter undergoes a series of transformations from molecular and solid states to superdense relativistic plasma states.

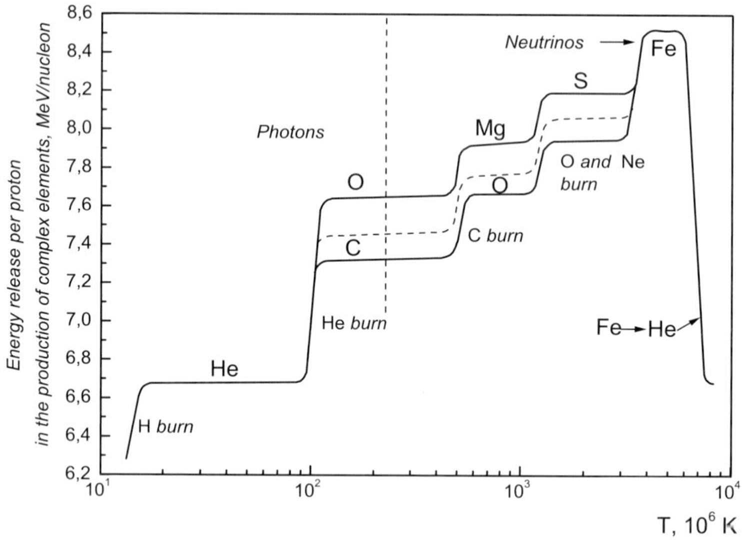


**Fig. 7.27** Energy liberation in the sequential burning of hydrogen, helium, carbon, oxygen, and neon. The upper curve represents the transformation of oxygen and the lower curve the transformation of carbon. The same takes place at other stages. It is not unlikely that the energy variation in real stars corresponds to the dashed curve [101].

The principal energy source in the subsequent stellar evolution is thermonuclear burn, which is responsible for the production of heavy elements and takes place in strongly compressed matter at extremely high temperatures (Fig. 7.27 [101]). In the course of this burning (Table 7.2 [180]), hydrogen transforms into helium (10 million K), which passes into carbon and oxygen (100–200 million K). Carbon at $\approx$ 800 million K yields neon, sodium, and magnesium, and oxygen at 2 billion K gives silicon and sulfur. A temperature of $\approx$4 billion K is required to transform silicon into iron and the neighboring elements. Here, the thermonuclear burn-up terminates, because energy input is required for the production of heavier elements. These elements (Cu, Sn, Ag, Pb, Th, U) are synthesized by neutron capture in the explosions of type II supernovae and red giants.

**Table 7.2** Main stages of thermonuclear stellar burn [180].

| Main component of a nuclear reaction | Central stellar temperature [K] | Central density [kg/m$^3$] | Stage duration |
|---|---|---|---|
| Hydrogen | $4\times10^7$ | $5\times10^3$ | $7\times10^6$ years |
| Helium | $2\times10^8$ | $7\times10^5$ | $5\times10^5$ years |
| Carbon | $6\times10^8$ | $2\times10^8$ | 600 years |
| Neon | $1.2\times10^9$ | $4\times10^9$ | 1 year |
| Oxygen | $1.5\times10^9$ | $10^{10}$ | 6 months |
| Silicon | $2.7\times10^9$ | $3\times10^{10}$ | 1 day |
| Core collapse | | | 0.2 seconds |
| Core explosion | | | |

Following [227], we now outline a brief picture of the evolution of single stars (Fig. 7.28).

*The mass of low-mass stars* $M < 0.08 M_\odot$ is too small for the onset of thermonuclear burning of hydrogen. Of significance at the center of such stars are the effects of electron degeneracy and Coulomb nonideality, and the object is an intermediate case between planets and stars. They are referred to as unsuccessful or "fusion-free" stars.

*Stars of mass* $0.08 M_\odot < M < 0.5\ M_\odot$ are entirely convective, which averts the subsequent ignition of the layer part upon hydrogen burnout in the core. Such stars do not become red giants, which heat up with time. The temperature of the helium core remains lower than the temperature of electron degeneracy removal, and eventually these stars would therefore be bound to form degenerate helium white dwarfs of mass $\sim 0.5 M_\odot$. However, since the hydrogen burn time for single stars with a mass smaller than $0.9 M_\odot$ exceeds the age of the universe, helium white dwarfs have not had time to evolve from single stars. Not managing to evolve during the cosmological age, the low-mass stars that have been formed in our galaxy are material for reconstructing the history of star formation. However, when a low-mass star is a component of a binary system, a part of its mass may flow to the accretion center, and the bare degenerate helium core of half the solar mass may turn into a white dwarf on cooling.

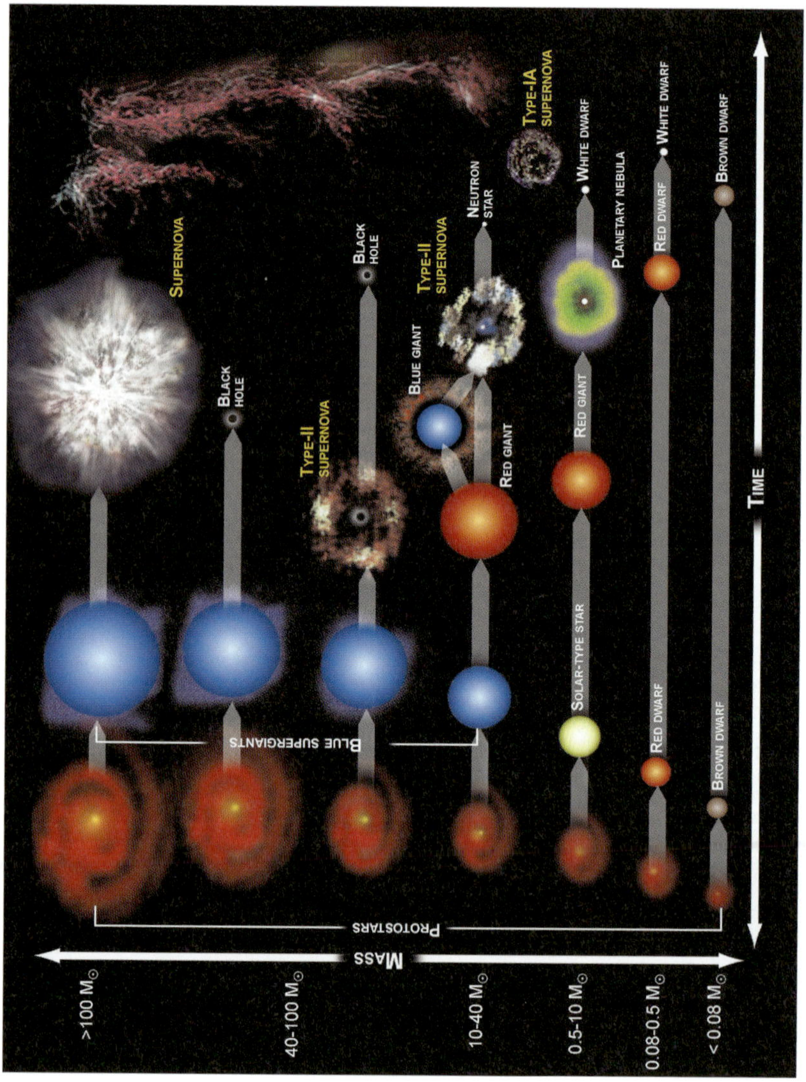


**Fig. 7.28** Diagram of stellar evolution [118]

*For stars of mass* $0.5M_\odot < M < 2.5M_\odot$, upon hydrogen burn there forms a degenerate helium core with a mass of about $0.5M_\odot$. At the red-giant stage thermonuclear burn of the hydrogen layer source occurs. The onset of the burning of helium and its transformation to carbon takes place under the conditions of plasma degeneracy and is accompanied by short-duration energy liberation (a helium flare). The stellar shell flies away and gradually disperses in space. The final product of the evolution is a carbon-oxygen white dwarf of mass $\sim 0.5M_\odot$ that gradually cools down.

Since our Sun is in precisely this stellar mass range, here we give two energy cycles of great importance in this case: the helium (Fig. 7.29) and carbon-nitrogen cycles (Fig. 7.30) [109].

The eventual result of the helium cycle: four protons transform to the nucleus of helium-4. In this case two neutrinos and $\gamma$-ray photons are produced, as well as two positrons, which subsequently combine with electrons to also give rise to $\gamma$-ray radiation. In the production of one $^4$He nucleus from four protons an energy of 26.7 MeV is realeased, which is equal to the difference between the energy of four protons and the energy of the resultant nucleus.

In the nuclear reactions considered above there emerge $\gamma$-ray photons, which travel through the solar medium. Along their path they interact with the atoms, ions, and electrons of the medium. Their mean free path is equal to 1 cm, while the solar radius is equal to $7 \times 10^7$ cm. Several hundred thousand years will pass before the "distant relatives" of the $\gamma$-ray photons born in the solar interior will manage to emerge to become detectable to the outside observer and make life on the Earth possible. According to recent ideas, the hydrogen burning in the solar interior proceeds primarily via the proton-proton cycle and only 1.6% of hydrogen burns in the reactions of the carbon-nitrogen cycle.

The temporal evolution of the Sun is represented in Fig. 7.31 [19], which provides the answer to the question of how long the Sun will shine: for about 5 billion more years. Following [19], we assume a crude estimate for the average lifetime is

$$t = 10 \text{ billion years} \left(\frac{M}{M_\odot}\right)^{-3},$$

which shows how rapidly the stellar lifetime decreases with the stellar mass. Since the brightness of the most massive stars $L \sim M$, with an increase in stellar mass the stellar lifetime ceases to shorten and tends to the value $\approx 3.5$ million years, which is extremely short on the space scale. If a massive star were born only 10 million years ago, today it is no longer in the main sequence, yet low-mass stars born even at the dawn of the evolution of the universe (which is approximately 14 billion years old) will live for a long time.

At the main-sequence stage there occurs a gradual, very slow increase of the stellar temperature, radius, and brightness. In the case of the Sun, for instance, during the nearly 5 billion years of its life the surface temperature has increased by only 2%, the radius by 10%, and the brightness by about 30%. During the next 4 billion years its temperature will hardly change, the radius will increase by a factor of 1.5,

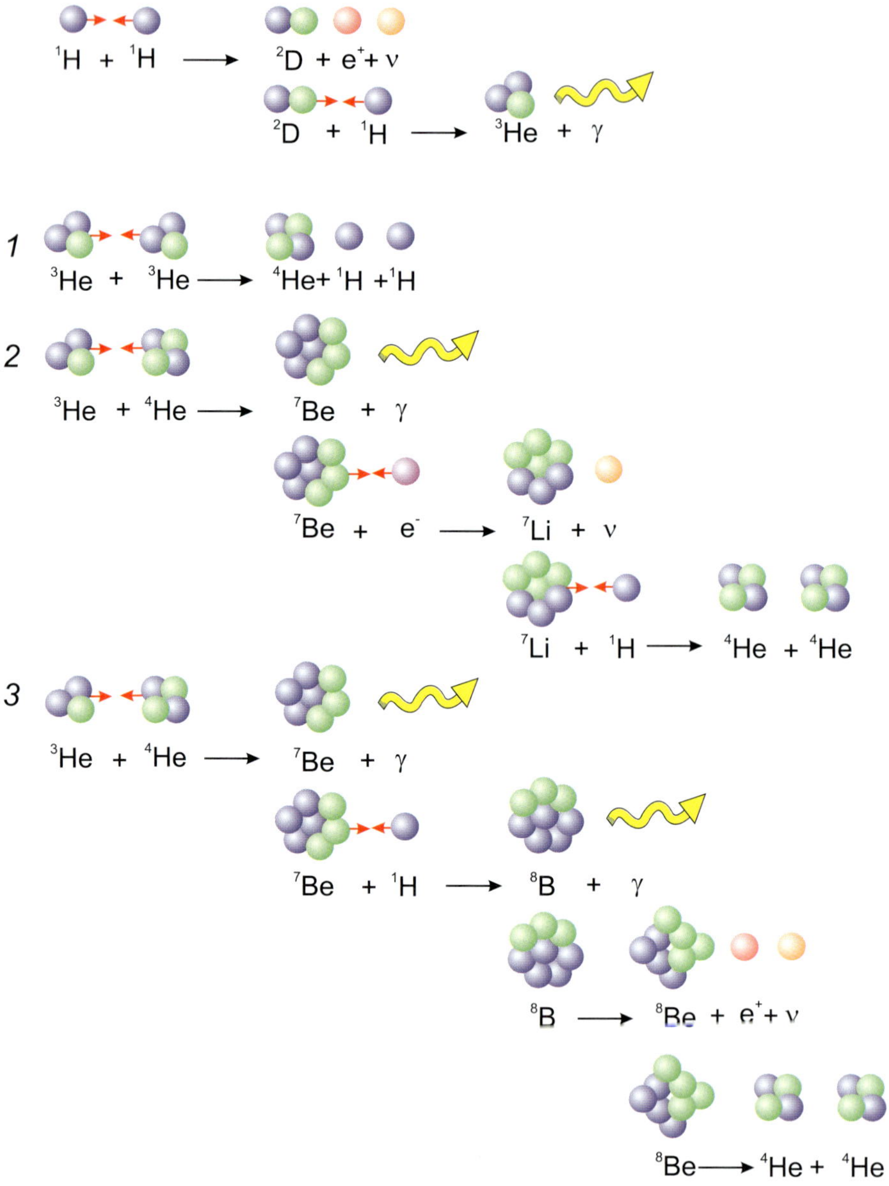


**Fig. 7.29** Proton-proton cycle [109].

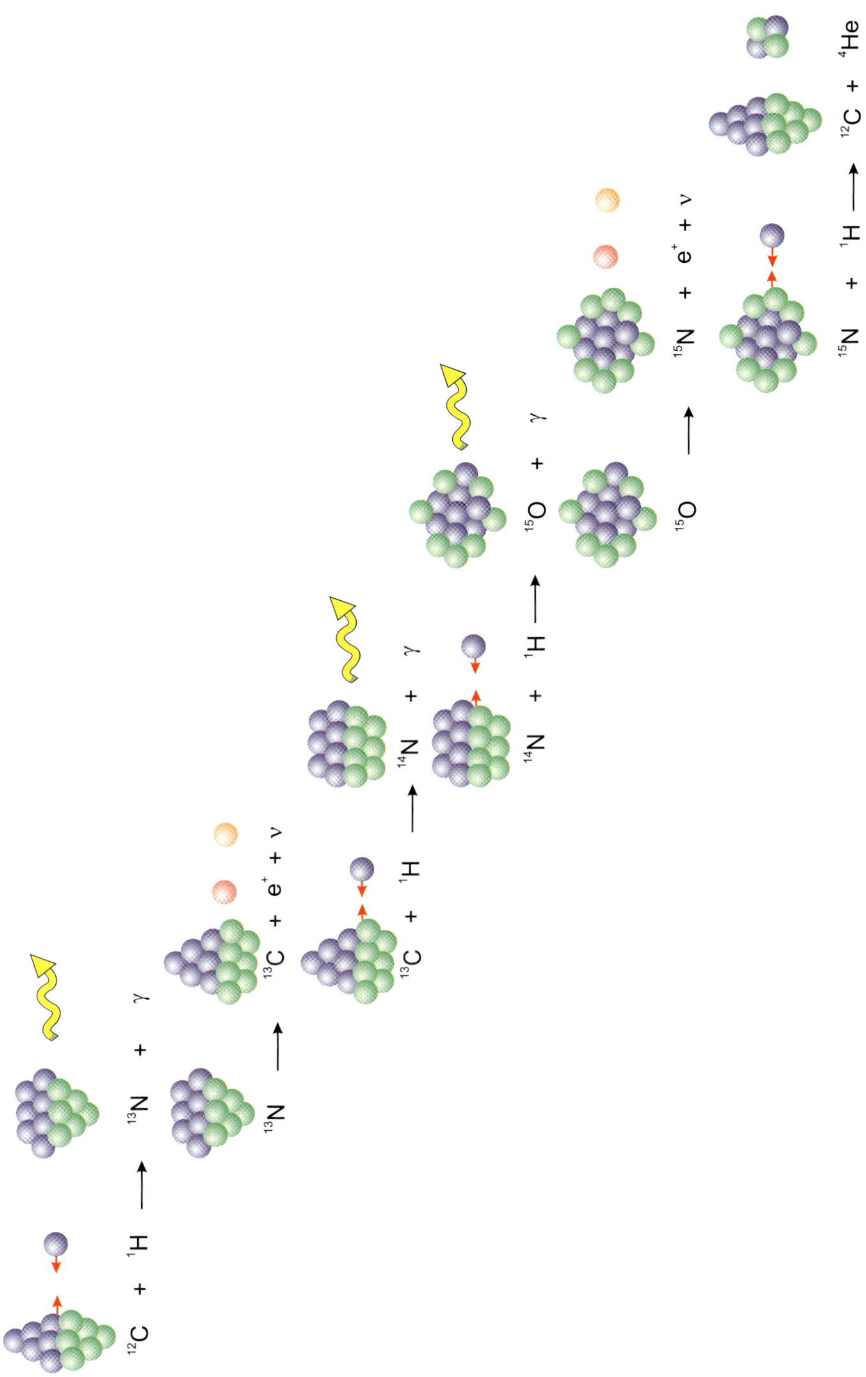


**Fig. 7.30** Carbon-nitrogen cycle [109].

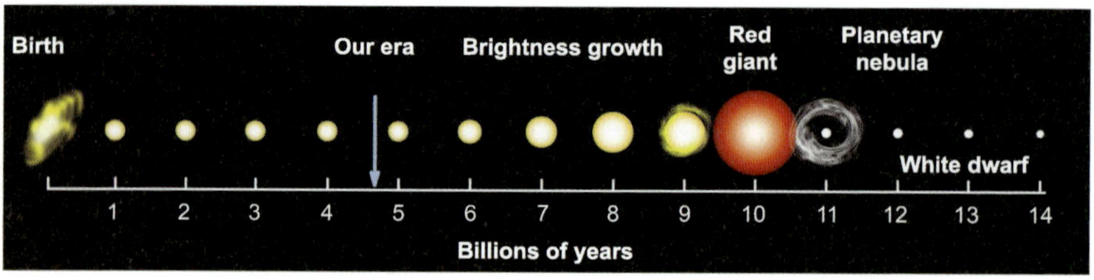


**Fig. 7.31** Solar evolution. Dimensions are not shown to scale. Reprinted, with permission, from [19].

and the brightness will rise two-fold. Stars pass through all other evolution stages – both the periods of stellar formation and those of decline – relatively quickly.

Despite the small changes in appearance, a star radically changes its internal structure during its lifetime: while initially it is a uniform sphere of hydrogen with an admixture of helium, by the end of its life the star has a core of practically pure helium surrounded by a shell of the initial hydrogen–helium mixture. In this case, a solar-type star burns about 10% of its initial mass during its lifetime.

Upon complete hydrogen burn-up in the core, if the conditions are insufficient for the ignition of helium ($10^8$ K, $10^5$ g/cm$^3$), nuclear reactions subside and the central stellar part begins to shrink dramatically and thereby heat up. When the temperature at the surface of the core in contact with hydrogen becomes as high as $10^7$ K, hydrogen begins to burn in the source layer around the core. As this takes place, more energy is released than in the burning of the hydrogen core before, and the star becomes brighter. The stellar radius increases, the temperature decreases, and the surface reddens. In this case, the star abandons the main sequence and finds itself in the branch of subgiants and then red giants. The source of energy at this stage is the burning of hydrogen in the constantly increasing source layer, which gives more and more energy to the star. The stellar radius and brightness continue to increase and the temperature remains almost invariable. for a solar-type star this process will take about 100 million years.

In the branch of giants, the helium core, which increases in mass, continues to contract. The layer heat source – the domain of hydrogen burning around the core – becomes progressively thicker and more powerful, while the stellar shell expands and becomes cooler. The zone of the shell entrained in convection becomes more and more lengthy. Finally the lower boundary of the convective zone descends to the hydrogen burn layer. The burn products (primarily carbon and nitrogen) are carried out to the surface and change the composition of the stellar atmosphere. The star continues to increase in radius and brightness, the temperature remaining moderate.

At the red giant stage, a star is unstable: its isothermal helium core continues to contract, the hydrogen-burn layer broadens, and its convective shell becomes progressively larger (Fig. 7.32). The radius of a red giant with a solar mass amounts to $200R_\odot$ at the phase of greatest extent. At this point in time, the solar surface will touch the orbit of the Earth. And though the solar surface temperature will de-

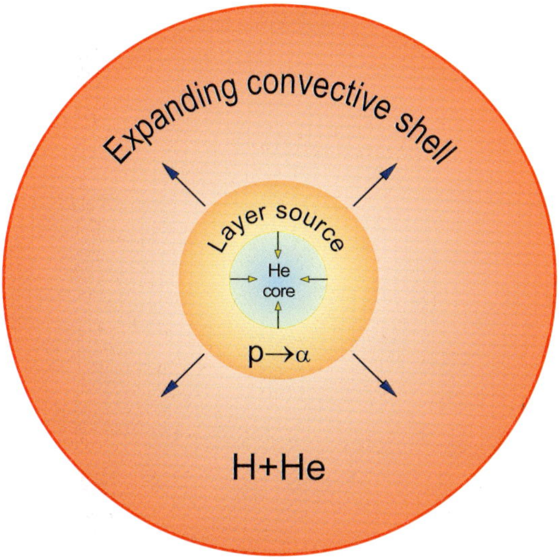


**Fig. 7.32** Structure of a red giant. At the center is a degenerate, gradually contracting core of helium produced in the course of hydrogen burn; it is void of nuclear reactions. Around the core is a progressively expanding layer energy source, where hydrogen burns. The extensive convective hydrogen-helium stellar shell expands. Reprinted, with permission, from [19].

crease to 3100 K during this period, the luminosity of this giant red ball will rise to 2400 times the present-day luminosity of the Sun. The evolution of life on Earth will terminate. However, in the vicinity of Pluto's orbit the conditions for life will become favorable. An analysis shows that life on the Earth may exist provided the solar mass is in the $1.6 \times 10^{30}$–$2.4 \times 10^{30}$ kg range. Outside of this range the terrestrial climate would be colder than the Martian or hotter than the Venusian one. Measurements of the solar mass yielded a value of $2 \times 10^{30}$ kg. The red giant stage is relatively short, and therefore only a relatively small fraction of the stars are observed at this stage – several percent of the total number.

*For more massive* $(2.5M_{\odot} < M < 8M_{\odot})$ *objects*, the helium core is nondegenerate upon hydrogen burn, and after the red giant stage there occurs the nondegenerate burning of helium with the production of carbon and oxygen. As a result, inside the red giant there forms a degenerate (C–O) core of mass $< 1.2M_{\odot}$. Due to the development of thermal instabilities, the shell is shed with the formation of a planetary nebula, which shines owing to the ultraviolet radiation of the hot ($T \approx 10^5$ K) core; the core gradually cools down and transforms to a relatively cold (C–O) white dwarf.

The model of a one-component plasma of C and O ions against the neutralizing degenerate electron background applies to these objects quite well. The characteristic parameters of white dwarfs are $\rho \approx 10^6$ g/cm$^3$ and $T \approx 10^6 - 10^8$ K, which leads to a strong plasma nonideality $\Gamma \approx$5–500. This plasma is liable to Coulomb crystallization. According to astroseismological observations, about 90% ($5 \times 10^{32}$ g) of the star BMP 37093 is in the crystalline state ("diamond" core; Fig. 7.33).

**Fig. 7.33** White dwarf BMP 37093 with a plasma crystal core.

**Fig. 7.34** New star explosion dynamics [118, 143]

*In the narrow mass range* ($8M_{\odot} < M < 10$–$12M_{\odot}$) the thermonuclear burn proceeds to give a mixture of oxygen, neon, and magnesium. Further reactions are not realized because the stellar shell disperses in the form of a planetary nebula. The result of evolution after the shell shedding is an (O–Ne–Mg) white dwarf with a mass close to the Chandrasekhar limit ($\sim 1.2M_{\odot}$).

Figure 7.34 shows the explosion dynamics of a new star [118, 143], when the Chandrasekhar limit is not reached by way of accretion, resulting in a weaker explosion than in the case of a supernova.

*The thermonuclear evolution in the core of a massive star* (10–12 $M_{\odot} < M <$ 30–40$M_{\odot}$) proceeds under nondegenerate conditions to produce the elements Fe, Co, and Ni (Figs. 7.35, 7.36).

Massive supergiants ($M > 8M_{\odot}$) that are at the latest stages of their evolution have a complex structure. In the stellar core new elements are synthesized, their complexity increasing with depth, up to iron. Such a supergiant has an inert iron core surrounded by burning shells of silicon, neon, oxygen, helium, and hydrogen. The stellar core of Fe, Co, and Ni with a mass of 1.5–2 $M_{\odot}$ is subject to several instabilities and collapses to form a neutron star. In this case, a certain amount of energy is carried away by neutrinos, and the shock wave emanating from the neutron core blows up the star from the inside. The process is accompanied by a type-II supernova burst (if an extended hydrogen shell has been retained) or Ib/c. In this case, the brightness of such a supernova is extremely high and may exceed the total brightness of all the rest stars of the galaxy. Later on, for several tens of thousands of years, the remnants of the dumped shell persist in the form of a luminous nebula – a supernova remnant.

Supernovae (Fig. 7.37 [51]) are among the most interesting and striking astronomical phenomena (a scheme of their birth is given in Fig. 7.38). The brightest of the supernovae was recorded on May 1, 1006, although chronicles note the occurrence of bursts of lower brightness in 185, 386, and 393. Having existed for several years, in brightness it exceeded Venus and was a rival to the Moon, because it was located at a relatively small distance of $\approx$7000 light years from the Solar System.

The apparent brightness–time curves (the so-called light curves) for type-Ia supernovae are quite similar. During the first 10–20 days the stellar brightness rises several-fold to attain its peak and then decreases at the same rate during the same period of time. Later on, the brightness decreases exponentially and the star vanishes in 1–3 years. Type-Ia supernovae exhibit a small spread in absolute peak brightness, which permits their use as "cosmic or standard candles" – sources of light for measuring distances in space. Type-II supernova brightness curves differ from those of type-Ia supernovae in that they exhibit a several-fold lower peak intensity, a longer decay time, and a different spectral composition of the radiation.

Anomalous optical effects observed in remote supernova bursts underpin a radical revision of the existing notions of the physical properties of empty space and of the structure and evolution of the universe [145, 41]. Owing to their extremely high brightness, supernovae are observed at great, truly cosmological distances, making it possible to estimate the apparent brightness of these sources as a function of their distance from the Earth. The brightness was shown to decrease faster than would be expected from cosmological models. This corresponds to the situation whereby the cosmological expansion proceeds with acceleration rather than with moderation, as was considered before. This discovery radically changes our understanding of the modern stage of the cosmological evolution of the universe and calls for a search for other manifestations of this effect, perhaps in the area of high-energy-density experimental physics.

Supernovae are subdivided into two types according to the nature of their spectra. Those with bright hydrogen lines are labeled type-II supernovae and those with a

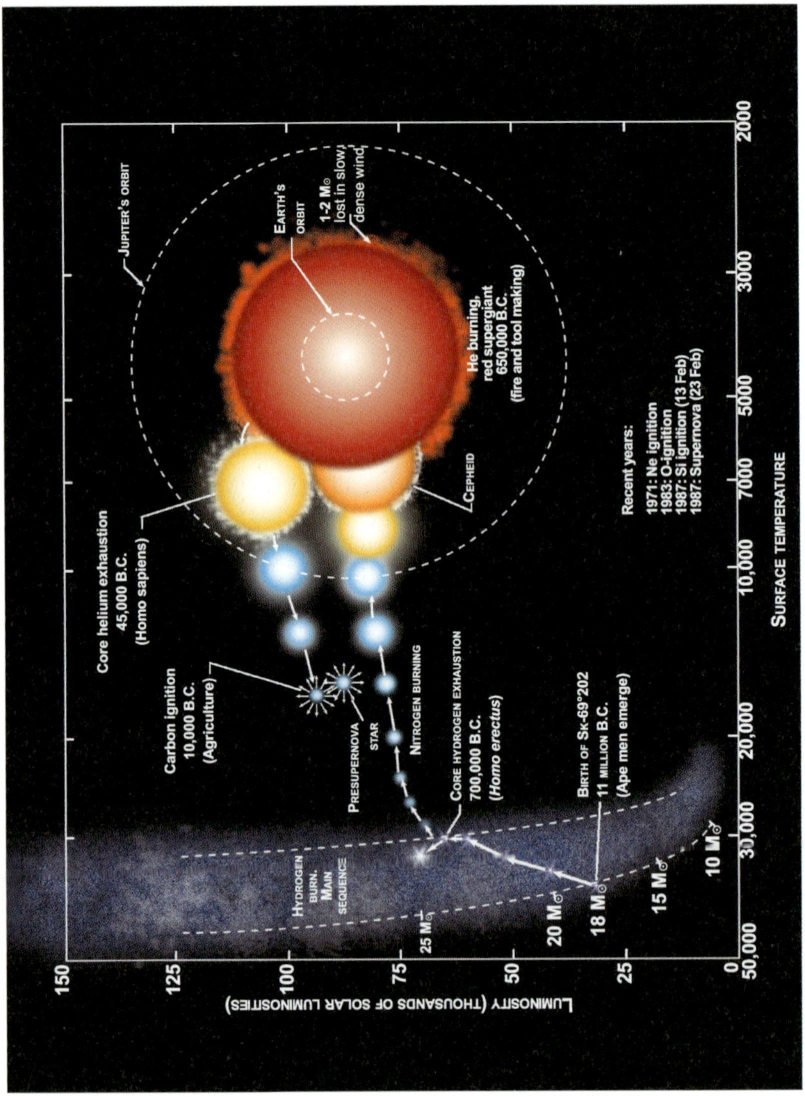


**Fig. 7.35** Evolution of the supernova SN 1987A in comparison with human evolution.

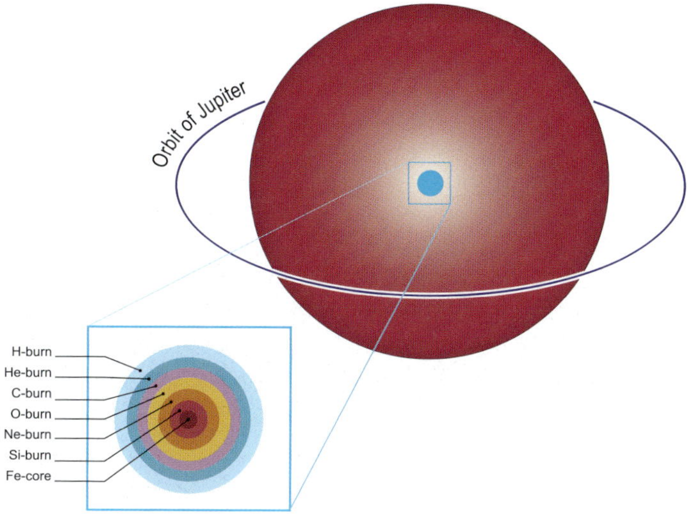


**Fig. 7.36** Structure of a red supergiant. Reprinted, with permission, from [19].

**Fig. 7.37** The supernova that flared up in the Galaxy NGC 4526 in 1994. The image was obtained by the Hubble Space Telescope [51]

deficit of hydrogen, type I. As a rule, type-I supernovae are brighter. They are in turn subdivided into subtypes Ia and Ib: the spectra of the former exhibit clear absorption lines of silicon, while the spectra of the latter exhibit those of helium. Most likely, type-Ia supernovae emerge in the catastrophic thermonuclear explosion of a carbon-

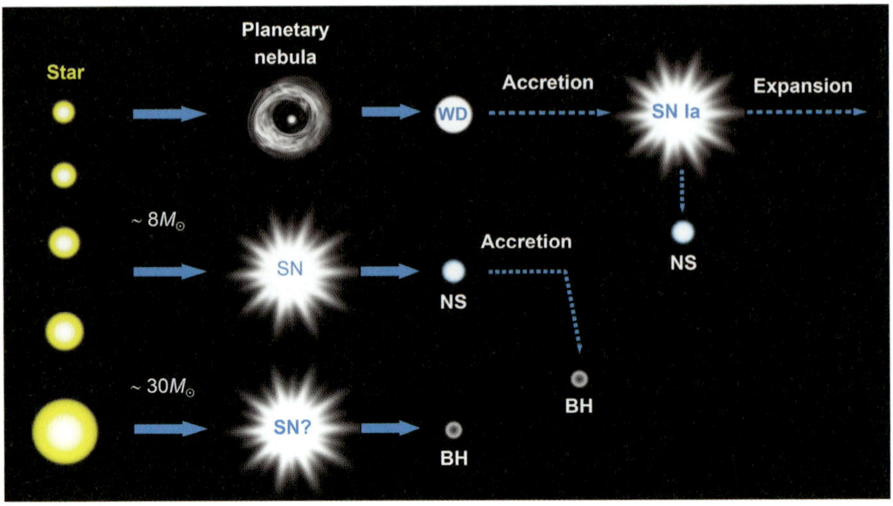


**Fig. 7.38** Scheme of stellar evolution that gives rise to supernovae. Low-mass stars transform to white dwarfs (WDs). A planetary nebula may emerge in this case. When a white dwarf is part of a close binary system, it may gain enough mass to explode as a type-Ia supernova (SN). The matter could disperse completely, but a neutron star may be formed (NS). Higher-mass stars transform to neutron stars upon supernova explosion. This compact object may also increase its mass in a close binary system and collapse (most likely without an explosion) into a black hole (BH). Most massive stars form black holes. It is still unknown whether this process is accompanied by a supernova explosion.

oxygen white dwarf. Other supernova types are associated with the gravitational collapse of the cores of supermassive stars.

In cosmological investigations, advantage is taken of type-Ia supernovae, because they are thought to be suited to cosmological observations better than other supernovae. First, type-Ia supernovae are very bright (their stellar brightness is $-19$) and rank below only the biggest galaxies ($-22$) and quasars ($-25$) in this respect. Second, their own luminosity at maximum light may be reconstructed from the slope of its observed light curve (i.e., the temporal dependence of stellar brightness). Third, there are grounds to believe that the bursts occurring at different cosmological times should not be much different (i.e., the cosmological evolution of the population of these objects should not be highly significant). Lastly, supernovae of this type have been theoretically studied rather well. This all permits observers to employ type-Ia supernovae as bright radiation standards in deep space.

In supernovae, owing to their high temperature $T \approx 30$ MeV, there exists an ultra-relativistic ($kT \gg mc^2$) plasma of massless particles. For $T \approx 10$ MeV, the energy density of this plasma is $\approx 4 \times 10^{23}$ J/cm$^3$; photons account for one third of this energy density; the plasma frequency $\omega_{\rm pl} \approx 1.5 \times 10^{21}$ Hz and the Debye radius $\lambda_D \approx 10^{-11}$ cm.

It is not unlikely that supermassive stars ($M > 30$–$40M_\odot$) collapse to form black holes with a mass of about $10M_\odot$. So far, reliable models of this process have not been introduced, although the data of astronomical observations about the existence of black holes of stellar masses appear to be rather convincing.

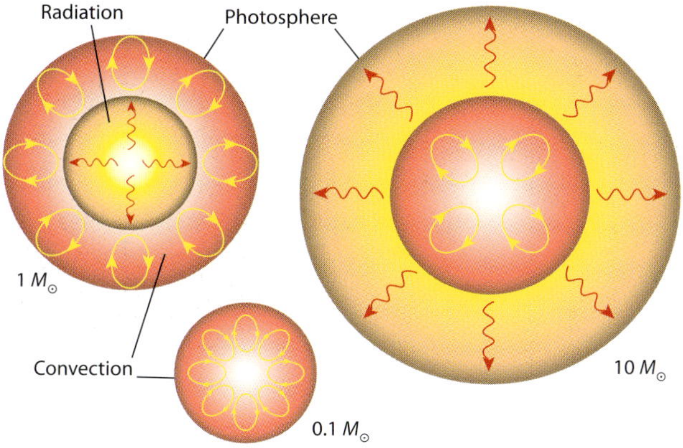


**Fig. 7.39** Mechanisms of energy transfer in main-sequence stars [19].

Depending on the stellar mass (dimension), in stars there operate different mechanisms of energy transfer (Fig. 7.39) from the thermonuclear burn domain in the compressed plasma of the central region to the stellar periphery with subsequent energy radiation into open space [19]. In the core of the Sun ($1\,M_\odot$) the energy is transferred by radiation, while in its relatively opaque shell the energy is transferred by convection. Stars of substantially lower mass (red dwarfs of $0.1\,M_\odot$) are completely convective. In stars significantly more massive than the Sun, in the hot and relatively transparent shell the energy is transferred by radiation, but in the core radiation cannot cope with this task, with the consequential development of convection.

To conclude this discussion, we give Table 7.3, from [19], which summarizes the schemes of single-star evolution.

What is the greatest mass of stellar objects? From the observations of binary system components it has been ascertained that there exist stars ranging in mass from 0.05–0.1 $M_\odot$ to 80–90 $M_\odot$ [204]. With an increase in stellar mass there occurs a sharp increase in stellar temperature and radiation pressure on external layers, which results in a buildup of radial oscillations, a loss of stability, and the shedding of the stellar shell. This mechanism imposes a mass limit (the Ledoux–Schwarzschild–Harm limit) of 60 $M_\odot$. With a more detailed inclusion of oscillation energy dissipation in shock waves and several other dissipation effects, the stellar mass limit increases to 100 $M_\odot$. The majority of stars satisfy this criterion, although there are thought to exist objects of greater mass as well: the hot supergiant $\eta$Car

**Table 7.3** Evolution of a single star [19].

<table>
<tr><th rowspan="2">Low mass (0.08–0.5$M_\odot$ )</th><th colspan="2">Intermediate mass (0.5–8$M_\odot$ )</th><th colspan="2">High mass (8–100$M_\odot$ )</th></tr>
<tr><th>0.5–3$M_\odot$</th><th>3–8$M_\odot$</th><th>8–10$M_\odot$</th><th>> 10$M_\odot$</th></tr>
<tr><td colspan="5">Burning of hydrogen in the core</td></tr>
<tr><td>Helium white dwarf</td><td colspan="4">Burning of hydrogen in the core</td></tr>
<tr><td rowspan="10"></td><td>Degenerate helium core</td><td colspan="3" rowspan="2">Nondegenerate helium core</td></tr>
<tr><td>Helium flare</td></tr>
<tr><td colspan="4">Quiet burning of helium in the core</td></tr>
<tr><td colspan="4">Burning of hydrogen and helium in layer sources</td></tr>
<tr><td colspan="2">Planetary nebula</td><td>Degenerate carbon-oxygen core</td><td rowspan="2">Nondegenerate carbon-oxygen core</td></tr>
<tr><td colspan="2">Carbon-oxygen white dwarf</td><td>Carbon detonation</td></tr>
<tr><td colspan="2" rowspan="4"></td><td colspan="2">Burning of carbon and subsequent elements (O, Ne, Si, Fe, Ni ...) in the core.</td></tr>
<tr><td colspan="2">Burning of carbon and subsequent elements (O, Ne, Si, Fe, Ni ...) in layer sources</td></tr>
<tr><td>Planetary nebula</td><td>Type-II supernova explosion</td></tr>
<tr><td>O, Ne, Mg... white dwarf or neutron star</td><td>Black hole</td></tr>
</table>

in the nebula NGC3372 with a luminosity $5 \times 10^6$ times the solar luminosity must have a mass of $\sim 200 M_\odot$ and the intense supernova SN 1961v in the galaxy NGC1058 possibly has a mass of $\sim 2000 M_\odot$ [204].

The lower stellar-mass bounds are quite diffuse, because small stars are hard to observe owing to their weak glow: a red dwarf of mass 0.06 $M_\odot$ has a temperature of only 2000 K. Examples of light stellar objects are provided by the binary system Ross 614 with components of mass 0.059 and 0.051 $M_\odot$, and the system LHS 1047 with the lowest-mass component with a weight of only 0.055 $M_\odot$ [204]. Calculations suggest that in stars with a mass below 0.07–0.1 $M_\odot$ the temperature is low and insufficient for thermonuclear burning, and gravitational compression is their energy source. The term "star" is scarcely applicable to low-mass objects without a thermonuclear energy source. Adjacent to this limit are objects of mass 0.02–0.04 $M_\odot$, referred to as "brown" dwarfs, which is indicative of their infrared radiation. By indirect techniques, astronomers have succeeded in detecting even lower-mass objects such as satellites of mass 0.009 $M_\odot$ in the system BD 68°946, which are transition objects intermediate in mass between stars and planets [204].

Therefore, in nature there exist and constantly come into being stellar objects in the $\sim$(100–0.05)$M_\odot$ mass range admissible by contemporary physical models, although a small number of objects lying outside of this range are also observed.

We have ascertained that the lifetime of a star is determined by its mass [180]. The greater the mass, the higher the temperature, density, and pressure inside and the higher the efficiency of the nuclear burn of the main fuel – hydrogen. So, large stars live for a shorter time, but shine brighter (Table 7.4).

**Table 7.4** Dependence of the stellar lifetime on the stellar mass [180]

| Stellar mass [$M_\odot$] | Lifetime [years] |
|---|---|
| 50 | 3–5 million |
| 10 | 30 million |
| 1.5 | 3 billion |
| 1.0 | 10 billion |
| 0.1 | 1 trillion |

One can see that small stars, which use energy economically, live much longer. Were our Sun only 1.5 times more massive, it would have ceased to exist long ago and the human mind would not have had time to come into existence.

The chemical composition of stars and the interstellar medium is of paramount importance to the physics of these objects. In particular, the oxygen-group elements (C, N, O) are catalysts of nuclear reactions and the iron-group elements determine the optical thickness and luminosity of stars. The commonly accepted atomic composition of the interstellar medium and the Sun is as follows: for every 1000 hydrogen atoms there are 100 helium atoms and two or three atoms of heavier elements. In this case, half of the atoms of the interstellar medium are combined in molecules and many atoms of heavy elements are in dust particles or molecules. Of course, these are only approximate values of the composition. In a number of stars, the content of heavy components may be 3–4 orders of magnitude below the value specified above [204].

White dwarfs [112], whose accretion may lead to type-Ia supernovae, are quite interesting astrophysical objects from the standpoint of realization of extreme states. In the evolution of stars of mass 8–10 $M_{\odot}$, thermonuclear burning terminates at the stage of a helium or carbon-oxygen degenerate core. This thermonuclear burning in the degenerate core is inherently explosive, and the increased temperature may partly remove the degeneracy and moderate energy liberation. That is why the outer shell of a red giant may come off: a thermal instability may develop at the interface between the layer source and the degenerate core, with the subsequent formation of a planetary nebula. Interestingly, positive ions here form a crystal lattice [227] – a kind of a phase transition in a nonideal plasma [68, 61, 60, 24, 69, 49, 189, 192, 163, 183, 140, 195, 64].

A different kind of stellar activity will emerge [227] when a white dwarf is part of a binary system, in which matter flows from the neighboring star to the white dwarf under the action of gravitational forces, thereby increasing its mass. As the Chandrasekhar limit [227] is approached, thermonuclear burning that is explosive in character (type-Ia supernova model) develops at the center of the white dwarf. A highly important problem arises in the development of models of this explosion: a deflagration burn wave mode is realized early in the explosion. Eventually it passes to the detonation mode of C-O burning with a transition to the iron-group elements with prevalence of the famous $^{56}$Ni nucleus, which furnishes, in its radioactive decay to $^{56}$Co and then to $^{56}$Fe, the entire energetics of the light curve of a type-Ia supernova [97]. However, as in the case of O–Ne–Mg white dwarfs, this does not necessarily occur, because the neutronization of matter may commence prior to the thermonuclear burning, and a collapse to a neutron star may then occur.

The presence of white dwarfs in binary stellar systems manifests itself in a broad class of variable stars, which are referred to as explosive variables [227]. Their main distinguishing feature consists in the occurrence of periodic or irregular flares of different amplitude, and the typical dimension of such a systems falls within the range between a fraction of the solar radius to several solar radii.

Stars in which the explosions are relatively small and take place in near-surface layers are termed nova-like stars. Also known are stars with higher-power flares, when explosions involve deeper layers of the stellar interior (several percent of the radius) as well. These stars are referred to as novae. Lastly, when the explosion encompasses a substantial part of a star, we are dealing with a so-called supernova.

The energy parameters of explosions also vary greatly. For the "weakest" astronomical explosions – solar flares – the time scale is about $10^3$ s and about $10^{32}$ erg is released in this case. For novae, $t \approx 10^8$ s and the corresponding energy $E \approx 10^{45}$ erg. For supernovae, $t \approx 10^{10}$ s and $E \approx 10^{50}$ erg. In galactic nuclei, it is likely that $t \approx 10^{15}$ s and $E \approx 10^{65}$ erg, etc.

The physical reasons for the flares in different kinds of explosive variables are different. Single high-power flares, which are typical for nova stars, are caused by the thermonuclear explosion of the matter accumulated on the surface of a white dwarf in the accretion from the neighboring main-sequence star or from a subgiant of approximately solar mass that had undergone a minor evolution. Calculations suggest that single flares are possible only in a specific range of masses of white dwarfs and of rates of matter accretion on their surface. For very low flow rates, the matter gradually degenerates and joins the white dwarf. For very high rates, the matter remains nondegenerate and the burning may proceed in a slack quasi-stationary mode. For the new star to be stable, degeneracy should take place in the matter incident on the surface and the temperature growth should not be accompanied by a pressure increase and a shell expansion, resulting in a thermonuclear explosion. Since the gravitational energy of the matter under these conditions is almost a hundred times lower than the caloricity of thermonuclear energy liberation, under an explosive thermonuclear burn the particles acquire a far greater velocity than the "escape velocity" on the surface of the white dwarf, and the exploded shell disperses in the interstellar medium.

The kinetics of white-dwarf evolution involves comprehensive data about the equation of state of the plasma, about its optical and transport properties, and is employed for the temporal analysis of the corresponding galactic domains. Unfortunately, only the outer regions of these objects are presently accessible to our experimental observations.

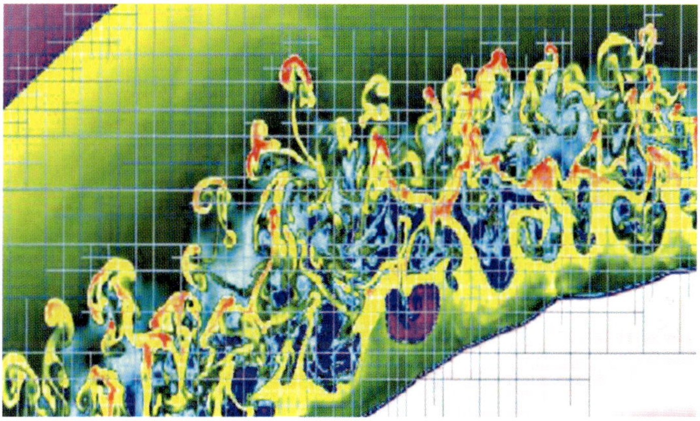

**Fig. 7.40** Two-dimensional numerical simulation of the SN1987A supernova explosion [104, 47]

At the same time, modern computer codes make it possible to carry out conceptual numerical modeling of supernova explosions [47]. Figure 7.40 [104, 47] shows the data of the two-dimensional simulation of a type-II supernova explosion, which clearly shows the development of hydrodynamic instabilities in the course of plasma expansion.

In several scenarios of stellar evolution (a supernova, a nova), thermonuclear burning emerges in the inner degenerate plasma domains and is then transferred to the outer regions by way of convection. In this case, there develop convective instabilities, resulting in the explosion of the object [133]. Similar convective processes, though without local thermonuclear energy liberation, are thought to take place in the outer regions of white dwarfs as well.

Unfortunately, convective effects in degenerate plasmas have not been adequately studied [145], which invites appropriate laboratory experiments. The available high-pressure techniques – diamond anvils, intense shock waves driven by light-gas guns, electrodynamic facilities, chemical and nuclear explosives, lasers, and high-current pinches – enable (see Chap. 3) studies of the equations of state and the optical and transport properties of strongly compressed plasmas up to a record high pressure of 4 Gbar [12, 219, 11, 212], as well as pressures up to 10 Gbar to be attained inside laser microtargets, so far without quantitative measurements of the plasma properties. In several cases [108, 147, 72, 68] it is possible to effect quasi-adiabatic compression and strongly suppress the effects of material heating.

It is hoped that the future NIF and LMJ facilities will radically broaden the attainable parameter range and enable conditions to be reached that are typical for terrestrial group planets, exoplanets, giant planets, brown dwarfs, and intermediate-mass stars, as well as for the outer layers of white dwarfs.

In the strongly compressed plasma of astrophysical objects, the static and dynamic charge screening lowers the repulsive potential barrier for ions, thereby increasing the rate of thermonuclear plasma burning [185, 92, 221]. In the solar plasma, this increase in reaction rate amounts to several percent, while for a supernova it may be million-fold.

That is why the study of thermonuclear burning in stationary and transient regimes under physical conditions close to the astrophysical ones (see Fig. 2.1) could become an important line in laboratory astrophysical plasma research. In almost 99% of the stars visible in the sky, hydrogen and/or helium are burning under parameters attainable by the laser systems OMEGA, NIF, LMJ, and by petawatt lasers (Chap. 4).

## 7.2 Superextreme States: Neutron Stars, Black Holes, Magnetars, and Wormholes

Depending on the initial mass of a star of solar chemical composition, three types of compact remnants may emerge in the stellar interior upon completion of thermonuclear evolution: white dwarfs, neutron stars, and black holes [105, 227].

Neutron stars are perhaps the most exotic astronomical objects, in which a broad spectrum of superextreme states of matter occur [226, 84] that are virtually unattainable for laboratory research. That is why neutron stars in a sense fulfill the function of a "cosmic laboratory", as their observable manifestations permit one to judge the behavior of matter under superextreme conditions: at supernuclear densities, in superstrong magnetic fields, with regard to the superfluidity of the baryonic component and intensive nuclear transformations of ultracompressed matter. It is reasonable that the observable manifestations of these processes are also highly diversified:

these are radio and X-ray pulsars, X-ray flare sources, $\gamma$–sources, X-ray transients, etc. [226, 193, 227].

Neutron stars are the smallest observable stars in the galaxy [96]. Their radii $R$ are of the order of 10 km, which is $10^{-5}$ times the size of an ordinary star. However, the neutron star masses $M$ are of the order of the solar mass $M_\odot$ and group about the value $1.4M_\odot$. The average matter density $\overline{\rho} = 3M/4nR^3 = 7 \times 10^{14}$ g/cm$^3$ of neutron stars is several times higher than the conventional nuclear density $\rho_0 = 2.8 \times 10^{14}$ g/cm$^3$. For convenience a neutron star may therefore be seen as a huge atomic nucleus 10 km in size. At the center of the star, the density may exceed the nuclear density by a factor of 10–20. At these densities, the condensation of pions, hyperons, and kaons is possible at the center of a neutron star. The possibility that strange quarks are produced is also under discussion.

The stellar body consists of the crust – inner and outer – in which there occurs matter neutronization, and the core – also inner and outer. The number of protons and electrons in the inner crust and outer core is equal to several percent of the number of neutrons. The gravitational energy of a neutron star amounts to a significant fraction ($0.2Mc^2$) of the rest energy of the star.

The existence of neutron stars was predicted by Baade and Zwicky [14] in 1934, two years after the discovery of neutrons. Despite their small size, neutron stars are among the most active stars: they radiate energy throughout the electromagnetic spectrum, from radio waves to ultrahigh-energy photons beyond 1 TeV. Rapidly gyrating neutron stars lose substantially greater amounts of energy. In particular, the radiation power of the pulsar in the Crab Nebula is $10^{38}$ erg/s, which exceeds the solar radiation power by many orders of magnitude. The energy radiated in the radio frequency band accounts for only a small fraction $10^{-5}$–$10^{-6}$ of the energy loss. The highest-power radio pulsars also emit in other ranges – optical, X-ray, and gamma-ray ranges.

Neutron stars that are short-period radiation sources (pulsars) are the final stage of the evolution of ordinary stars with $M > 8M_\odot$, when gravitational forces compress their matter to nuclear densities to produce nuclear matter [105, 106, 227], which was predicted by L.D. Landau back in 1932. Neutron stars exist due to the mutual repulsion between neutrons (and protons), and not between electrons, which emerges as a result of the Pauli principle.

Apart from radio pulsars, neutron stars are also sources [96] of high-power X-ray radiation (X-ray pulsars), $\gamma$- and X-ray bursts (magnetars), and constant X-ray radiation emanating from the centers of supernova explosions, very weak optical stars. Studying neutron stars permits several basic physical problems to be solved. First and foremost, this is an investigation of the equation of state of superdense matter, $\rho > \rho_0$.

The thermonuclear burning of $^{32}$Si with the production of the iron isotopes $^{56}$Fe, $^{58}$Fe, $^{60}$Fe, and $^{62}$Ni, etc. concludes the chain of thermonuclear reactions in the nondegenerate core of a massive star, because further thermonuclear fusion is possible only with the absorption of energy. The density at its center amounts to $\approx 3 \times 10^9$ g/cm$^3$, $T \approx 8 \times 10^9$ K for a core mass of 1.5–2 $M_\odot$ [227]. An important process stimulating the gravitational collapse is the photodissociation of an iron nucleus into 13 alpha particles, $\gamma + {}^{56}_{23}\mathrm{Fe} \rightarrow 13\,{}^{2}_{4}\mathrm{He} + 4n$, and the neutronization of matter.

Following Kirzhnits [106], according to L.D. Landau's predictions, the constituent particles of an atomic nucleus (protons and neutrons) are capable of making up another stable system apart from an atomic nucleus: neutron matter [105, 106]. This consists primarily of neutrons with a small admixture (several percent) of protons and the same number of electrons. Due to Pauli's principle, neutron matter is stable against neutron decay by the scheme $n \rightarrow p + e^- + \bar{\nu}$, because the energy level of the electron that might be emitted in the decay is already occupied by other electrons that make up the composition of the matter.

Neutron matter is produced under superstrong compression of ordinary matter consisting of electrons and nuclei, which takes place in the gravitational collapse of a star passing through the supernova explosion stage. Under this compression, when the matter density amounts to $10^{11}$ g/cm$^3$, the density, and hence the electron energy, becomes so high that inverse $\beta$-decay begins $p + e^- \rightarrow n + \bar{\nu}$: the electron energy is sufficient to overcome the neutron–proton mass difference. As a result, as the compression becomes stronger there occurs a progressive speeding-up of the process involving electron capture by nuclei with the transformation of protons to neutrons. Eventually, for a density only a little lower than the nuclear density, the state of neutron matter occurs, which fills the inner part of neutron stars – pulsars.

A specific feature of neutron stars is their superhigh (nuclear) density of $\approx 2.8 \times 10^{14}$ g/cm$^3$ or 0.17 baryon/fm$^3$. However, unlike an atomic nucleus, in which nucleons are held together by a strong interaction (quarks), nucleons in a neutron star are held by gravitational forces, while neutron $\beta$-decay is suppressed by the strong electron degeneracy in the compressed matter.

The elasticity of degenerate matter is lowered in the neutronization [227], because the electron density decreases while the baryon density is retained; the matter becomes "softer". The pressure growth with increase in density slows down, and the effective adiabatic exponent of the matter $\gamma = d\log P/d\log\rho$ lowers from 5/3 to 4/3, leading to mechanical instability of the object in accordance with the virial theorem [227]. The neutronization of matter is therefore one of the main physical processes responsible for the collapse of the cores of massive stars in the late stages of their evolution.

The neutronization of matter is extremely hard to model in laboratories even with the use of superhigh compression in laser fusion targets, yet it may be possible to employ its equivalent process of absorption of electron antineutrinos from, say, high-power sources with reactor strontium [94].

An additional cause of the loss of hydrostatic stellar stability lies in the effects of general relativity: when the matter pressure makes a contribution to the attractive force to increase the force tending to compress the matter [105, 106]. In this case, the collapse of the core of a massive star is accompanied by a type II or Ib/c supernova burst.

In the neutronization of matter a star quite quickly loses stability: the loss of elasticity leads to compression and heating, but the negative heat capacity of ordinary stars does work in this case, because the pressure of degenerate gas, which opposes compression, depends only slightly on the temperature. In addition, most of the energy released in the gravitational compression is carried away by the neutrinos produced in the neutronization, and even if the temperature growth in the collapse removes degeneracy and increases the pressure of the electron gas, the energy continues to be carried away by antineutrinos in the course of the $\beta$-decay of neutron-oversaturated nuclei.

The collapse of the stellar core terminates at densities on the order of the atomic nucleus density $\approx 2.8 \times 10^{14}$ g/cm$^3$, whereby the effects of neutron degeneracy become significant, and the matter compressibility becomes capable of resisting the action of gravitational forces. For a solar mass, the radius of a uniform neutron star with a density equal to the nuclear one is equal to about 12 km. The freely falling outer stellar layers strike against the rigid compressed core and "bounce off", which might be thought to be the cause of shell shedding and the observed effect of a supernova. However, numerical simulations have shown that this is insufficient for the explosion of the supernova shell. At present, it is universally recognized that the causen has to do with non-unidimensional effects such as magnetic field, supersonic convection, rotation of the stellar core, etc. [47, 104, 95, 93].

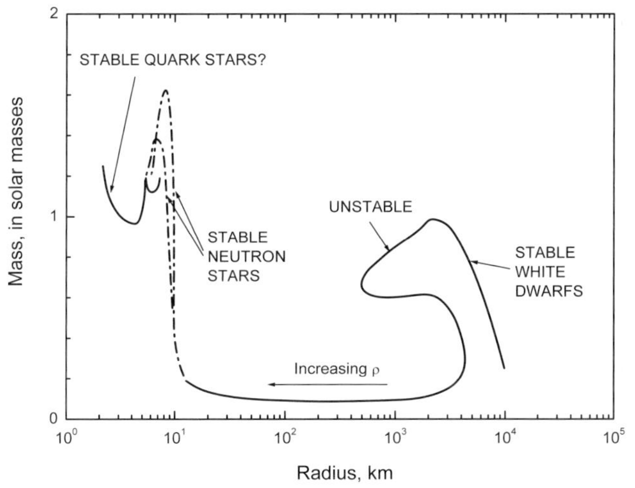


**Fig. 7.41** Transition to neutron and quark stars [83].

The transformation of stars from dense white dwarfs to superdense neutron and quark stars shows (Fig. 7.41) that an instability stage occurs in this case. Eventually there forms a compact star with $M \sim M_\odot$, only about $\approx$10 km in size (Fig. 7.42), with an initial temperature of $\approx 10^{11}$ K, and a core density of $\sim 1.5$–$15\rho_0$. The star possesses strong magnetic ($B \approx 10^{11}$–$10^{16}$ Gs) and gravitational fields (gravitational acceleration of $\approx$(2–3)$\times 10^{14}$ cm/s$^2$), which necessitates the use of general relativity theory for its description. Neutron stars rotating with a period of 0.0016–1 s (radio

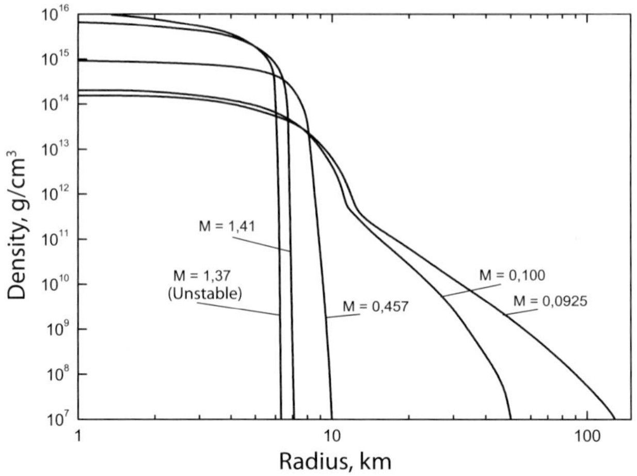


**Fig. 7.42** Density distribution in neutron stars of different mass (in solar mass units $M_\odot$) [20].

pulsars) are the only astrophysical objects in which the mechanism of rotation moderation (and hence their evolution) is defined by electrodynamic forces. The neutron stars that are part of abinary system manifest themselves as X-ray sources [227].

Despite the small size of neutron stars, the spectrum of matter states (Fig. 7.43) and the physical processes (Figs. 7.43, 7.44) in them are immensely diversified. The atmosphere of a neutron star ranges from tens of centimeters to several millimeters in thickness and has a density of 0.1–100 g/cm$^3$; it consists of a nonideal plasma with $T \leq 10^6$ K and has a tremendous magnetic field.

The atmosphere may consist of iron-group elements formed at the stage of star formation or of light elements such as H and He as a result of their accretion. Owing to the strong gravity, the stellar atmosphere is very thin (0.1–10 cm) and dense (0.1–100 g/cm$^3$), so that the matter is strongly nonideal from the standpoint of interparticle interaction. When the stellar temperature is not too high, light atoms, molecules, metallic droplets, and clusters may be formed in the crust. In the case of a strong magnetic field, the crust may be condensed, with a small amount of gas above it. Superstrong magnetic fields may lead to quantum electrodynamic effects (such as vacuum polarization), which are important for radiative processes. All these problems seriously hinder the calculation of the composition, the optical properties of the atmosphere, and hence the radiative properties of the star itself and are therefore of great interest for theoretical astrophysics and nonideal plasma physics [226, 193].

The outer crust, which several hundred meters in thickness [226], consists of a dense plasma, in which the electrons pass from the Boltzmann state to the degenerate state with depth, and then (for $\rho \gg 10^6$ g/cm$^3$) to a degenerate relativistic gas. For $\rho \geq 10^4$ g/cm$^3$ there occurs complete pressure-induced plasma ionization. On further compression there occurs $\beta$-capture and matter neutronization.

At the interface with the core of a neutron star ($2\rho_0 < \rho < 20\rho$), atomic nuclei vanish and the neutrons in the inner crust may be superfluid, which is reflected in the

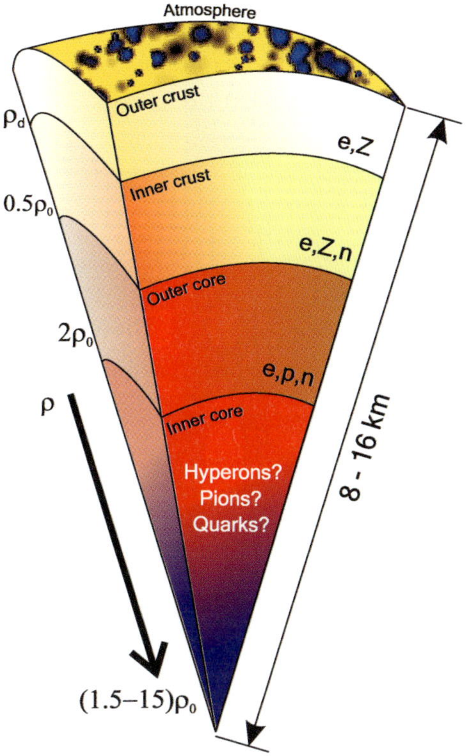


**Fig. 7.43** A sectional view of a neutron star of mass $1.4\,M_\odot$. The stellar parameters depend greatly on the equation of state of its layers [227, 226]; $\rho_0 = 2.8 \times 10^{14}$ g/cm$^3$.

dynamics of cooling and in the neutron luminosity of the object. The superfluidity of neutron matter may be observed in the breakdown of stellar rotation in the course of deceleration. In some cases, the breakdown dynamics is adequately explained by superfluid vortex separation from the stellar crust. The evolution of a magnetic field "frozen" into the star allows conclusions to be drawn about the superconductivity of the stellar core. Since the internal stellar temperature depends crucially on the critical temperature of nucleon transition to the superfluid state, Page and Applegate [155] came up with the idea of using a neutron star as a "thermometer" for measuring the critical temperatures of nucleons in asymmetrical nuclear matter, which would yield indirect information about the equation of state of the nuclear matter.

The inner core composition of a neutron star is not exactly known owing to insufficient knowledge of the physics of strong interactions in superdense matter [105, 227]. It is not unlikely that the core consists of nucleon–hyperon matter, pion condensate, quark–gluon plasma, or some other exotic states. According to Yakovlev et al. [226], if the properties of the neutron star crust ($\rho < 0.5\rho_0$) are

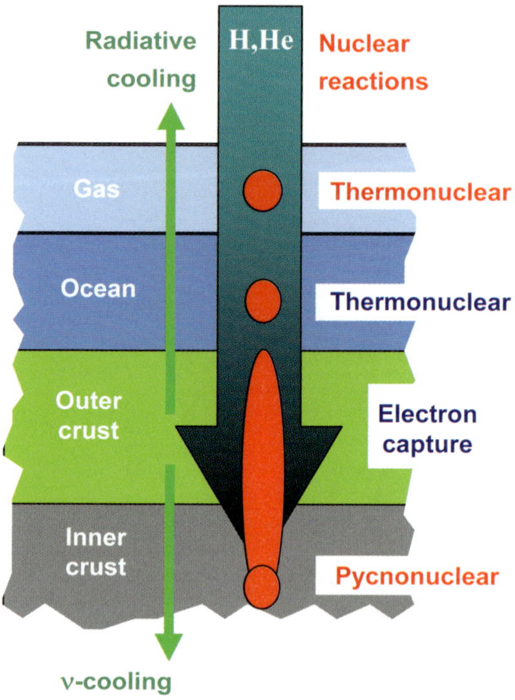


**Fig. 7.44** Physical processes in a neutron star.

described by nonideal plasma models, for $\rho \geq \rho_0$ the description of supernuclear-density matter is severely hampered by the incompleteness of laboratory data and the absence of a complete theory of supernuclear-density matter [151].

Many neutron stars are parts of binary systems and manifest themselves as X-ray pulsars when matter flows to a neutron star with a high magnetic ($> 10^{10}$ Gs) field. At lower magnetic fields, the matter accumulates on the stellar surface until some critical value is reached, when there occurs a thermonuclear explosion, which reveals itself as regular X-ray bursts. Unlike the explosions on the surface of a white dwarf, the intense gravitational field of a neutron star prevents the escape of explosion products and brings them back.

Quite unusual is the structure of matter under the high magnetic fields $B$ inherent in neutron stars, $B > 10^{12}$ Gs [96]. Under such fields, the cyclotron radius of atomic electrons is smaller than the Bohr radius, the atom is strongly compressed in the direction perpendicular to the magnetic field, and is needle-shaped. The properties of matter formed by such atoms may be inferred from the interaction of the surface of a neutron star, where the matter density is $10^5$ g/cm$^3$, with its magnetosphere.

A very important achievement which follows from observations of neutron stars that manifest themselves as radio pulsars is the verification of the general theory of relativity, i.e. the measurement of post-Newtonian corrections to the dynamics of motion of two neutron stars that make up a close binary. In particular, the shortening of the orbital period arising from the lowering of the binding energy of the stars due to their emission of gravitational waves was measured for the pulsar B1913+16. The parameters of interstellar medium – electron density, magnetic field, parameter nonuniformity – are determined from the delay of the arrival of the radio signal and its polarization. The timing of radio pulsars permits the cosmic background of gravitational waves to be studied.

Finally, studying active neutron stars makes it possible to investigate the electrodynamic processes occurring in superhigh magnetic fields, $B > 10^{12}$ Gs, inherent in neutron stars.

When the magnetic field of a neutron star in a close binary system is extremely high ($10^{12}$–$10^{14}$ Gs), a special kind of accretion on the neutron star may occur, whereby the matter of the normal star together with the magnetic field frozen in it are incident along the induction lines in the regions of the magnetic poles. In this case, the excess angular momentum is transferred to the star by magnetic field rather than hydrodynamic motion. The velocities of falling on the surface of the neutron star can reach up to several hundred thousand kilometers per second, and small (hundreds of square meters) polar regions of the surface experience tremendous fluxes of matter and energy, which generate a plasma temperature of $\approx 10^9$–$10^{10}$ K. The released energy is radiated in the form of hard photons emanating from two hot "X-ray" spots (Fig. 7.45). The strong magnetic field makes the radiation of these spots anisotropic.

Since the magnetic axis in the general case does not coincide with the axis of mechanical rotation, an observer will record one or two X-ray radiation pulses during one revolution of the neutron star about its axis. Such sources are referred to as X-ray pulsars [227] (Fig. 7.45). In the case of gamma radiation we talk about $\gamma$-ray bursts. When the magnetic field is captured in the course of evolution, the neutron star shows up as a magnetar [113], a star possessing a superhigh magnetic field (Fig. 7.46, [113]).

A magnetic intensity on the order of $10^{18}$ Gs, where the energy of the magnetic field is comparable to the gravitational energy $E_s$, is believed to be the limiting magnitude of the magnetic field of a neutron star [96].

Observations suggest [96] that the energy lost by a gyrating neutron star (radio pulsar) goes primarily to form a relativistic particle stream, which is referred to as the pulsar wind. This particle flux amounts to $10^{40}$ particles per second.

As mentioned earlier, there is a class consisting of sparse and very active neutron stars that produce bright *gamma*- and X-ray bursts and which possess magnetic fields substantially higher than those of pulsars. These stars (magnetars) gyrate relatively slowly, period $P \approx$5–10 s, but decelerate considerably more strongly, $dP/dt \approx 10^{-10}$–$10^{-12}$ s/s. Their energy is due not to gyration but to their magnetic field. The X-ray radiation flux from such a star is $W_X \approx 10^{35}$–$10^{36}$ erg/s, which far exceeds the gyration energy lost by the star. This is an indication that the source of

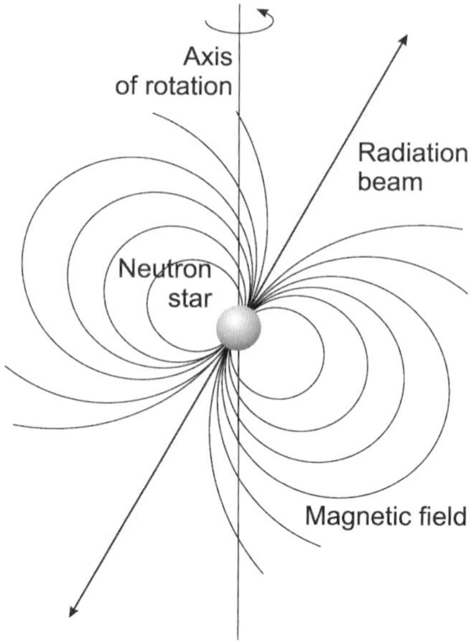


**Fig. 7.45** Neutron star with a magnetosphere. Shown are the magnetic lines of force of the neutron star, with the magnetic dipole axis inclined to the axis of stellar rotation. The pulsar radiation in one of the models emanates from two cones (the conic beams in the drawing) coaxial with the magnetic dipole [164].

activity of the star is its magnetic field rather than the stellar rotation, as for radio pulsars. Such stars are therefore referred to as magnetars.

A strong magnetic field in the magnetosphere of a neutron star sets the stage for the generation of plasma and the production of wind – the stream of relativistic electrons and positrons emitted by active stars.

Efficient particle production begins at magnetic intensities $B_c = m^2c^3/(e\hbar) = 4.4 \times 10^{13}$ Gs, where the spacing between neighboring Landau levels is equal to the electron rest energy, $\hbar\omega_c = mc^2$. The vacuum becomes unstable under this magnetic field and gives birth to electron–positron pairs, which make a significant contribution to the radiation dynamics of these exotic objects.

As we saw in Sect. 5.2, the quark–gluon plasma (QGP) is an exotic superdense state of matter. Such a plasma has been recently produced in the laboratory: it consists of quarks, antiquarks, and gluons [85, 150, 177, 77]. The QGP possesses the highest density and may emerge at the center of a neutron star or in the collapse of an ordinary star (Fig. 7.47). In this instance they are referred to as quark stars or hybrid stars and consist of quark–gluon matter at the center and a hadronic shell [191]. In this case, quark stars should be smaller than neutron ones owing to the higher compressibility of the QGP (Fig. 7.48). A candidate for a quark-gluon star or a hybrid star is the star RXJ1856 with a radius of more than 16 km, which was detected by the Newton and Chandra Space Telescopes in the X-ray range.

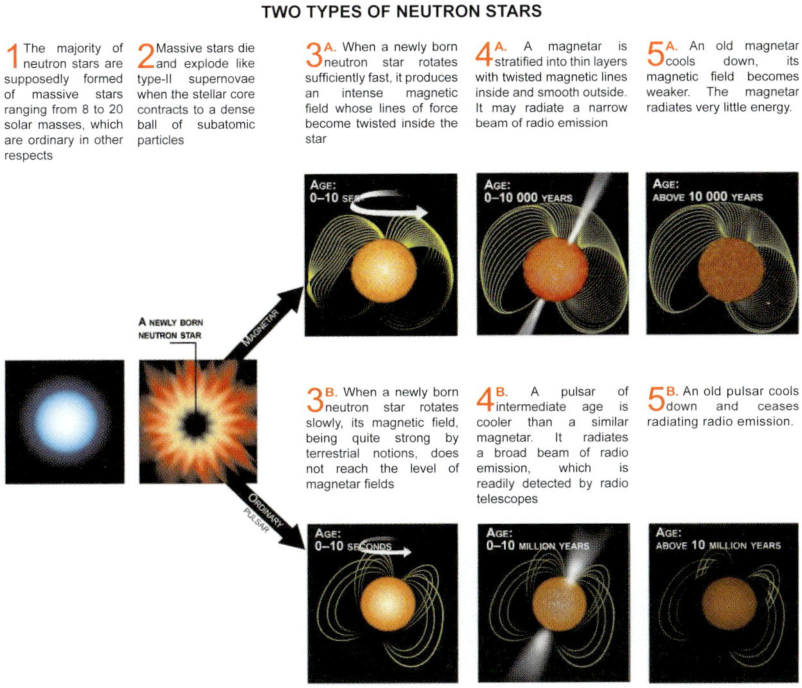


**Fig. 7.46** Two types of neutron stars. Reprinted, with permission, from [113].

Superdense states of matter and, in particular, QGPs may also show up in black holes – objects predicted by general relativity theory wherein the gravitational field is so high that the escape velocity is equal to the velocity of light [36, 214]. However, the idea of "black holes" was supposedly first suggested by the Cambridge lecturer J. Micheli in 1783 and somewhat later by Laplace from the corpuscular theory of light [89]. A black hole is a space-time domain with a distinct event horizon – a surface of dimension $r_e = 2GM/c^2 \sim 3M/M_\odot$ km, from under which no information can come out and become available to an external observer. The black hole's surface acts as a kind of valve, which transmits matter in one direction only – into the object – while light is unable to come out. This unique property of black holes leads to a unique fact: an external observer perceives it as a hot body that is a source of thermal radiation.

Three types of black holes are known [38]:

– Stellar mass black hole $M = (3-50)M_\odot$ forming at late stages of massive stellar evolution. Stellar evolution ends up with the formation of a white dwarf (if the

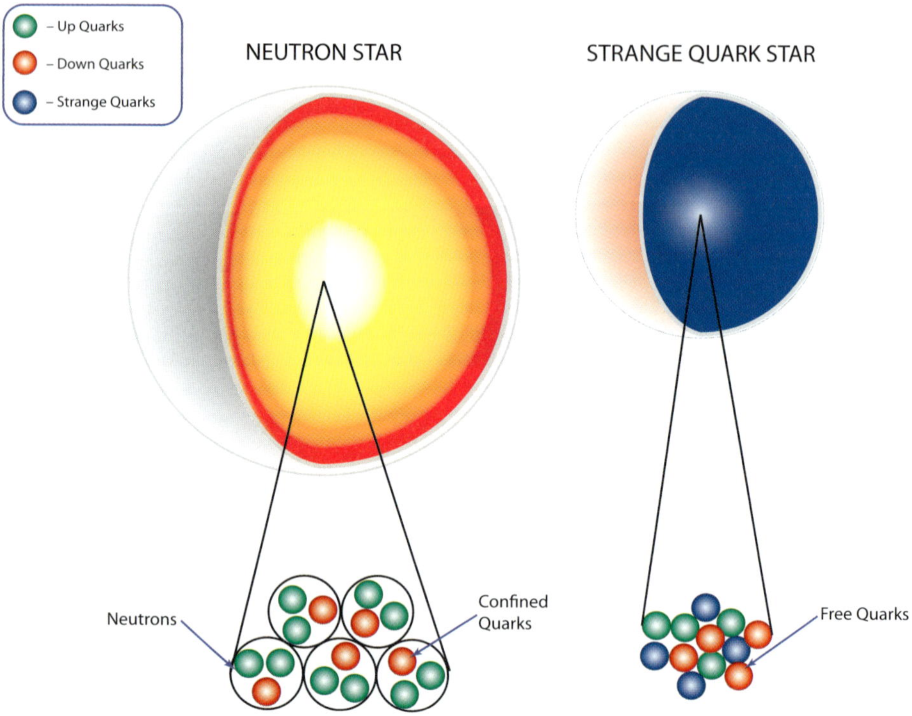


**Fig. 7.47** Neutron and quark stars.

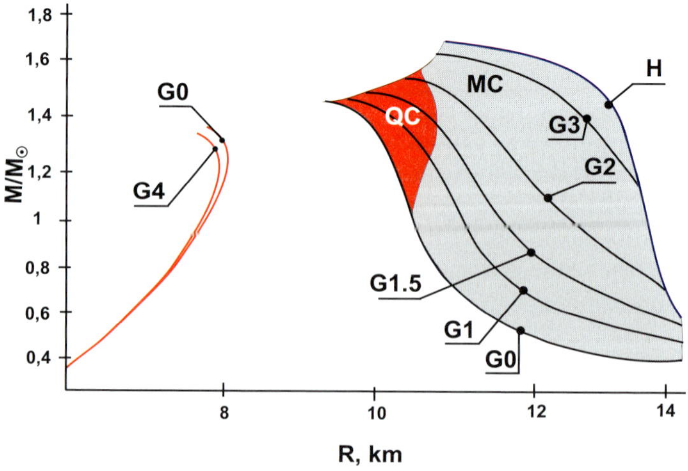


**Fig. 7.48** Dimensions of quark and hybrid stars [191].

mass of the evolved stellar core is $M_c \leq (1.2 - 1.4) M_\odot$), a neutron star (if $M_c \leqslant 3M_\odot$), and a block hole (if $M_c \geq 3 M_\odot$).
- Supermassive black hole in galactic nuclei $M = (10^6 - 10^{10}) M_\odot$;
- Primordial black hole formed at the early stages of the universe. Only sufficiently massive primordial black holes with $M > 10^{15}$ g could survive to the present time because of quantum evaporation of black holes (proposed by Hawking).

The prerequisite to the occurrence of black holes is ultrastrong matter compression: $r_{bh} \approx 3$ km for the Sun, $r_{bh} \approx 1$ cm for the Earth, and $r_{bh} \approx 0.12$ mm for the Moon. These conditions correspond to the compression of matter to some critical density $\rho_{cr} = 2 \times 10^{16} (M_\odot/M)^2$ g/cm$^3$, which is inversely proportional to the mass of an object. For a typical black hole of stellar mass $M = 10 M_\odot$), the gravitational radius is equal to 30 km and the critical density amounts to $2 \times 10^{14}$ g/cm$^3$ and is equal to the density of an atomic nucleus.

For a black hole in the galactic nucleus ($M = 10^{10} M_\odot$), the gravitational radius is equal to $3 \times 10^{15}$ cm = 200 AU, which is five times the Sun–Pluto distance. The critical density in this case is $0.2 \times 10^{-3}$ g/cm$^3$, which is several times lower than the density of the air ($1.3 \times 10^{-3}$g/cm$^3$).

For the Earth ($M = 3 \times 10^{-6} M_\odot$), the critical density is extremely high and equal to $2 \times 10^{27}$ g/cm$^3$, which is higher than the density of an atomic nucleus by 13 orders of magnitude.

It is known [38] that for a black hole forming at our time the event horizon has not yet been formed due to the relativistic time retardation near the horizon for a remote observer. So the surfaces of "modern" black holes lie very close to the event horizon and approach it for an infinitely long time from the viewpoint of a remote observer. For such an observer, all processes near these surfaces are infinitely stretched in time and hence unobservable.

Cherepashchuk [38] notes that modern observations, taking into account observational selection effects, suggest that in our galaxy, comprising about $10^{11}$ stars, there are $\approx 10^{10}$ white dwarfs, $\approx 10^8$ neutron stars, and $\approx 10^7$ black holes.

It is possible to indirectly infer the presence of these objects from their gravitational action (Fig. 7.49) on neighboring objects: in our galaxy there are about 20 candidates for black holes of stellar mass, while supermassive black holes (millions of times more massive than the Sun) are probably present in 300 galaxies [205]. However, there are even more of these objects. During their long history, many stars should collapse on using up their fuel, so that the number of black holes may be comparable to the number of visible stars and, according to Hawking [89], their mass may explain the high rates of galactic rotation (see Figs. 7.85 and 7.86).

Einstein's general theory of relativity endows black holes with surprising properties. According to Cherepashchuk [37], on the event horizon the course of time terminates in the view of an external observer. A spacecraft sent to a black hole would never intercept the event horizon from the standpoint of a distant observer but would continuously slow down as the horizon is approached. An astronaut in his spacecraft is in principle capable of penetrating beyond the event horizon, but he will be unable to transmit any information to the external observer. In this case,

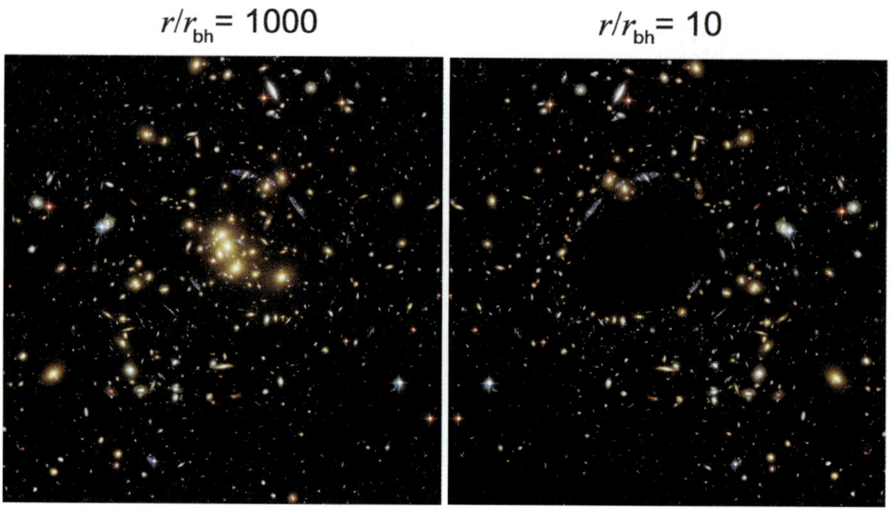


**Fig. 7.49** Black hole against the stellar sky background. At a distance of 1000 $r_{\text{bh}}$ its presence is hard to discern (the left drawing); at a distance of 10 $r_{\text{bh}}$ (the right drawing) it distorts the appearance of the stellar sky [164].

the astronaut falling freely below the event horizon would probably see a different universe and even his own future. This is due to the fact that the spatial and time coordinates change places inside the black hole and travel in space is replaced here by travel in time [37].

The properties of rotating black holes are even more extraordinary. Their event horizon is smaller in radius and is embedded in an ergosphere – a space-time domain in which bodies are bound to move continuously, being entrained by the vertical gravitational field of the rotating black hole [37].

Black holes of stellar mass may be produced either in the accretion-induced collapse of neutron stars in binary systems or in the collapse of the nuclei of massive single stars [227]. The necessary conditions for the production of black holes emerge at the final stage of evolution of stars with masses greater than approximately $3M_{\odot}$ (Fig. 7.50 [39]). The uncontrollable compression of such a star – its collapse – eventually leads to the fulfillment of the condition $R < r_{\text{bh}}$ and the production of a black hole of mass between 3 and 100 $M_{\odot}$.

Heavier, massive and supermassive, black holes with masses up to $10^9 M_{\odot}$ emerge due to the collapse of huge masses of gas at the centers of globular star clusters, in galactic nuclei, and in quasars. Light black holes with masses below $3M_{\odot}$ might be formed due to the growth of density fluctuations of the supercompressed matter of the early universe (primary black holes).

Astrophysicists relate black holes to compact X-ray sources in our galaxy and the nearest one. Most of them are X-ray binary systems, in which the optical star – the donor – delivers matter to the neighboring relativistic object: a neutron star or a black hole (Fig. 7.51 [39]).

**Fig. 7.50** The binary system GRO J1655-40 in the Scorpio constellation consists of a normal star of mass $2M_\odot$ (at the right) and a black hole of mass $7M_\odot$. The drawing was made using the data of the Chandra X-ray Observatory (by M. Weiss; the image is taken from the site http://chandra.harvard.edu). The matter of the normal star flows to the black hole. "Wind" blows from the accretion disk of the black hole under the action of magnetic forces, while magnetic forces in the inner region force the matter to approach the black hole [39].

The accretion of this matter is accompanied by tremendous energy liberation in the X-ray range with a luminosity of $\approx 10^{36}$–$10^{39}$ erg/s, which is millions of times higher than the bolometric luminosity of the Sun. This means that the optical star plays the role of a probing body in the gravitational field of a relativistic object. In this case, the black hole candidates differ from neutron stars not only in their greater mass (by more than a factor of three) and the absence of a magnetic field, but also in observable manifestations – these high-power X-ray radiators are neither X-ray pulsars nor type I X-ray bursters [36, 77, 75]. The radiation intensity of these objects ranges up to gigantic values of $10^{27}$ W/cm$^2$, which exceeds the record intensities for focused laser radiation by six orders of magnitude (Sect. 3.2, Tables 3.2 and 3.3). For a typical black hole with a mass $M = 10M_\odot$ and a supernuclear density $\rho \approx 10^{16}$ g/cm$^3$, available models suggest that the pressure is at a level of 2–$3\times10^{24}$ Mbar.

Quantum mechanics introduces some corrections to the models of black holes: with retention of the event horizon, a black hole ceases to be "black" but becomes a radiation source [89]. The nature of this radiation is the same as for electron–

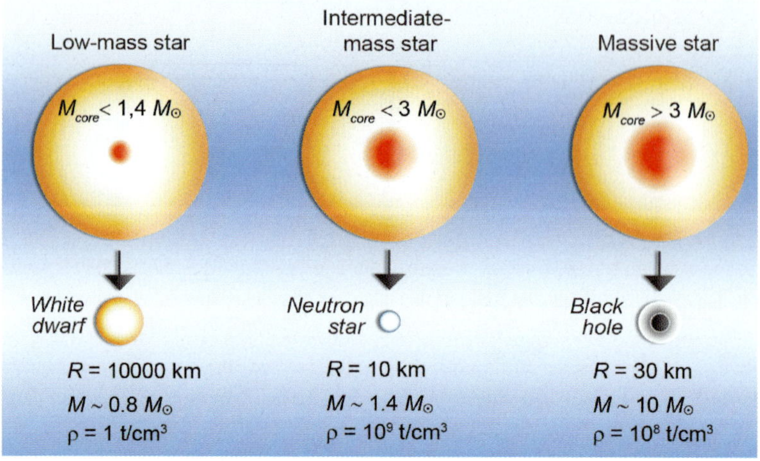


**Fig. 7.51** Schematic of the occurrence of a black hole compared to the evolution of stars of different mass [39].

positron pairs produced by a strong electric field, which increases the energy of virtual (short-lived) pairs in vacuum, making them real. The pairs in the strong gravitational field of a black hole are produced in a similar way. These pairs emerge in the "empty" space domain located in front of the event horizon. One of the components of this pair becomes a real particle outside of the event horizon and, having a positive energy, may escape to infinity; the other particle emerges inside the horizon and falls, with a negative energy, into the black hole. Eventually the black hole becomes a source of a continuous stream of particles going to infinity. No particle intersects the event horizon in the formation of this radiation. In the analysis of this picture, Hawking discovered [89] that the properties of this radiation are the same as for the radiation of a hot black body of radius $r_{\mathrm{bh}}$ heated to a temperature (in kelvin)

$$T \approx \frac{\hbar c^3}{8\pi kGM} \approx 0.5 \times 10^{-7} M_{\odot}/M.$$

In this case it was shown [89] that the smaller the black hole, the higher its brightness temperature, and the easier the black hole is to notice. Like all hot bodies, for the description of which the second law of thermodynamics was formulated, black holes possess a large amount of entropy and may emit particles (vaporize). A single black hole with a mass equal to about one million solar masses (like the one that is thought to be located at the center of our galaxy) possesses an entropy that exceeds one hundred-fold the entropy of all the particles in the observable universe.

In the course of radiation, the mass of the black hole decreases and its temperature rises, which enhances the emission and thereby speeds up the mass loss. That is why the black hole "flares up" with time, its temperature rising rapidly, and the black hole ceases to exist within a finite time on exhausting its mass.

One scenario of the universe's evolution (see Sect. 7.6) implies that it will eternally expand. Distant galaxies will vanish from our view. Those which remain close to us will transform into black holes, which will vaporize into the surrounding medium. It is not unlikely that billions of years later the universe would be empty. Then and only then would it possess the maximum possible entropy. The universe would come to a state of equilibrium, and from then on nothing would ever happen in it.

It is significant that the last moments prior to the disappearance of a black hole will occur in the regime of a high-power explosion with an energy liberation of $10^{30}$ erg in a time of about 0.1 s. Such explosions might be observable at a long distance from the Earth. This does not apply to stellar objects and even more so to massive and supermassive black holes: even for a mass equal to the solar one, the temperature is equal to an insignificant fraction of a degree and the lifetime is much longer than the life of the universe. Only primary black holes with a mass of about $10^{15}$ g are therefore capable of exploding in our epoch. Unfortunately, such explosions have never been observed.

An interesting radiation-induced change of accretion dynamics is related to the so-called Eddington luminosity limit [227], above which the light pressure exceeds the gravitational attraction. A classical example of a binary system in which a supercritical accretion to a compact object is realized is SS433 (Fig. 7.5, [65]), a massive binary star, which emanates directed matter jets traveling at a velocity of $\approx 0.26c$, where $c$ is the velocity of light. Binary systems with relativistic jets have come to be known as galactic microquasars. In most of them, the compact objects are greater than $3M_{\odot}$ in mass and are reliable black hole candidates. It is likely that the near-critical accretion to a black hole is always accompanied by the production of relativistic jets; however, the mechanism of their formation is as yet imperfectly understood.

It is likely [227] that supermassive (with $M > 10^5$–$10^{10} M_{\odot}$) black holes are the central elements of a number of galaxies. The matter falling freely on a black hole is accelerated to velocities close to the speed of light, and near the black hole this stream spins into a dense and extremely hot accretion disk due to the existence of the initial angular momentum. The energy of this "spinning top" is the principal energy reservoir of the active galactic nucleus. It is there and not in the black hole itself that high-power short-wavelength radiation fluxes are generated, and due to its magnetic field there occurs acceleration of protons and electrons to very high energies; these protons and electrons are ejected in the form of well-collimated jets perpendicular to the accretion disk plane. In some cases, the jets emanating from galactic centers measure hundreds of kiloparsecs in length and end in huge clouds [205].

By and large, this picture resembles that which is realized in the matter accretion to a black hole or a neutron star, but the scale of energy in this case is many orders of magnitude greater. The bulk of the energy of the active nucleus is the energy of matter released in its fall into a deep gravitational well produced by a supermassive black hole. As in the case of accretion onto compact stellar remnants, in the accretion disk around this object the energy of the falling gas is partly converted into thermal energy and partly into disk rotation energy. Integrally, the released energy turns out to be tens of times higher than the nuclear energy contained in the nuclei of the falling matter due to its acceleration to velocities close to the speed of light in the gravitational field of the black hole.

Supermassive black holes may be elements of quasars [38], which are very active nuclei of rather distant giant galaxies (with diameters of 50 kpc). Quasars are the most powerful stationary emitting objects in the universe. The total luminosity of quasars, including radio, IR, X-ray, and gamma-ray bands, is $10^{47}$–$10^{48}$ erg/s, which exceeds the luminosity of the host galaxy by three orders of magnitude.

It is not unlikely that at the centers of quasars there are black holes with masses of $\sim 10^8 M_\odot$, which accounts for the high-power radiation emanating from the polar regions of these objects [89].

Although nowadays there are a great number of registered phenomena attributed to black holes, there is still no definite confirmation of their existence in the universe [38]. Observational studies of black holes are carried out in two directions:

(1) Searches for massive compact objects – black hole candidates. Up to the present time, great progress has been made: the number of known black hole candidates (in X-ray binary systems or galactic nuclei) is approaching 100.

(2) Searches for sufficient criteria allowing for the unique identification of the discovered black hole candidate as a real black hole. There are a great many difficulties here, but there is some progress and much hope is placed on future space X-ray, interferometric, and gravitational wave experiments.

Arkani-Hamed et al. [6] and Bleicher [26] came up with the interesting idea of producing mini black holes in terrestrial conditions (Fig. 7.52). In this formulation, two exawatt lasers generate colliding ion streams with energies $\gtrsim 1$ TeV by irradiating gold targets, which should give rise to a black hole with a radius of $\approx 10^{-4}$ fm.

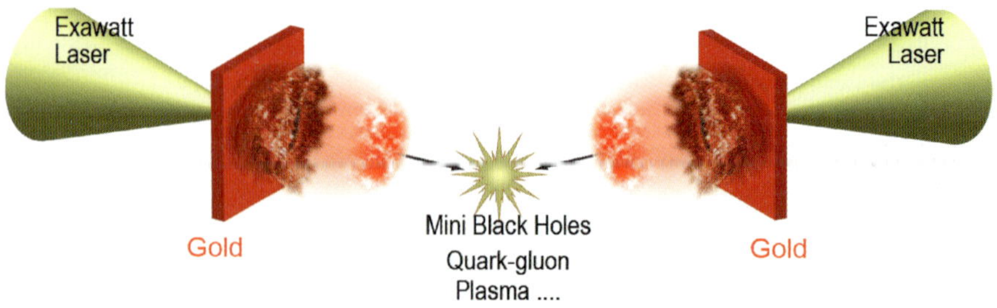


**Fig. 7.52** Laser-assisted production of a QGP and a mini black hole [6, 26].

An important feature inherent in relativistic astrophysical objects is the existence of gigantic magnetic fields, which determine to a large measure the dynamics of their motion and radiative characteristics.

The recorded radiation of a neutron star is primarily soft X-rays; this radiation becomes accessible $10^5$–$10^6$ years after the stellar birth. Recently recorded was a new class of neutron stars – magnetars [124, 156] – which possess an ultraintense magnetic field up to $10^{15}$ Gs; this field affects their gamma-ray radiation formed together with the thermal radiation of the surface [227, 124].

An example of such an object is provided by the brightest gamma-ray burst recorded in our galaxy to date; its source is the neutron star SGR-1806-20, which belongs to the class of magnetars and is at a distance of $\approx$50 000 light years from the Earth. The burst lasted for only $\approx 0.1$ s, and the amount of energy released within several seconds after the burst was greater than the energy radiated by our Sun in 250 000 years.

The extremely short duration of the burst testifies [124] to the fact that the source of its energy is the dissipation of magnetic field stored in the magnetosphere and not in the core of the neutron star. The resultant expansion of the "magnetic cloud" with a small fraction of the baryon component (with $M < E/c^2$) is determined by the magnetic field. It is strongly relativistic and anisotropic, and retains these properties for several weeks after the burst.

In magnetars there occur extremely strong magnetic bursts. In this case, the stellar surface cracks under the action of the Lorentz force and protons escape from these cracks, which interact with the magnetic field and radiate energy. The stellar magnetic field is determined from the amount of this energy [124, 156]. Zasov and Surdin [228] came up with the following model of magnetic magnetar bursts.

A magnetar is calm most of the time, but the stress induced by the magnetic field in its solid crust gradually increases (phase 1 in Fig. 7.53). At some point in time, the stress in the crust exceeds its ultimate strength, and the crust breaks into a multitude of small fragments (phase 2). This "starquake" gives rise to a pulsating electric current, which rapidly decays to leave an incandescent plasma ball (phase 3). The plasma ball radiates X-rays from its surface, cools down, and vaporizes within a few seconds (phase 4).

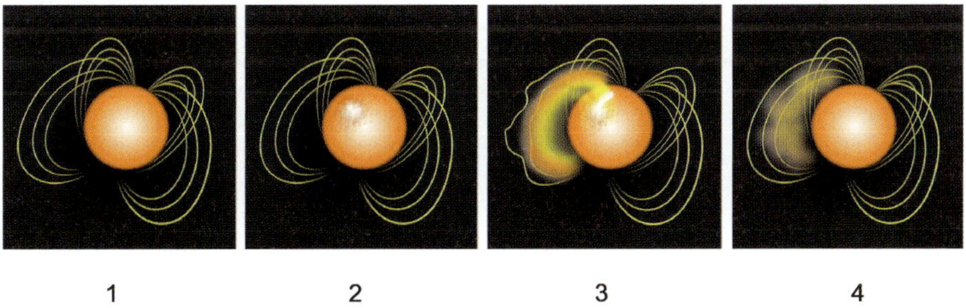


**Fig. 7.53** Magnetar burst formation [113].

For the SGR 1806-20 magnetar – the highest-power magnet in the universe – the measured magnetic field turned out to be the record breaking $\approx 10^{15}$ Gs, which corresponds to a density of $\approx 10^{13}$ g/cm$^3$, approaching the density of nuclear matter. For comparison, the induction of the solar magnetic field varies from 1 to 5 Gs, while the terrestrial magnetic field amounts to only 0.31–0.62 Gs. If SGR 1806-20 were in the place of the Moon, its magnetic field would change the arrangement of molecules in all living organisms. However, fortunately there are no such extreme stellar objects in the solar neighborhood (the nearest magnetar is $\approx$ 13 000 light years distant). The magnetic field of SGR 1806-20 is several hundred times that of ordinary radio pulsars. In this case, the stellar surface temperature is as high as $\approx 10^6$ K. Due to this superintense magnetic field alone, a piece of iron at the surface of SGR 1806-20 would experience a force 150 million times stronger than terrestrial gravity. Only 14 magnetars have been discovered to date, although it is assumed that our galaxy should contain at least several million of them.

It is pertinent to note that the existence of magnetars is sometimes associated [89] with the hypothetic possibility of traveling in time and space with a transition to a different domain of the universe via a "wormhole", provided the development of instabilities does not make this process unfeasible.

In the analysis of observational data it is very important to have ideas about the physical properties of the partially ionized dense plasma of the outer layers of a neutron star in the presence of intense ($B \approx 10^{11}$–$10^{16}$ Gs) magnetic fields [193, 84].

It is common knowledge that strong magnetic fields radically change the properties of matter. In particular, for $H \approx 150$ MOe the equation of state becomes four times more rigid. For $B \approx 10^{12}$ Gs, the pressure of the magnetic field at the surface of a neutron star amounts to about 40 Pbar (petabars). For

$$B \gg B_0 = m_e^2 e^3 c/\varepsilon^2 \hbar^3 \approx 2.35 \times 10^9 \text{ Gs}$$

($\varepsilon$ is the permittivity), the cyclotron electron energy,

$$\hbar\omega_e = \hbar(eB/m_e c) = 11.58B/10^{12} \text{ keV},$$

is much higher than the electron–nucleus Coulomb interaction energy. Under the conditions of a neutron star, the Coulomb force is a weak perturbation with respect to the magnetic one, and for typical stellar temperatures the electrons are in the Landau level [75, 84]. Since the motion is limited in the direction perpendicular to the magnetic field, atoms and ions assume a cylindrical shape and have high binding energies [103, 165]. Such ovoid atoms make up thread-like molecular chains with covalent bonds, which in turn may form a three-dimensional condensate. Furthermore, when an atom moves in a constant magnetic field, in its frame of reference, due to the Lorentz transformation, there emerges an electric field, which may ionize the system [103, 165]. Much is still unclear about the structure as well as the electronic and mechanical properties of such an exotic state of strongly magnetized matter [145, 84].

The central element in the understanding of the structure and evolution of stellar objects (especially of the structure of neutron stars), the collapse of stars, supernova explosions, etc. is the equation of state as well as the composition and optical properties of the nonideal plasmas of stellar matter [105, 62, 71, 84, 63]. Their equation of state includes the important condition for nuclear statistical equilibrium, but it is applicable only for extremely high temperatures ($\approx 5 \times 10^{10}$ K). The corresponding equations of state are formulated for an arbitrary collection of atomic nuclei (including unstable ones) with the inclusion of the strong Coulomb nonideality effects of the matter, which is sometimes referred to as a Coulomb liquid [141]. Apart from the great complexities of describing strongly nonideal plasmas, there are additional complications arising from the inclusion of a magnetic field, which hinders the correct description of the relation between the nuclear center-of-mass motion and the electronic structure. Only the first steps have been made along this path. A phase transition of a nondegenerate gas to the state of a macroscopic condensate is predicted to occur at relatively moderate fields and temperatures [62, 71, 68, 64]. With increasing field and falling temperature, the saturation vapor density of this condensate will become lower, lengthening the optical free paths and thereby making it easier for the thermal radiation to escape from the stellar center occupied by the condensate of metallic liquid [145].

Describing the radiation transfer in ionized plasmas with $T > 10^6$ K and magnetic fields $B > 10^{14}$ Gs involves certain difficulties. One of them is related to the ion cyclotron resonance at a proton energy $0.63\,(Z/A)\,B/10^{14}$ keV.

Another problem is related to the polarization of the vacuum and virtual electron–positron pair production at ultrahigh magnetic fields,

$$B > B_{\mathrm{cr}} = m_e^2 c^3/e\hbar = 4.4 \times 10^{13}\ \mathrm{Gs},$$

which change the dielectric properties of the medium, polarize the proton motion, and thereby change the radiative plasma properties. In these fields, the orbitals of bound electrons are strongly perturbed by the magnetic field: $h\omega_c/E_{\mathrm{c}} > 1$, where $h\omega_{\mathrm{c}}$ is the electron energy at the cyclotron frequency and $E_{\mathrm{c}}$ is the Coulomb energy [99].

Poorly studied is the effect of "vacuum resonance", whereby the influence of plasma and vacuum on the linear polarization becomes weaker, giving rise to resonance effects in absorption and scattering [145]. This effect manifests itself at proton energies

$$E_v(\rho) \approx 1.0\,(Y_e\rho)^{1/2}\left(B/10^{14}\right)^{-1} f(B)\ \mathrm{keV},$$

where $\rho$ is the density in g/cm$^3$, $Y_e = Z/A$, and $f(B)$ is a weakly varying function of the field induction $B$.

Already today, new-generation superhigh-power short-pulse lasers generate fields of $\approx 10^9$ Gs in laser-produced plasmas (see [46, 138] and Chap. 4). The advancement of this line of research may provide new information about the radiation capacity and spectra of astrophysical plasmas in intense magnetic fields.

An ingenious way of studying physical effects in superhigh magnetic fields is discussed in [98]. For an electron–hole plasma (liquid) in semiconductors, the fields lowered by a factor $(\varepsilon m_0/m)^2 \approx 10^4$–$10^3$ (where $m_0$ is the proton mass and $m$ is the hole mass) compared to an electron-proton plasma will still be "superhigh". The electron–hole plasma is therefore a unique condensed object accessible for laboratory research in superhigh magnetic fields.

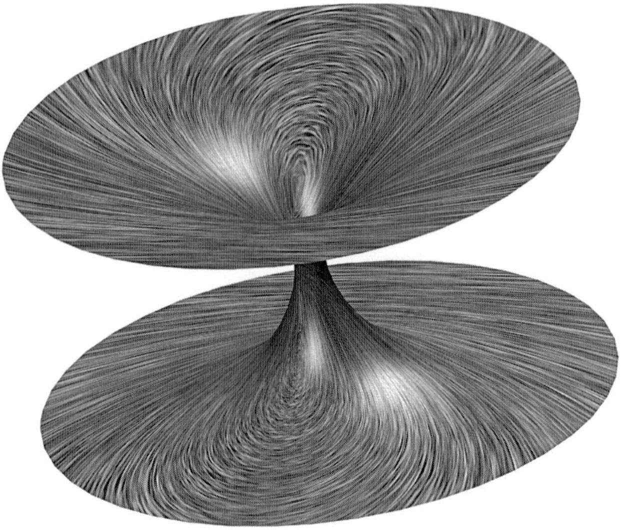

**Fig. 7.54** Wormhole connecting two spaces (two universes). Reprinted, with permission, from [180].

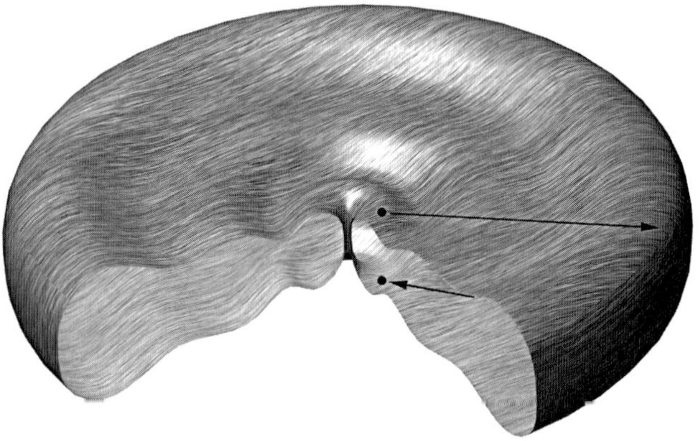

**Fig. 7.55** Wormhole connecting two local regions of the same space. The path between them in ordinary space may be far longer than the path through the wormhole. For obviousness, our space is curved. Reprinted, with permission, from [180].

More recently, one more mechanism of the origination of black holes related to the collapse of so-called "magnetic tunnels" or "wormholes" [199, 122] has come under discussion [102]. These highly exotic objects supposedly emerged at early stages after the Big Bang and have persisted for various reasons until the present time. According to the chaotic inflation model, which underlies modern cosmology [102], apart from our universe there may exist an infinite number of other universes, which originate in a scalar field in different spatial domains and at different points in time to make up a "spatio-temporal foam" [199, 122]. Some parts of our and other universes may be connected by traversable spatio-temporal tunnels – "wormholes" (Figs. 7.54 and 7.55, [180]) – which existed in the initial scalar field and persisted after inflation. There emerges an intriguing possibility of studying the multielement universe and searching for specific manifestations of extraordinary-type astrophysical objects – entrances to the tunnels.

Kardashev et al. [102] considered different models of a "wormhole" sustained by the electromagnetic field penetrating through these holes and by the combination of magnetic field and "phantom" energy, magnetic field and "phantom" matter, which was introduced to account for the accelerated universe expansion discovered recently [150, 177]. In this case, the "phantom" energy and "phantom" matter should possess extraordinary equations of state close to the vacuum one, $P/E < -1$, and be isotropic or anisotropic. Shatskii et al. [196] investigated the dynamics of a spherical model of a wormhole filled with matter consisting of magnetic field and dust with a negative mass density.

The entrance to such a magnetic tunnel, which is a magnetic monopole of macroscopic dimensions, turns out to be amenable to observation, and the accretion of the ordinary matter onto it may give rise to a black hole with a radial magnetic field. Kardashev et al. [102] discuss the likelihood that some active galactic nuclei and other objects of the galaxy are or were entrances to these magnetic holes, which range in mass from 2 kg to 2 billion solar masses. The long cosmological lifetime of these objects is attributable to the circumstance that the Hawking evaporation effect [88] is not operable for them, because they lack the event visibility horizon. This is the difference between wormholes and primary black holes, which have a lower mass bound of $10^{15}$ g. A wormhole may manifest itself as a binary system formed by the entrance to a spatial throat, which produces multiply connected topology, emitting nonthermal radiation and gamma-ray bursts.

Table 7.5, taken from Ref. [102], shows the throat parameters ($\nu_G$ is the oscillation frequency of a probe particle and $\nu_H$ is the gyrofrequency) of a magnetic "wormhole" estimated for a quasar nucleus, objects with the critical field for electron–positron pair production, objects of terrestrial and solar mass, objects with the critical magnetic field for monopole–antimonopole pair production, and an object of Planck's mass.

One can see that "wormhole" parameters correspond to ultrahigh energy densities, which may be comparable to those in the immediate vicinity of the Big Bang. Attention is drawn to the extremely high magnetic fields of these objects. For fields $B = m_e^2 c^3/e\hbar \approx 4.4 \times 10^{13}$ Gs a level's Landau excitation energy exceeds the rest energy of the electron. Positronium atoms acquire stability at fields above $10^{24}$ Gs; they are spontaneously produced in the vacuum to fill the entire medium. The breakdown of the vacuum and monopole pair production take place at fields of $\approx 2.6 \times 10^{49}$ Gs, which leads to a "wormhole" mass of only 1.8 kg, which is supposedly the lower bound for these objects.

**Table 7.5** Parameters of wormholes of different mass [102]

| $M_\infty = 2M_0$ | $r_0$ [cm] | $B_0$ [Gs] | $\rho\,(r_0)$ [g/cm$^3$] | $\nu_G$ [Hz] | $\nu_H$ [Hz] |
|---|---|---|---|---|---|
| $6\times10^{42}$ g $= 3\times10^9\,M_\odot$ | $4.5\times10^{14}$ | $7.8\times10^{9}$ | $2.7\times10^{-3}$ | $7.6\times10^{-6}$ (1.5 days) | $2.2\times10^{16}$ |
| $10^{39}$ g $= 5\times10^5\,M_\odot$ (pair production $e^\pm$) | $7.4\times10^{10}$ | $4.4\times10^{13}$ | $9.7\times10^{4}$ | 0.045 (22 c) | $1.3\times10^{20}$ |
| $2\times10^{33}$ g $= M_\odot$ (the Sun) | $1.5\times10^{5}$ | $2.3\times10^{19}$ | $2.4\times10^{16}$ | $2.4\times10^{4}$ | $6.6\times10^{25}$ |
| $6\times10^{27}$ g $= M_\oplus$ (the Earth) | 0.45 | $7.8\times10^{24}$ | $2.7\times10^{27}$ | $7.6\times10^{9}$ | $2.2\times10^{31}$ |
| $5\times10^{10}$ g (positronium) | $3.5\times10^{-18}$ | $10^{42}$ | $4.4\times10^{61}$ | $9.7\times10^{26}$ | $2.7\times10^{48}$ |
| $1.8\times10^{3}$ g (pair production $\mu^\pm$) | $1.3\times10^{-25}$ | $2.6\times10^{49}$ | $3\times10^{76}$ | $2.6\times10^{34}$ | $7.3\times10^{55}$ |
| $2\times10^{-5}$ g (Planck's mass) | $1.5\times10^{-33}$ | $2.3\times10^{57}$ | $2.4\times10^{92}$ | $2.3\times10^{42}$ | $6.6\times10^{63}$ |

Also discussed is the feasibility of observing – against the background of cosmic objects – entrances to magnetic tunnels or black holes produced by way of accretion from relict black holes [102]. The difference in structure and rate of spatial decay of these objects from the structure of ordinary black holes may serve as an indication. Furthermore, the rotation of a monopole excites a dipole electric field, resulting in relativistic particle acceleration. In this case, electrons are accelerated toward one pole and protons toward the other. Such unidirectional jets are observed to emanate from some sources, for instance, the 3C273 quasar [102]. A black hole may be distinguished from the entrance to a "wormhole" by the absence of an event horizon in the latter: a luminous object incident on the "wormhole" would be continuously observed with an alternating red or blue shift. Observations of the gravitationally lensed quasar Q09577+561 supposedly lend credence to this model [102]. A distinctive feature of the "wormhole" as a strongly relativistic object without an event horizon is the possibility of occurrence of periodic oscillations of the stellar mass relative to the throat (see Table 7.5) accompanied by the corresponding variation of the recorded red shift. In the motion of radiating sources along circular orbits around the entrance to the tunnel, a compact source will also change its flux and red shift. Lastly, an external observer may discover radiation at the gyrofrequencies and observe effects related to $e^{\pm}$ and $\mu^{\pm}$ pair production [102].

Binary systems involving entrances to magnetic tunnels as well as the approach of two entrances to the same magnetic tunnel may also be accompanied by observable radiative effects. In particular, the second of these processes ends with the transformation to a black hole without magnetic field and with the radiation of an external magnetic field. The electromagnetic pulse accelerates electron–positron pairs, which may explain the observed gamma-ray bursts.

Kardashev et al. [102] also considered other possibilities for the observable manifestations of new types of primary cosmological objects – entrances to "wormholes" and specific black holes resulting from them. In any case, the discovery of "wormholes" opens up intriguing possibilities for investigating the multielement universe and even paves the way to semifantastic ideas of traveling via "wormholes" into the past by a "time machine" in violation of the causality principle and to other universes before these tunnels collapse to black holes [199, 122].

Figure 7.56 [157] depicts the Hillas diagram with the characteristic magnitudes of magnetic fields and dimensions of astrophysical objects. It also demonstrates their capabilities as particle accelerators. Solid lines represent estimates of dimensions (in parsecs) and magnitudes of magnetic field (in gauss) of the objects capable of accelerating protons to energies of $10^{20}$ and $10^{21}$ eV for a shock wave velocity 1/300 the speed of light. The dotted line represents the same for iron nuclei. Dark spots indicate the observed dimensions and magnetic fields of different astrophysical objects. It seems that in the universe known to us there are no, under the assumptions made about the particle acceleration mechanism (Fermi) itself, obvious candidates for the role of a zevatron accelerator – an accelerator of observable-energy cosmic particles. Also shown for comparison here are the characteristics of the Tevatron accelerator and the LHC operating on the Earth.

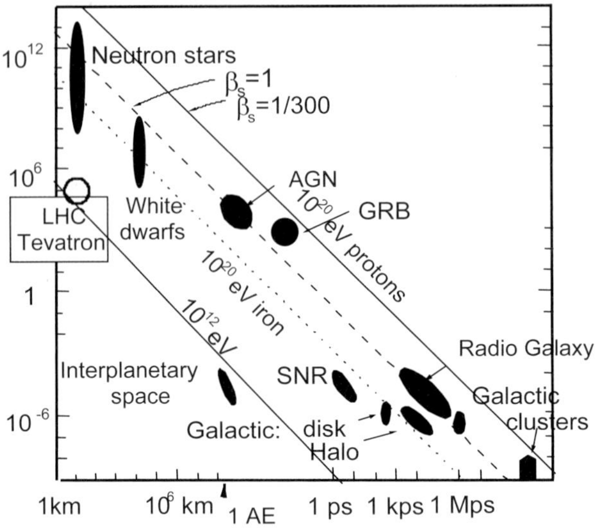


**Fig. 7.56** Hillas diagram showing the characteristic magnitudes of magnetic fields and dimensions of astrophysical objects [157].

The gravitational accretion of matter is the principal source of mass and energy for many astrophysical objects, for example binary stars, binary X-ray sources, quasars, active galactic nuclei, and jet ejections [227]. Although accretion theory has long been known, it gained a new impetus with the discovery of the first X-ray sources. As noted above, the accretion of matter onto stars (including neutron stars) may take place on their inner surface or on the outer surface of the magnetosphere of a strongly magnetized star. In this case, the gravitational energy of the falling matter is converted to heat and is reradiated into open space.

A different situation arises in the accretion to black holes (Fig. 7.57), which are possibly present in the binary X-ray sources of our galaxy or in active galactic nuclei. In this case, as discussed earlier, the matter transits the "event horizon" of a black hole, while the corresponding radiation cannot come out. Unlike the accretion onto stars, the radiation capacity of accretion to black holes is a priori unknown and depends on several factors, such as the angular momentum of the falling matter and the matter-entrained magnetic field. For the spherical accretion of nonmagnetized gas, the efficiency of energy conversion to radiation may amount to $10^{-8}$, while the presence of a magnetic field increases its magnitude to 10% [145, 150, 177, 193]. The small value of this quantity is of significance in explaining the extremely low luminosity of the neighborhood of black holes at the center of our galaxy. Both jets and emanating plasma streams make a contribution of their own to the low luminosity (Fig. 7.58).

Viscosity is a special problem in the theory of matter accretion [145]. Normally viscosity lowers the effect of the angular momentum to enhance the accretion. For

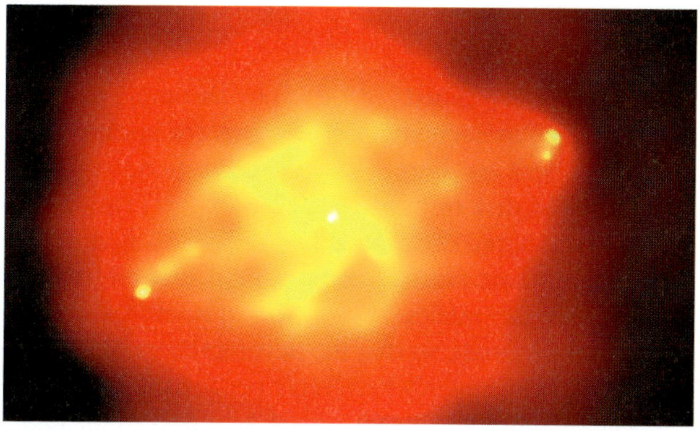

**Fig. 7.57** Accretion of matter onto a black hole [47] in the galaxy Cygnus A. One can see cosmic jets responsible for high-power X-ray and radio radiation. Under deceleration due to the ambient medium, they produce "hot spots" and gigantic cavities. The radio emission power of the object Cygnus A is $\approx 10^{45}$ erg/s, which is approximately 10 million times the emission power of the Andromeda Galaxy.

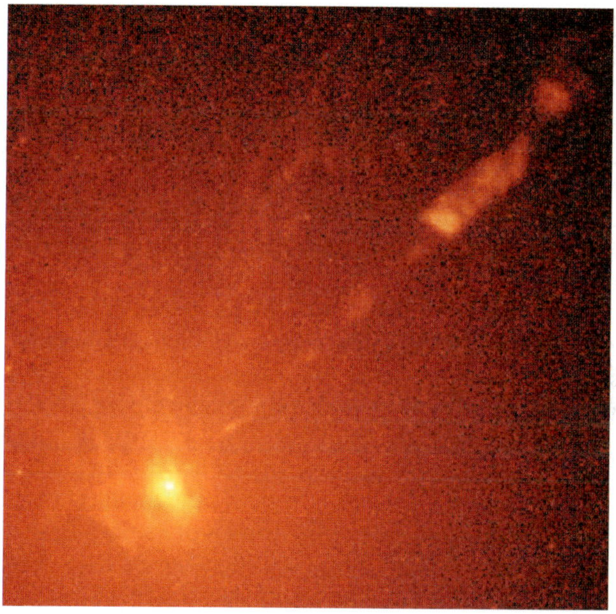

**Fig. 7.58** Jet formation in the accretion of matter onto a massive object [47] at the center of the galaxy M87. The jet consists of high-velocity particles and exhibits a structure of inhomogeneities (clusters) measuring $\approx 10$ light years.

a laminar viscous flow, molecular viscosity is low and a stationary accretion disk should be massive and thick, and the accretion rate should be low in this case. This does not correspond to the observations of high accretion rates for binary X-ray sources.

The discrepancy may be explained [145, 235] by assuming that the real viscosity coefficient far exceeds the "molecular-kinetic" value, for instance, by assuming the accretion disk to be turbulent and introducing the corresponding turbulent viscosity.

Explaining the accretion disk turbulence [217, 216, 201] calls for special consideration [145]. In the linear approximation, a Keplerian disk is stable (after Rayleigh [56]) against axially symmetric perturbations, which conserve the angular momentum. For a very high Reynolds number, $Re = \rho V\ell/\eta \approx 10^{10}$, the amplitude of nonlinear perturbations will be small, so that the development of turbulence from nonlinear perturbations is possible even when linear perturbations are stable.

At the present time, magnetohydrodynamic instabilities as the cause of accretion disk turbulization are being studied intensively. The linear (magnetorotational) instability of differentially rotating disks (the Alfvén velocity is below the velocity of sound) may develop into a turbulent regime, which it has been possible to model employing sophisticated three-dimensional magnetohydrodynamic (MHD) codes.

Laboratory experiments in this area involve rotating liquid metals, though for lower-than-required values of the magnetic Reynolds number $Re_{\rm m}$ owing to the substantial magnetic diffusion in metals [201]. Another type of laboratory experiment might involve laser-produced plasma, in which it is possible to generate $\approx 100$ kT fields and high accelerations.

Apart from the magnetorotational instability [217, 216], Rossby wave instability [145, 56] may be significant in accretion problems involving an unmagnetized flow with a nonmonotonic profile of the density, pressure, or entropy [56]. Two-dimensional simulations of such situations revealed the emergence of 3–5 long-lived vortices, appreciably changing the accretion dynamics, which is of importance in studying the formation of planets around young stars.

At present, as we see, considerable attention is being given to binary systems in which one of the objects is a collapsing object such as a neutron star, a black hole, or a white dwarf. Active galactic nuclei and quasars are at the other extreme. Their anomalous luminosity results from falling-matter energy conversion to a massive black hole located at the galactic center. The corresponding high-intensity X-ray radiation from these objects ionizes the matter in the falling accretion disk and transforms it to the plasma state, making significant the effects of magnetic field, radiation, and Coulomb interaction on the dynamics of motion. To study the physics of these processes, advantage is taken of the high-energy X-ray cosmic observatories Chandra and XMM. In this case, many questions [145] concerning plasma photoionization in equilibrium and kinetic modes remain to be answered.

High-current Z-pinches, lasers, and relativistic heavy-ion accelerators (see Chaps. 3 and 4) can to a certain extent sufficiently reproduce the photoionization effects in plasmas, providing a clearer insight into the dynamics of matter accretion to black holes.

## 7.3 Cosmic Jets, Radiative Shock Waves, and Molecular Clouds

At a certain stage of the evolution of a protostar, the excess parent gas flows off the protostar (Fig. 7.59 [206]). Massive protostars are typically characterized by a spherically symmetric outflow, because any unidirectional structures are destroyed here by radiation and the stellar wind. Typical for low-mass protostars and young stars are unidirectional gas flows, which are evidently due to the conversion of rotational energy to the kinetic energy of a jet [204]. The reason for the collimation of these jets and flows as well as one of their energy sources probably is circumstellar disks, which are about 1 pc in diameter and up to hundreds of $M_{\odot}$ in mass.

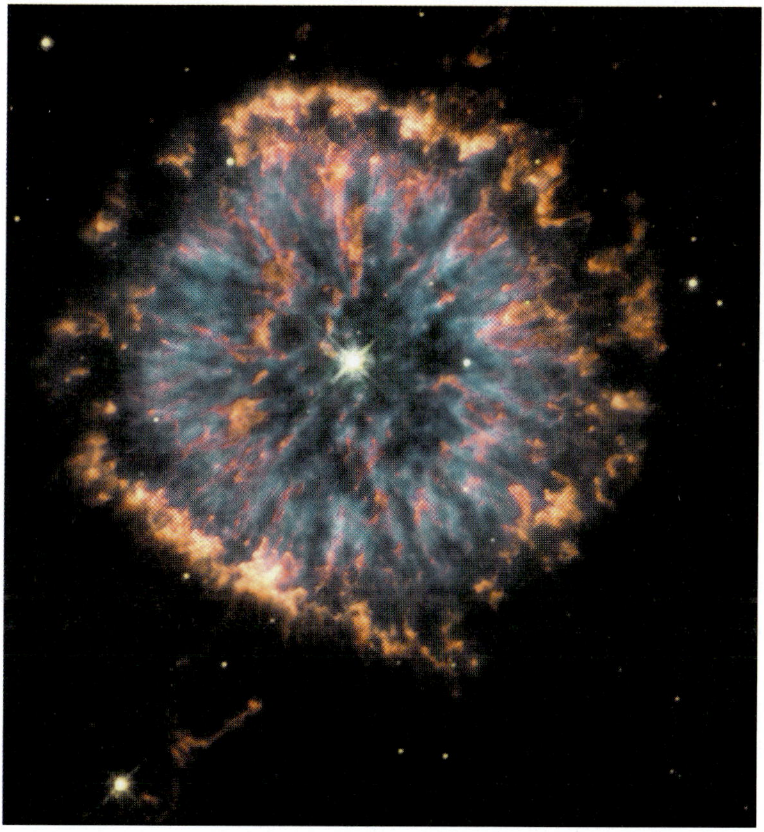

**Fig. 7.59** Planetary nebula NGC 6751 in the Aquila constellation. One can see the streams of gas escaping from the central star. Photo from the Hubble Space Telescope [206].

Axially symmetric flows [227] have long been discussed in connection with numerous astrophysical problems [227, 214, 145, 231]. Among the flows of this class are the accretion to ordinary stars and black holes [193, 28, 231], axially symmetric stellar (solar) wind [158, 25, 210, 136], jet ejections from young stellar objects [116], and the ejection of particles from the magnetospheres of rotating neutron stars [135, 23]. Magnetohydrodynamic (MHD) models are actively being developed also in connection with the theory of the structure of the magnetospheres of supermassive black holes, which are believed to be the "central engine" of active galactic nuclei and quasars [162, 21]. The interrelation [214] of different cosmic objects is illustrated in Fig. 7.60.

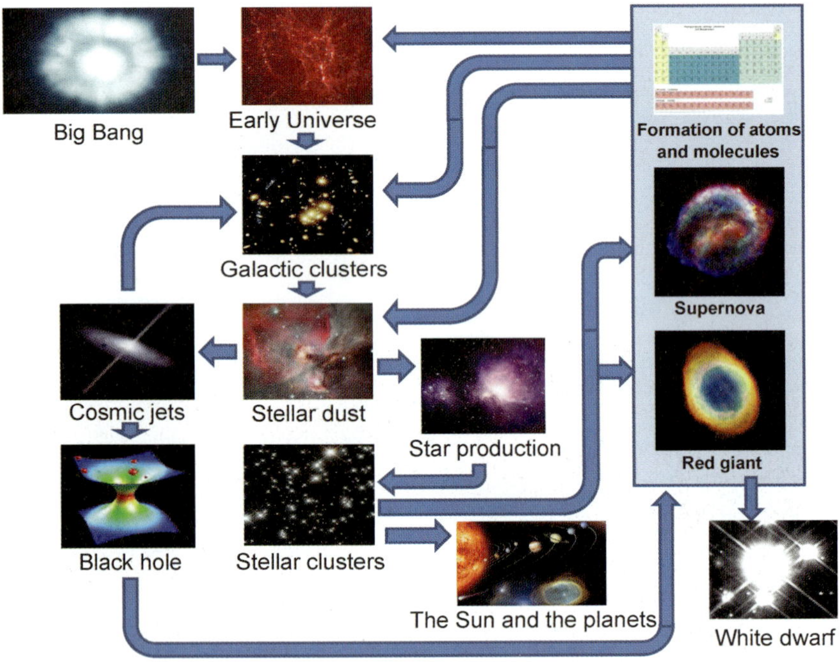


**Fig. 7.60** Schematic of the interrelation between different cosmic objects [214].

High-power ($\approx 10^{39}$ W), long (0.01–2 pc long), and well-collimated (with divergence angles of 5°–20° and even less than 1°) differently directed molecular and plasma relativistic jets (with velocities up to $0.9c$) are observed in many active galaxies, quasars, and old compact stars [145, 227, 214, 57] (Figs. 7.5, 7.58, 7.61 and 7.62). In the case of the most active quasars, these jets are often accompanied by electromagnetic radiation with a wide spectrum – from radio-frequency emission $\approx 10^8$ Hz to superhard gamma-ray radiation with a frequency of $10^{27}$ Hz. Well-collimated jets with velocities of 100–300 km/s are also observed for many young stars.

Observations and theoretical analysis suggest that the acceleration and focusing of these plasma jets is effected by the magnetic fields rotating together with accre-

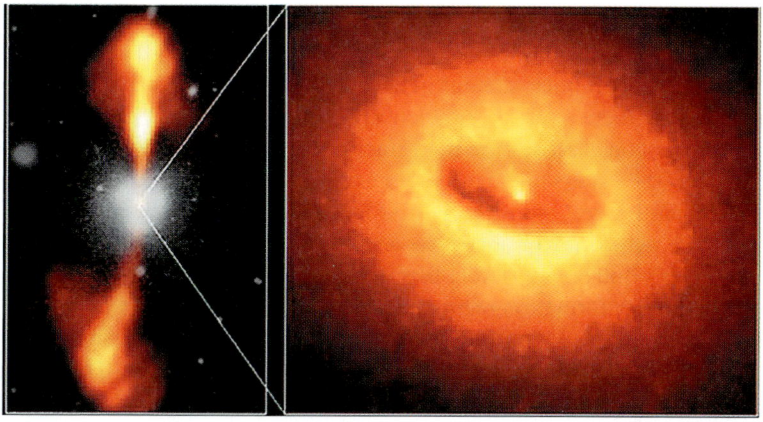

**Fig. 7.61** Active elliptical galaxy NGC 4261 in the Virgo constellation is the high-power radio source 3C 270. Left panel: General view (at the frame center) with the image of elongated radio emission regions superimposed on it. Each of the radio emission streams has moved away from the galactic center by 20 kpc. Right panel: Photo of the central galactic region obtained from the Hubble Space Telescope. The galactic nucleus is surrounded by a massive gas-dust disk about 300 kpc in diameter [228].

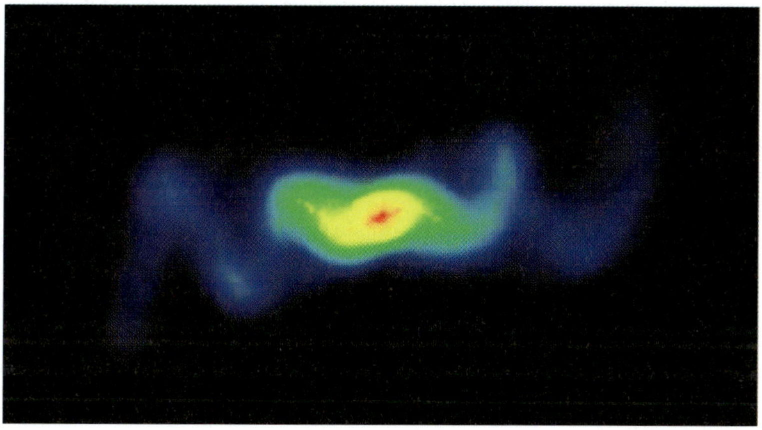

**Fig. 7.62** Radio image of the jets from the black-hole candidate SS4333, which is part of a binary system with an ordinary star $M \sim 10M_{\odot}$, which emanates a strong stellar wind [214].

tion disks, the fields taking the energy from the accretion masses or from the rotating black holes.

For many stellar objects the hydrodynamic [57, 56] jet dynamics mode is realized, where the jet energy and angular momentum are generated by the electromagnetic field and kinetic energy of the ambient matter [145].

For extragalactic, microquasar, and gamma-burst jets, a different (Poynting) mechanism is thought to operate, in which the electromagnetic field is the main source. In any case, the jet velocities exceed the "escape velocity" sufficiently for the jets to leave the accretion disk.

The occurrence of bipolar streams in the evolution of a protostar appears to somewhat paradoxical: there occur simultaneously both accretion and jet efflux of matter. However, this is considered to be the principal mechanism by which the protostar can rid itself of the excess angular momentum and magnetic flux during collapse.

Of significant interest are recently discovered structures in the radio continuum and 21 cm line, which tower above the accretion disk, are referred to as "worms", and measure 100–300 pc. It is likely that they owe their origin to hot plasma streams emanating from star formation regions [204].

Laboratory experiments on plasma jets in collisional and collisionless regimes may be carried out in high-current Z-pinch facilities (Sect. 3.7). The energy exchange and the generation of radiation and shock waves may be studied in the collisional regime. By lowering the density of targets it is possible to model the penetration of the jets into the targets, their dynamics, and thermalization.

A comparison of radiative hydrodynamic processes in the laboratory and in astrophysics was made by Falize et al. [58], who found similarity criteria for the reproducibility conditions of radiative hydrodynamic phenomena.

In any case, the use of Z-pinches and lasers [47] has made it possible to obtain hydrodynamic plasma jets with Mach number $\mathrm{M} \approx 5$–10, where $\mathrm{M} = v/c_s$ and $c_s$ is the velocity of sound, and strongly radiating jets with $\mathrm{M} \approx 50$–60 [145, 223] in laboratory conditions. A new class of high-energy, electroneutral, and well-collimated proton jets with an energy up to 100 MeV was produced employing picosecond terawatt lasers [139, 167, 43, 127, 125, 100, 111].

The Poynting acceleration mechanism is close in geometry to the geometry of a magnetically shielded diode system with a potential $m_e c^2/e$ applied to it, which leads to the generation of intense charged-particle streams. The same pulsed high-current accelerators with an operating voltage of $\approx 10^6$ V [131, 213] are employed for the simulation of plasma jets from the accretion regions.

Gigantic pulsed energy releases in astrophysical objects are inseparably linked with the generation of intense shock waves, solitons, and contact surfaces in cosmic plasmas [145, 227, 47, 214] (Fig. 7.63), which may be caused by a supernova explosion, stellar wind, a galactic spiral wave, or the mutual collision of clouds and stars, among others.

When an explosion wave leaves a dense medium, its radiative cooling time becomes shorter than its hydrodynamic time. Such radiative explosion waves may emerge at the final stages of supernovae evolution, at the stage of the afterglow of gamma-ray bursts and supernovae.

In dense molecular clouds [145, 227, 214], the temperatures of shock-compressed plasmas range up to tens of millions of degrees. These waves are subject to radial oscillations if the cooling time increases slowly with temperature. Furthermore, dense cooled domains are sometimes longitudinally unstable. Extensive computer calculations have shown that the propagation rate of such radiative waves in the interstellar medium is $n = (d\ln R/d\ln t) \approx 1/3$, where $R$ is the radius of the radiation waves $(R(t) \sim t^n)$, $t$, is time and $n$ the propagation rate, which is below the Sedov–Taylor

**Fig. 7.63** Cat's Eye Nebula (NGC 6543) [143] located 1 kpc away from the Sun. Seen at the center is the remnant of the star that shed its plasma shells. Photo from the Hubble Space Telescope.

rate of 2/5, but is higher than the simple self-similar limit of 2/7 [56]. That is why the study of radiative explosion waves in astrophysics is of significance, connecting complex hydrodynamic and radiative processes in plasmas (Fig. 7.64) [134, 218].

The study of radiative explosion waves (Fig. 7.65) in a laboratory requires the production of a strongly radiating, but optically thin domain of shock-compressed gas [47, 148]. Experiments at Lawrence Livermore National Laboratory with the Falcon short-pulse laser irradiating a xenon jet made it possible to produce and investigate radiative waves and to analyze their attenuation rate [224].

Nicolai et al. [148] investigated the structure, dynamics, and main parameters of laser-produced jets. By varying the gas parameters and its chemical composition they studied regimes ranging from the adiabatic to strongly radiating. It was possible to reproduce the basic features of astrophysical jets, such as the formation of detached shock waves, the region of shock-compressed plasma, a contact surface, and the Mach disk.

Interstellar space is not empty or uniform (Figs. 7.66 and 7.67). It is filled with low-density matter, radiation, and a magnetic field with the corresponding characteristic energy density of $\approx 1$ eV/cm$^3$. On average, one cubic centimeter of interstellar space contains no more than one hydrogen atom and far fewer other chemical elements. In one cubic kilometer of this space there are no more than ten dust particles, measuring 1 $\mu$m each. The matter temperature varies over a wide range – from $\approx$ 10 K to $10^6$ K. Dense gas and dust domains have come to be known as clouds (or nebulae), Figs. 7.68 and 7.69, of size 100–300 pc, up to $10^7 M_\odot$ in mass,

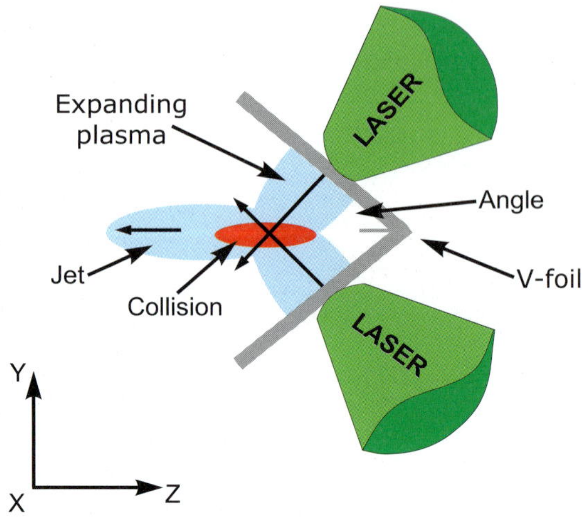


**Fig. 7.64** Schematic of experiment [218] on laser generation of plasma jets. Two lasers irradiate thin foils of thickness 3–5 $\mu$m each, with the effect that the intersection of two backside jets produces a longitudinal jet.

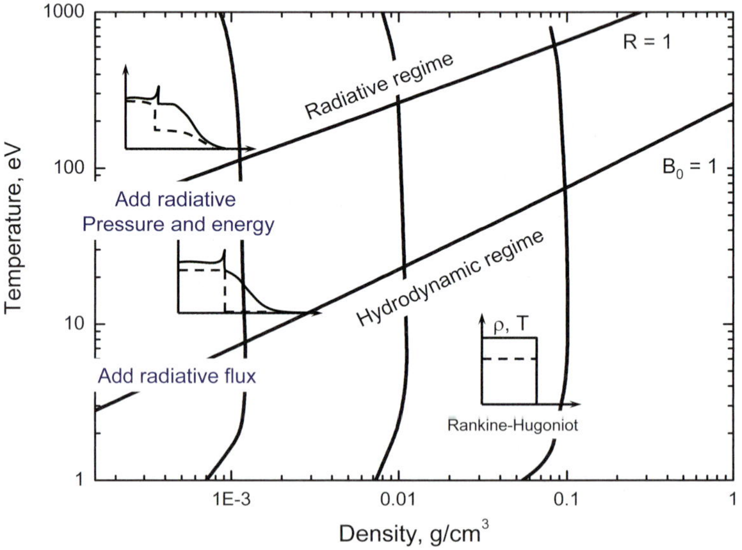


**Fig. 7.65** Radiative shock waves in xenon. $R = E_{\text{th}}/E_{\text{rad}} = \dfrac{\frac{3}{2}\rho_0 T/m}{\frac{4\sigma}{3c}T^4}$. Here $B_0 = \dfrac{\rho_0 C_v T V_s}{\sigma T^4}$.

and are classed as diffuse ($T \approx 10^2$–$10^3$ K, $n \approx 1$–$10^2$ cm$^{-3}$), dark (10–$10^2$ K, $10^2$–$10^4$ cm$^{-3}$), molecular (5–50 K, $4 \times 10^2$–$10^6$ cm$^{-3}$), or globular (10–30 K, $10^3$–$10^6$ cm$^{-3}$). It is likely that some of them came into being in the regions of active star

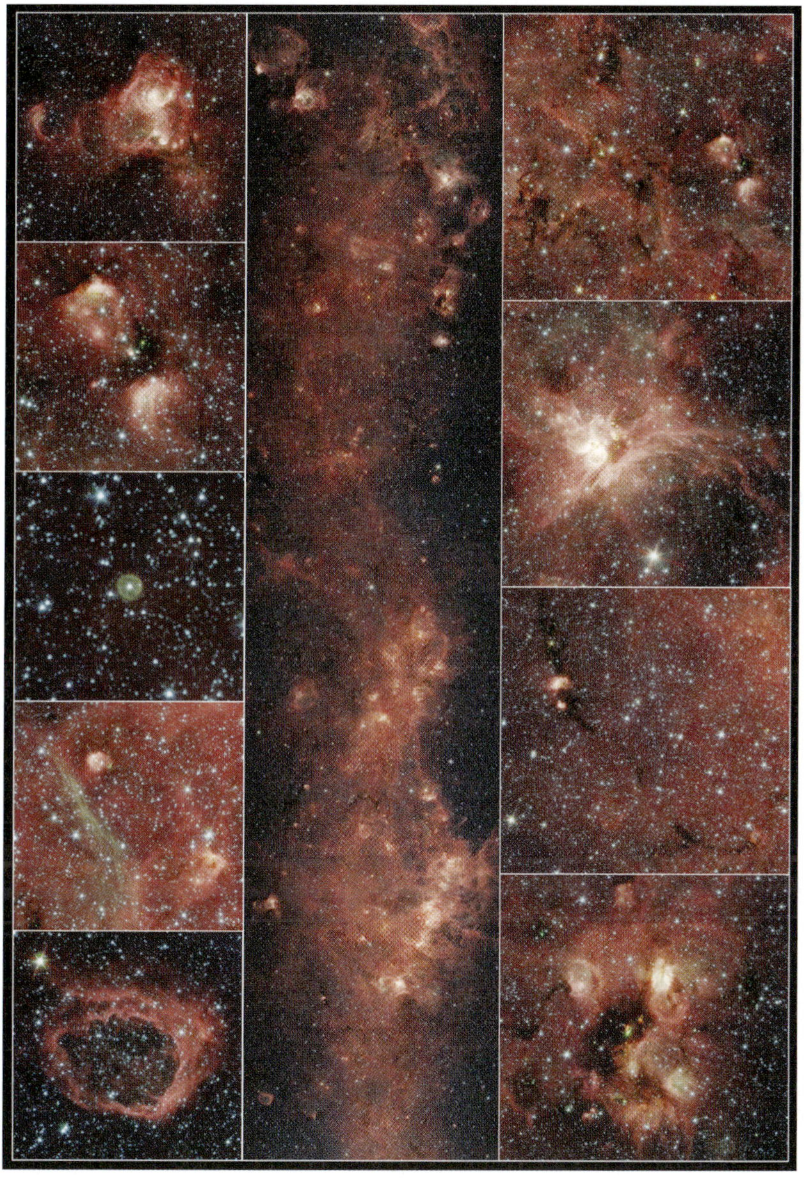

**Fig. 7.66** Milky Way in the direction of the Norma constellation in the infrared range. The central strip: a 9″-long and 2″-wide part of the Milky Way. Top and bottom rows: Enlarged fragments of the central image that show typical objects and structures of the Milky Way – star formation regions, dense and diffuse interstellar clouds, and planetary nebulae. The photos were made by the NASA Spitzer Space Telescope in the 3.6–8.0 $\mu m$ infrared range in 2005 [144].

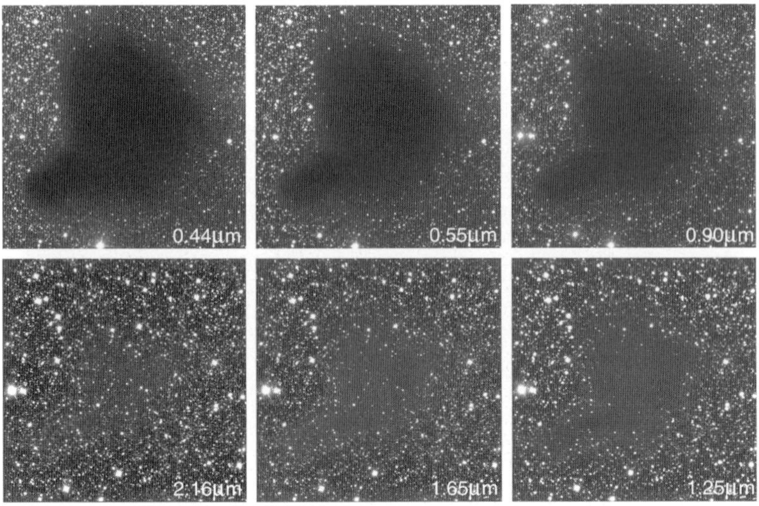


**Fig. 7.67** Gas-dust cloud B 68 measuring 0.2 pc in the Ophiuchus constellation located 150 pc away from the Earth. No stars are seen behind it in the visible range (the first photo). However, infrared rays experience weaker absorption by the interstellar dust, and the cloud becomes progressively more transparent with increasing wavelength (indicated in micrometers). Photo from the ESO VLT Telescope [206, 54].

formation, and the emergence of molecules in turn is an indication that the medium has cooled down, become denser, and is ready for star formation [204]. A wide spectrum of molecules has been discovered, up to structures containing 13 atoms (cyano-decapeptide), which are produced in gas-phase ion reactions and on the surface of dust particles possessing catalytic properties. Surdin [204] showed that the simplest molecules are formed in $10^4$–$10^5$ years, while it takes tens of millions of years to form a 10–13 atom molecule.

Dust particles ranging from 0.01 $\mu$m to 0.2–0.3 $\mu$m in size are present in the interstellar medium and participate in the catalytic synthesis of molecules, entrap stellar radiation and transfer its momentum to the interstellar gas, thereby participating in the radiation balance, and are, of course, an important factor in star and planet formation. Although the dust mass is only 0.03% of the total mass of the galaxy, its luminosity amounts to 30% of stellar luminosity and completely determines the IR radiation of the galaxy (see Fig. 7.1). The radiation-induced expulsion of dust from the atmospheres of red giants as well as nova and supernova explosions are the main dust generators in space [(3–4) $\times 10^{-3} M_\odot$ per year].

The equilibrium in giant molecular clouds is maintained by gravitational forces and the dynamic pressure of large-scale matter streams, such as collimated jets up to $10^{37}$ erg/s in power, shock waves, collisions with other clouds, and star formation [204]. The last-named process eventually ruins the molecular clouds, which last up to $10^7$–$10^8$ years on the average. On the other hand, massive interstellar clouds

**Fig. 7.68** Cone Nebula (NGC 2264) [143].

themselves may exert a destructive effect on stellar clusters (for more details, see [204]).

Several models are discussed in the analysis of interstellar cloud formation: agglomeration in accidental collisions, the Parker (or magnetic Rayleigh–Taylor) instability, the gravitational (Jeans) instability, and gas densification by the expanding gas shells around star formation domains.

The amount of interstellar gas in galaxies depends on many factors [227] – star formation, gas shedding by post-evolution stars (Figs. 7.63 and 7.70), accretion onto a galaxy, etc. Atomic and molecular gas accounts for the bulk of gas mass in disk galaxies (Fig. 7.71 [220]). About 97% of the mass is accounted for by hydrogen (70–75%) and helium (20–25%). The interstellar gas medium is nonuniform (Fig. 7.71 [220]) on many scales due to density waves, shock waves, solitons, and hot or, conversely, cold giant clouds. The hot, or coronal, gas has a temperature of $\approx 10^5$–$10^6$ K and dimensions of hundreds of parsecs. It owes its origin to the activity of young stars: stellar wind, star formation, supernova bursts, etc. The hot gas in giant galaxies results from the ejection of the gas shell by red giants in the course of their evolution or collision (for instance, with a white dwarf, Fig. 7.72 [194]). Since a substantial energy fraction of a supernova explosion is transferred to the ambient medium, there emerge expanding regions of hot gas (Fig. 7.73 [228]) and shock waves propagating

**Fig. 7.69** Omega/Swan Nebula (M17) [143].

**Fig. 7.70** Image of the result of the Tycho (SN1572) supernova explosion [47].

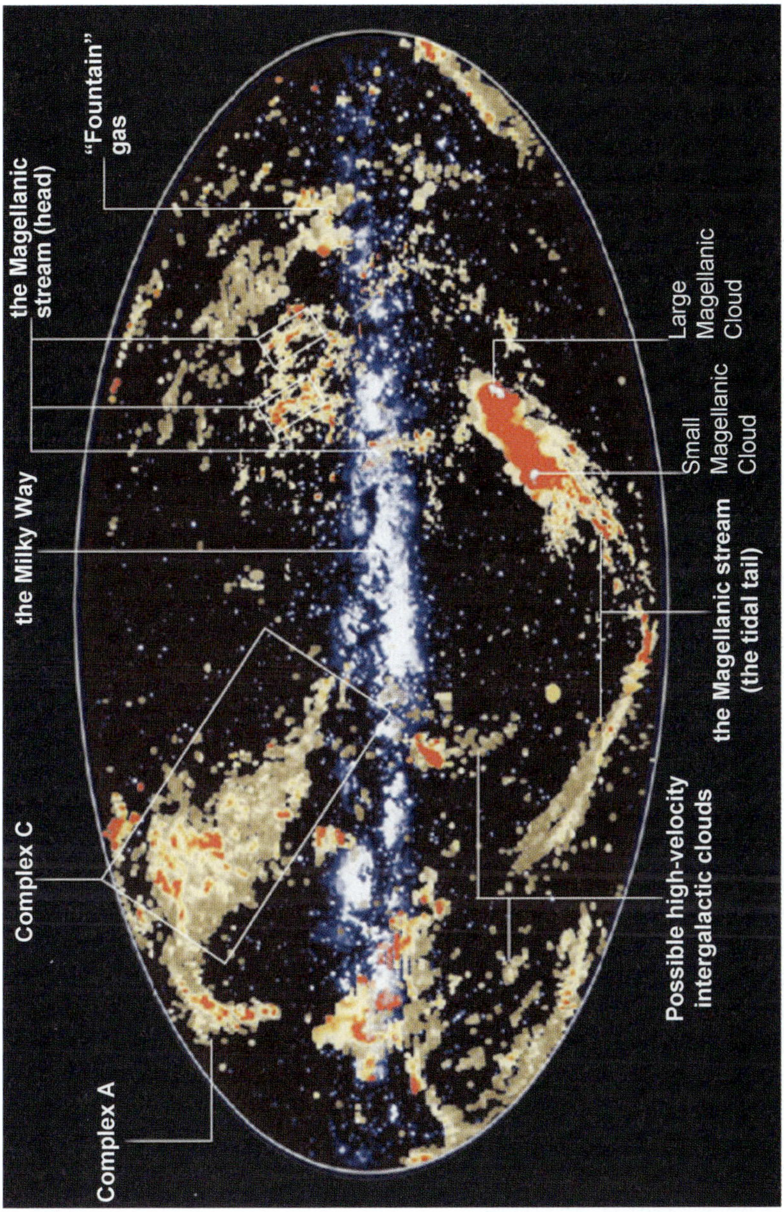


**Fig. 7.71** Galactic gas distribution constructed from the data on neutral hydrogen distribution (colored spots) and superimposed on the image of the galaxy in visible light (white color). The galactic disk is viewed from the side; the galactic nucleus is at the center. High-velocity hydrogen clouds (in particular, the complexes A and C) can be seen above and below the disk. Reprinted, with permission, from [220].

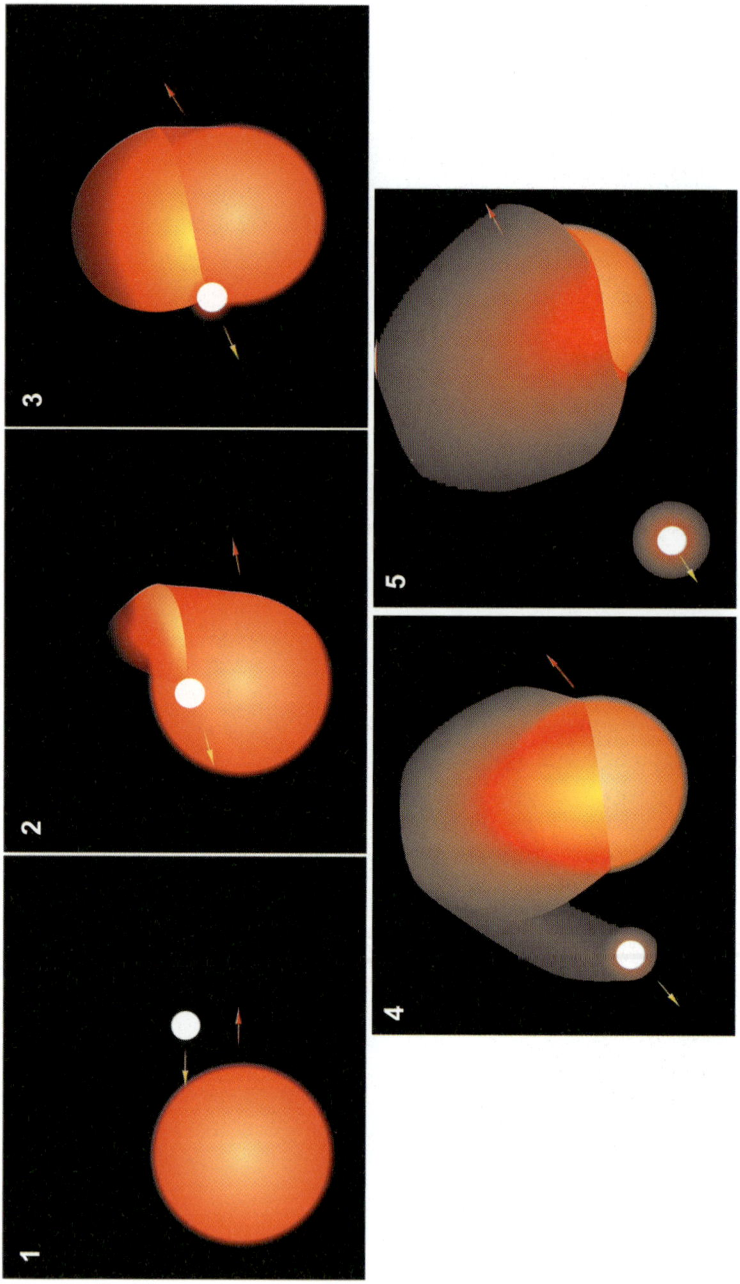


**Fig. 7.72** Collision of a white dwarf with an inflated red giant [194]. The dwarf remains intact on collision and entrains a part of the gas of the giant. The giant itself disintegrates, although its nucleus remains intact, and transforms into another white dwarf.

**Fig. 7.73** Galaxies M81 (bottom) and M82 (top) viewed in the ultraviolet spectral range. The explosive nature of star formation in M82 is demonstrated by the ejection of gas heated by the stars of the central part of the galaxy [228].

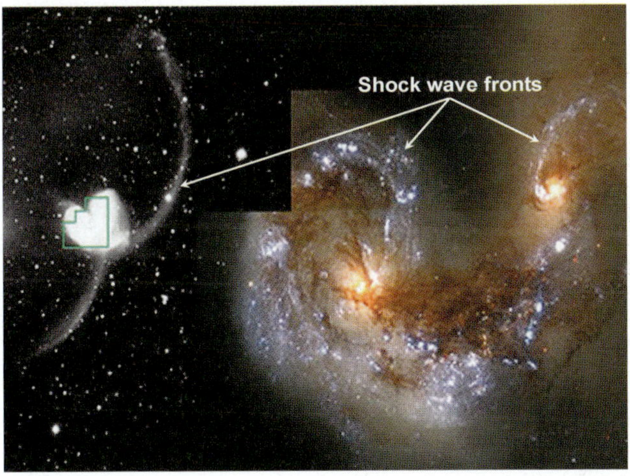


**Fig. 7.74** The interacting Antennae Galaxies (NGC4038/9). At the right: A photo from the Hubble Space Telescope shows active star formation in the central regions of the merging galaxies [228], including shock wave fronts.

**Fig. 7.75** The Red Rectangle Nebula, which surrounds the star HD 44179. One can see the ejected gas jets [17].

**Fig. 7.76** Newborn stellar cluster (region N90) surrounded by the residual gas of which it was formed [206].

**Fig. 7.77** At the top right of the photo is the central star of the planetary nebula NGC 6302 – one of the hottest – surrounded by a dense, dusty, and carbon-saturated torus [17].

through the interstellar gas. They foster star formation by compressing dense gas regions, speeding up their cooling and further compression, thereby giving impetus to star formation (Fig. 7.74). The same role is played by the shock waves generated by strong tidal interactions in compact galactic systems. That is why cool and dense molecular clouds illuminated by bright and young nearby massive stars fulfill the role of a "stellar incubator" in the universe (Figs. 7.68, 7.69, 7.75, 7.76, 7.77).

The high-intensity stellar radiation incident on the clouds produces ablation pressure, which is supplemented by the pressure of the stellar wind. As a result, the cloud is compressed by shock waves and set in motion away from the radiating object. The Cone (Fig. 7.68), Omega (Fig. 7.69), Rosette, and NGC 3603 nebulae are good cases in point. The interest in dense molecular clouds located near bright stars arises from the hypothesis that they are the birthplace of new stars [227]. Great interest is aroused by the shape of extraordinary columns in nebulae (Fig. 7.68), which might have resulted from Rayleigh–Taylor type [56] instabilities of the ablation front. An alternative explanation relies on the "cometary" model, whereby the

**Fig. 7.78** Photo of the galaxy NGC 3079 obtained from the Hubble Space Telescope, which demonstrates the high activity of the nucleus. Hot gas ejections into intergalactic space provide the material for the production of young stars [214].

matter under irradiation is carried out from the dense regions to the clouds, as is the case in cometary tails.

Observations made with the Hubble Space Telescope [143] and the ground-based BIMA millimeter interferometer yielded valuable information about the velocity distribution and the matter dynamics inside the clouds, making it possible to verify many hypotheses for the radiative matter dynamics, the role of the cloud matter in the formation of stars, etc. (Figs. 7.78 and 7.79)

Not only are dust particles present in the interstellar medium and molecular clouds, but they can also form dust clouds of their own, which are detected by excess infrared radiation. Today more than 100 young planets are known with gas-dust disks that contain the matter of a planetary system under formation [204]. An analysis of the IR spectra of the dust disks suggests that the particle dimensions are rather large – from 0.5 $\mu$m to several centimeters.

The jet dynamics of dust clouds and dust disks is affected by the fact that the dust particles may be charged (owing to photoionization) and form a so-called dust plasma [69], which possesses unusual properties in many respects and, as evidenced by experiments, increased viscosity (for more details, see [69, 70]).

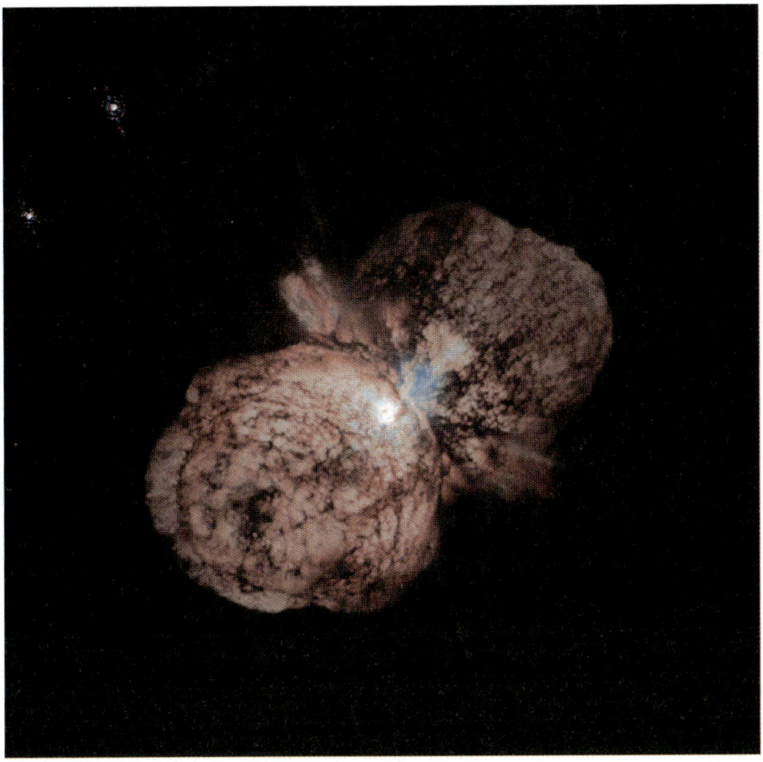

**Fig. 7.79** Huge gas and dust clouds from the supermassive star Eta Carinae are the material for new stellar objects [214, 143].

To describe the initial phase of spontaneous star formation, different models [204] are taken advantage of: the gravitational (Jeans) model generalized to the case of general relativity theory by Lifshits [120], the magnetohydrodynamic model, thermal instability, as well as the model of shock-wave compression of interstellar matter. Along with spontaneous star formation, also considered is induced formation stimulated by the explosion in the birth of a neighboring star, which gives rise to compression waves and a "chain" reaction of star formation [120]. In support of this stimulated mechanism we mention the domain around the emission nebula Omega (M17). On one side of this nebula there is a gas-free space with a group of young stars (in the emission nebula itself there are even younger stars), and on the other side there is a molecular cloud with indications of star formation.

Imshennik and coworkers [97] also provide other examples of stimulated star formation waves traveling along one molecular cloud on a $\approx 100$ pc scale. Conceivably, in this case an analogy with the propagation of combustion and detonation waves through chemically active media is appropriate. The more so as Öpik [154] came up with the idea that the shock wave from a supernova explosion triggers

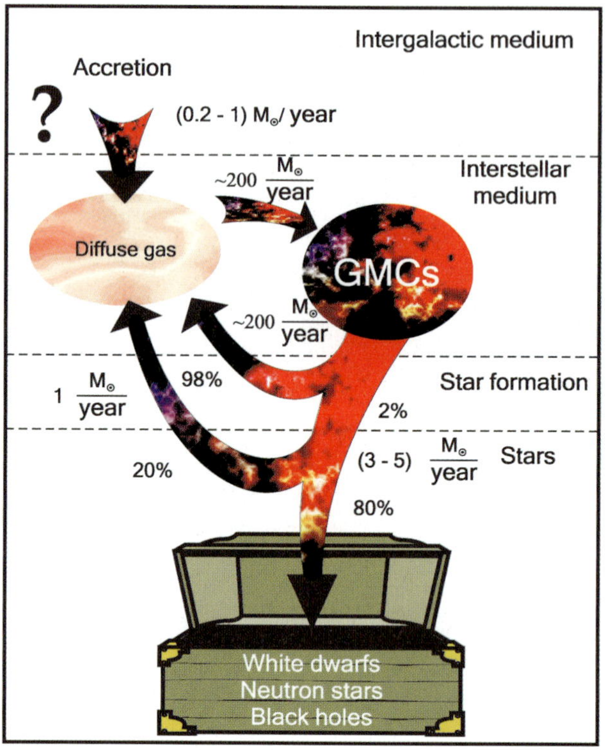


**Fig. 7.80** Matter exchange between the stars and the intergalactic medium [204].

star formation. The mechanisms of interstellar cloud compression by external radiation pressure and the idea of the instability of the gas shells compressed by the stellar wind stream were considered as the mechanism for the triggering of the self-propagating or epidemic wave of star formation. At the same time, the birth of a star and a stellar system in the parent cloud frequently leads to its disintegration and may lend impetus to the formation of new interstellar clouds.

The exchange of matter between stars and the interstellar medium is schematized in Fig. 7.80 [204]. The interstellar medium is constantly replenished with the material of stars in their explosion, while the stars themselves are produced from the interstellar medium (Fig. 7.81 [180], Fig. 7.82 [89]). In this case, from this turnover there steadily drop out the indestructible remnants of evolution: white dwarfs, neutron stars, black holes, and planets. By and large the stars of mass $1M_\odot$ return 40% of their matter to the interstellar medium and a star of mass $9M_\odot$ more than 90% [204]. The total stellar mass flux into the interstellar medium is $\sim 1M_\odot$per year, while the gas inside the interstellar medium passes from the diffuse medium to molecular clouds and back with a rate of $10^2M_\odot$ per year, the expenditure of material for star formation amounting to only $\approx 1$–3% of the latter figure.

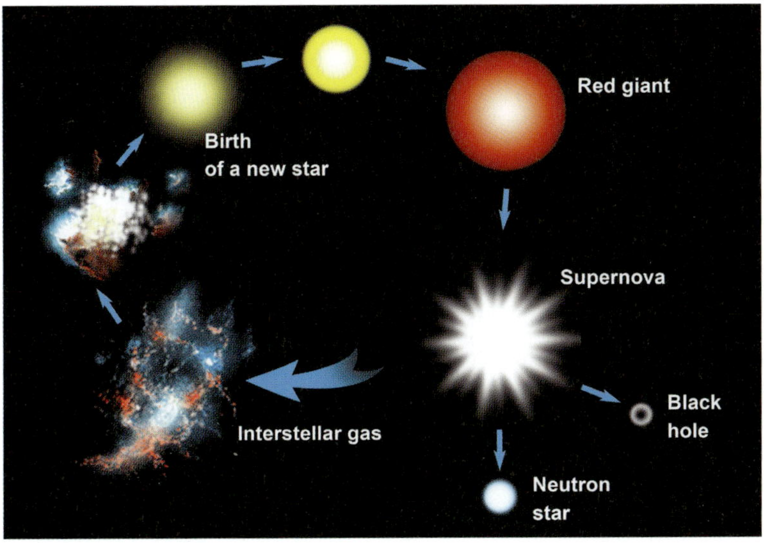


**Fig. 7.81** Matter turnover in galaxies. A typical young star comes into being in a gas cloud, is in a quasi-stationary state for a long time, as our Sun is now, and passes into the stage of a red giant upon exhaustion of hydrogen. A red giant explodes to deliver heavy elements into ambient space and replenish the gas clouds with the nuclei of heavy chemical elements. Reprinted, with permission, from [180].

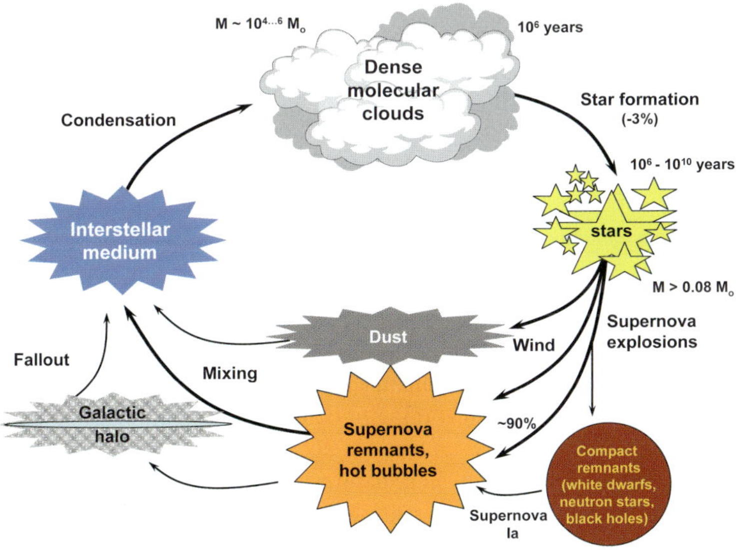


**Fig. 7.82** Dynamics of the formation of cosmic structures [89].

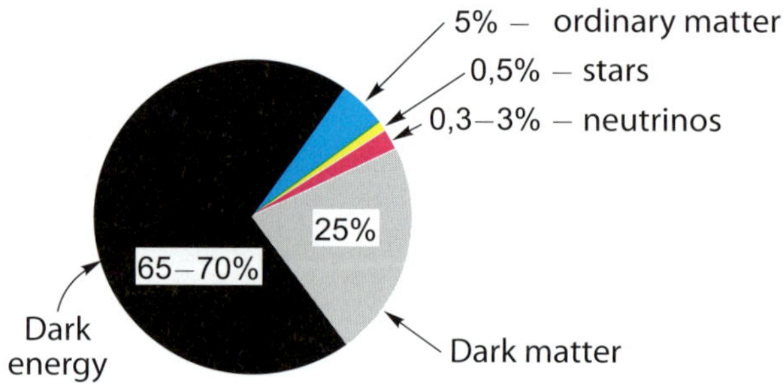


**Fig. 7.83** Energy balance in the modern universe [178].

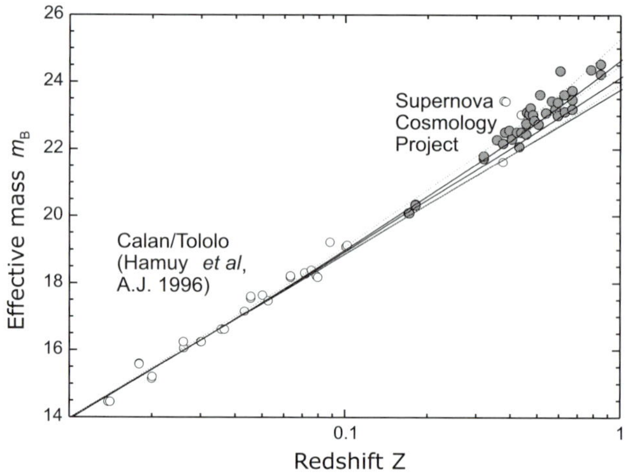


**Fig. 7.84** Stellar magnitude at the peak of brightness for type-Ia supernovae as a function of redshift $Z$ (data from [161]). The upward deviation, towards lower brightness of the farther supernovae, from the straight line (the linear Hubble law) is indicative of an accelerated expansion of the universe.

While on the subject of matter transformation in the universe, we cannot help mentioning the problem of "dark" matter and "dark" energy, which has intrigued researchers since 1998–1999. According to contemporary notions [178], about 70–75% of the mass–energy in the universe is accounted for by the so-called "dark" energy, Fig. 7.83 and Table 7.6, which manifests itself only via gravitational effects and is not directly related to known particles. It might be represented by Einstein's cosmological constant [42]. The main indication of the existence of this kind of energy is the acceleration of the universe's expansion, which follows from observations showing that distant supernovae turn out to be systematically weaker than would be expected from the linear Hubble law (Fig. 7.84). Consequently, in the past it expanded more slowly than today.

**Table 7.6** Composition of the universe [194].

| **Substance** | **Typical particles** | **Typical particle mass (energy) [eV]** | **Number of particles in the visually graspable universe** | **Probable contribution to the total mass of the universe [%]** | **Evidence for its existence** |
|---|---|---|---|---|---|
| Ordinary ("baryonic") matter | Protons, electrons | $10^6 \ldots 10^9$ | $10^{78}$ | 5 | Direct observations, estimate of element abundance |
| Radiation | Photons of the microwave cosmic background | $10^{-4}$ | $10^{87}$ | 0.005 | Observations with radio telescopes |
| Hot dark matter | Neutrinos | $\leq 1$ | $10^{87}$ | 0.3 | Neutrino measurements, models of galaxy formation |
| Cold dark matter | Supersymmetric particles | $10^{11}$ | $10^{77}$ | 25 | Estimate from galaxy dynamics |
| Dark matter | "Scalar" particles | $10^{-30}$ (assuming that dark energy comprises particles) | $10^{118}$ | 70 | Acceleration of universe expansion from the data of supernova observations |

This effect is associated [42] with antigravity – a new physical effect discovered at distances of 5–8 billion light years, at the edge of the expanding universe. Antigravity shows up as the cosmic repulsion of distant galaxies, which exceeds gravitational attraction. This antigravity is produced not by galaxies or other bodies but by the previously unknown form of energy–mass referred to as "dark" energy, which accounts for 70–80% of the entire energy–mass of the observable universe. This "dark" energy manifests itself only due to its property of producing antigravity. Otherwise it is invisible and imperceptible: it does not emit, absorb, or scatter light. It possesses negative pressure and its density is constant; its pressure–density ratio $w = P/\varepsilon_\Lambda = -0.97 \pm 0.09$. The value of the "dark" energy density is many orders of magnitude smaller than the magnitudes arising from dimensionality considerations for the fundamental interactions, i.e., strong, electroweak, and gravitational [179]:

$$\varepsilon_\Lambda \sim 10^{-46}\varepsilon_{\rm QCD} \sim 10^{-54}\varepsilon_{\rm EW} \sim 10^{-123}\varepsilon_{\rm grav},$$

while the energy scale itself ($M_\Lambda = \varepsilon_\Lambda^{1/4} \approx 10^{-3}$ eV) is much smaller than the energy scales of the known interactions $\Lambda_{\rm QCD} \approx 200$ MeV, $M_{\rm W} \approx 80$ GeV, and $M_{\rm Pl} \approx 10^{19}$ GeV (Pl stands for Planck).

Unlike Newtonian gravity, the force of antigravity increases linearly with distance [42], which has serious implications for the dynamics of the universe's motion after the Big Bang [42]. It turns out that zero acceleration occurred when the universe was 7–8 billion years old. Since that time the universe has begun to expand by an exponential law. The value of the cosmological constant is close to the critical density at that time and is equal to $\approx 10^{-56}$ cm$^{-2}$. Since the age of the modern universe is about 14 billion years, it turns out that the history of the universe is divided into two nearly equal parts: during its first part there prevailed the gravitation of dark matter, baryons, and radiation, while the second part was dominated by the antigravity of dark energy.

Another manifestation of dark energy is afforded by the WMAP spacecraft, which observed nonuniformities of the temperature distribution of the relic radiation background – "imprints" of the primary pregalactic structure of the universe, whose development later led to the emergence of galaxies. The evolution of these variations and their observable quantitative characteristics depend on the physical parameters of the universe as a whole and, in particular, on its geometry.

The most significant (and supposedly the most reliable) finding of WMAP is that the total world's energy density $\rho_0$ is close to the critical density $\rho_c = (3/8\pi G)H_0^2 = (1 \pm 0.1) \times 10^{-29}$ g/cm$^3$, where $H_0 = 72 \pm 0.4$ km/(s·Mpc) is the Hubble constant. In this case, the strict equality of these densities is not ruled out. These data are indicative of the high density of the universe, which exceeds the density of ordinary and dark matter.

According to the WMAP data, the matter density in the universe is equal to the critical value with a high degree of accuracy. This signifies that dark energy should account for 70–80% of the total density. The point is that the contribution of other energies (baryons, dark matter, and radiation) is severely bounded above to about 30–20% of the critical density, which is known in advance from a set of other independent cosmological requirements. Today it is agreed that 26% of matter is due to hidden mass and only 4% due to visible matter.

Such overestimation of the mass of the universe follows from nonuniformity of the distribution of galaxies and the analysis of mass evolution in accumulations of galaxies.

The dark energy density was estimated even in the first experiments and turned out to be quite low: $\approx 0.7 \times 10^{-29}$ g/cm$^3$ [42]. In a cubic meter of space, this dark energy mass is equal to the mass of three hydrogen atoms. For a medium with such a density, the force of antigravity is also vanishingly small: it is balanced by the gravitational attraction of two hydrogen atoms spaced 50 cm apart [42].

Also associated with dark matter are the hypothetical Weakly Interacting Massive Particles (WIMPs), which emerged at the early stages of formation of the universe. This class comprises neutrinos, neutralinos, photinos, gravitinos, axions, etc. It is significant that the fraction of the average density of the baryon matter component relative to the average density of the nonbaryon one (WIMPs) amounts to only $\approx$ 0.07. At the same time, the fraction of the average density of visible matter relative to the total average density of the universe amounts to only $\approx$0.003–0.007.

Among this list of candidates for "dark" mass carriers are more or less preferable objects; however, the final answer to the question as to what hidden mass consists of has to be provided by observations, probably the LHC-based ones.

An additional point in favor of the existence of "dark" energy is the solution of the so-called Hubble–Sandage paradox [42]. This consists in the following: for a strong nonuniformity and obvious randomness of matter distribution at a distance of up to 20 Mpc, there exists a regular matter expansion flow with direct proportionality between velocity and distance, which is valid for a uniform distribution. Indeed [42], in the range from several to several tens of megaparsecs cosmic matter is distributed quite nonuniformly. It conglomerated in separate bunches – galaxies, galactic groups and clusters, which are randomly strewn over space. Under these conditions the Friedman model is inapplicable: the assumption of matter distribution uniformity is not fulfilled.

However, detailed observations showed [187] that a regular expansion flow obeying the Hubble law $v = H_s R$ is traced with confidence in a broad spatial range from 4 to 200 Mpc. It turns out that the regular galactic kinematics is in no way impeded by the strong nonuniformity of their distribution inside a uniformity cell. In this case, the Hubble constant $H_s$ measured inside the uniformity cell is close to the value of the global quantity $H_0$ furnished by the WMAP observations.

It seems that the existence of dark energy resolves this paradox. The dark energy has a perfectly uniform density and dominates everywhere beyond heavy crowdings of matter, making almost the whole universe nearly uniform. The contradiction between the regularity of galactic motion and the irregularity of their spatial distribution inside a uniformity cell is thereby removed: in the presence of the dominant dark energy background, the total mass/energy distribution also turns out to be nearly regular [42].

Lukash and Rubakov [123] consider the measurement of peculiar galactic velocities in clusters and superclusters, gravitational cluster lensing, the measurement of galactic rotation curves, the determination of luminosity–mass relationships, the temperature measurement of X-ray clusters, etc. as arguments for existence of "dark" energy.

The theoretical models of "dark" energy effects are analyzed by Rubakov and Tinyakov [179]: the cause of accelerated expansion might be associated with new low-energy physics (the quintessence model [41]).

Another model of gravity modification at long range proceeds from the brane-world theory and the idea of high and infinite additional dimensions [176]. In such theories, ordinary matter is localized on a three-dimensional hypersurface (brane) embedded in a higher-dimensional space. Rubakov's idea [176] is that gravitons may propagate over "our" brane for a finite (though long) time to subsequently go to additional dimensions.

The remaining 30% of the mass–energy of the universe exists in the form of matter – particles possessing mass. The most common of them are protons, neutrons, and electrons, which account for one sixth of matter, i.e., for 4–5% of the entire universe. The major part of this mass is due to the energy of motion of quarks and gluons, which are parts of protons and neutrons. A smaller amount (0.3–1.0%) of the universe's mass is accounted for by three neutrino families. Yet another part – the fourth, cosmic energy, component – is "radiation", which is taken to mean relic photons (as well as perhaps gravitons); radiation accounts for no more than several hundredths of one percent of the total density. The remaining 20–25% of the universe's mass–energy is constituted by "dark" matter invisible to us, whose existence is evidenced by its gravitational effect on observed cosmic objects. It is thought to consist of massive particles of mass 10 GeV–1 TeV, which are not among the particles of the Standard model, for it makes up galaxy-sized clusters. Their lifetime should be comparable to the age of the modern universe of $\approx$14 billion years.

The particles of "dark" matter possess ordinary properties relative to gravitational interactions: they are capable of assembling in clusters, forming gravitational fields, etc., which is of consequence for the formation of structures in the universe – galaxies and their clusters. Studying these structures and the polarization of microwave relic radiation suggests that the "dark" matter particles were nonrelativistic even at the very early stages of the evolution of the universe. At the same time, the "dark" matter particles are void of electric charge and interact only weakly with matter, which hinders their detection. One of the early observable manifestations of "dark" matter is presented in Fig. 7.85.

According to Newtonian mechanics, the rotational velocity of gravitationally bound masses should be inversely proportional to the square root of the distance to the center of rotation. Data measurements presented in Fig. 7.85 [182, 181] show, however, that the remote galactic parts rotate with the same velocity as the parts nearest to the center and in doing so do not fly away due to the centrifugal force. This is an indication that the galactic mass increases linearly with radius, so that the greater part of the gravitating though invisible mass is located far beyond the limits of the visible galactic image [55].

A similar example of anomalous rotation of our galaxy is given in Fig. 7.86 [52], which shows the velocity of rotation as a function of distance to the galactic

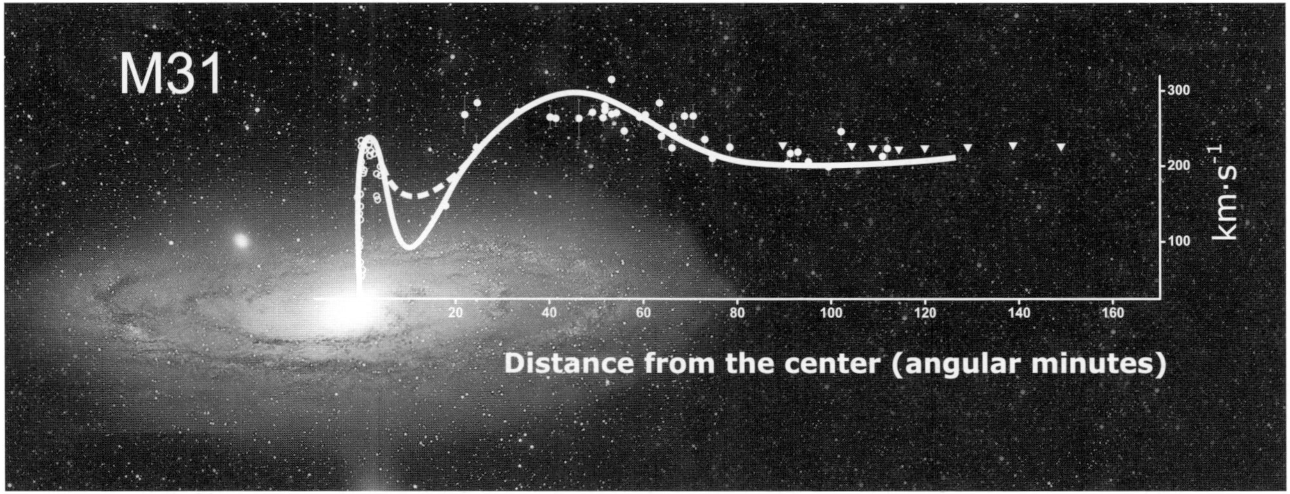


**Fig. 7.85** Photo of the Andromeda galaxy (M31) with radial velocity distribution superimposed on it. The data were obtained by light (circles) and radio measurements [181, 182]. The velocity of rotation remains invariable beyond the visible image, which is an indication that the mass of M31 increases linearly with radius.

center. In the very inner domains, the character of rotation is close to the solid-state one. Then, the velocity decreases slightly, but farther from the center, up to the most distant objects with measurable velocities, the velocity of rotation remains constant or slowly increases with distance. By the 1970s, this had been borne out by neutral hydrogen observations in our galaxy and in the majority of other galaxies. The almost universally accepted explanation of this strangeness is that, in addition to the visible halo of objects, the galaxies are surrounded by the far more extensive halo of gravitating and yet unobserved matter.

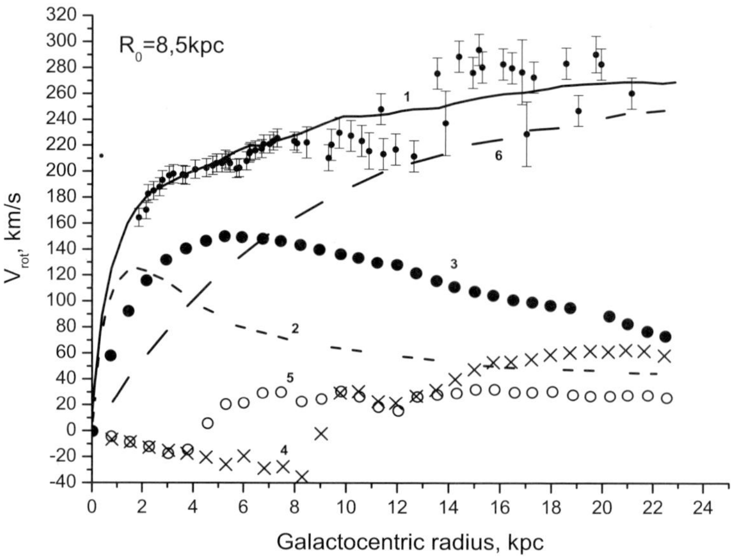


**Fig. 7.86** Summary curve of galaxy rotation (1) and its constituent rotation curves for individual galaxy constituents: the badge (2), the stellar disk (3), the neutron hydrogen layer (4), the molecular hydrogen layer (5), and the halo of unobservable dark matter (6), whose contribution determines the rotation curve at long distances from the center. From [153].

A great body of other observational data of different nature have been accumulated to date, which testify to the existence of "dark" mass in the universe [225]. Apart from cosmological data, measurements of the gravitational fields in galaxies or their clusters are also indicative of huge amounts of "dark" matter.

Figure 7.87 shows the gravitational "probing" data, when the gravitational field of a cluster curves the light rays emanating from the galaxy located behind the cluster. The gravitational field therefore acts as an optical lens. Several images of this remote galaxy are sometimes observed in this case. In the left half of Fig. 7.87 they are in blue. The curvature of light rays depends on the mass distribution in the cluster. When the lensing effect is taken into account, there results the image of the radiating galaxy shown in blue in the right half of Fig. 7.87. One can see that it is very different from the luminous ("visible") matter. The masses of galactic clusters determined in this way suggest that the "dark" mass accounts for about 25% of the total energy density in the universe. This corresponds to the estimates that follow from a comparison between observations and the theories of the formation of galaxies and clusters of galaxies.

**Fig. 7.87** Gravitational lensing of a remote galaxy. Photographs in the direction of the galactic cluster 0024+1654 obtained with the Hubble Space Telescope [143].

These are estimates of the virial or dynamic masses of galactic clusters, which are several times greater than the visible cluster masses obtained by direct summation of observable galactic masses that are estimated from the mass–brightness dependence. Galaxies in the clusters move much faster ($V \gg 1000$ km/s) than follows from the estimate of the visible matter mass.

Figure 7.88 [180] shows a photo of the UGC 10214 galaxy: a stream of matter flows from this galaxy towards an invisible gravitating center, which supposedly consists of dark matter.

The presence of hidden mass is also attested by the discovery of hot ($T = (3$–$10) \times 10^7$ K, $n_e > 0.001$ cm$^{-3}$) gas in galactic clusters and other observational data. To solve several theoretical problems (the problem of the formation of large-scale structure in the universe, cosmological problems involving the explanation of the recently discovered spatial fluctuations of the relic microwave background, etc.) it is also necessary to invoke the hidden mass.

The study of molecular and dust clouds in laboratory conditions may be carried out with the use of high-power lasers and high-current devices, which model the ablation dynamics of clouds, the stability of interfaces, and the effect of radiation on the structure and development of radiative hydrodynamic instabilities. Furthermore, the generation of shock waves in such media may give rise to substantial nonuniformities and turbulent processes. Research into the generation of laser-driven shock waves and of compression waves with Z-pinches is now in

**Fig. 7.88** Tadpole galaxy (UGC 10214). According to one hypothesis, the matter "tail" was formed due to the attraction of an invisible "dark" galaxy [180].

progress [47, 78, 200, 45, 169] and may yield much new information for the understanding of molecular clouds in space. The imposition of a magnetic field on the experimental volume makes the clouds "stiffer" and substantially alters their dynamics and structure.

The interstellar medium exhibits one more activity of a specific kind by generating maser radiation at the lines of the water hydroxyl molecule (OH), silicon dioxide, and methanol ($CH_3OH$). The maser effect consists in the fact that the linewidths correspond to a temperature of only $10^2$ K for a brightness temperature of $10^{16}$ K [204]. This laser radiation is highly variable: its polarization, intensity, lineshapes, etc. sometimes change radically over a day or even several minutes, which corresponds to the masing regions of size of order 1 AU. These sources range in luminosity from 1 to $10^3$ solar luminosities. They are located in the domains of active

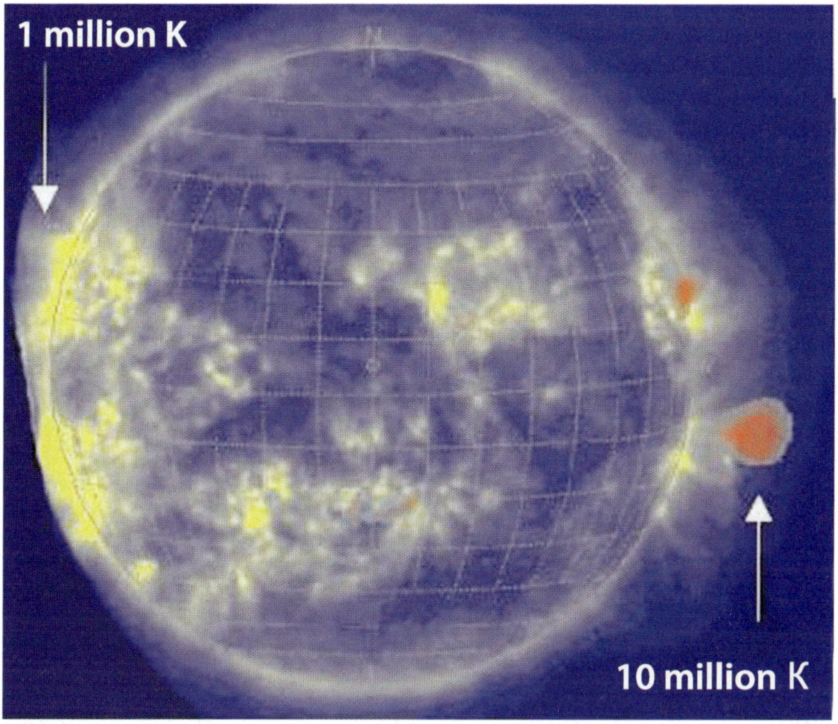


**Fig. 7.89** Solar flare; the flare temperature reaches 10 million K.

star formation, which are rich in gas and massive stars, as well as in galaxies with active nuclei. Chemical reactions, IR radiation, or supersonic gas expansion into vacuum may be the sources of population inversion of the cosmic masers, which is thought ot be realized in the atmospheres of protostars and old supergiants.

Magnetic fields exist throughout cosmic space, including the interior of the Earth and the interstellar and intergalactic media. The origin and structure of these fields are only partly understood, and this applies specifically to the fields of neutron stars and giant planets [145]. Most thoroughly elaborated is the model of terrestrial field generation (the "dynamo" model), which relies on detailed measurements and extensive magnetohydrodynamic calculations [201, 15].

Transient magnetodynamic phenomena, which determine stellar activity, present a special problem. For the Sun they are well known as the cycles of its magnetoactivity, coronal plasma ejections, sunspots, and other long-observed phenomena (Figs. 7.89 and 7.90), which result from the complex nonlinear interaction of magnetic fields and the solar plasma.

Magnetohydrodynamic phenomena on the Sun have been studied, of course, most thoroughly [9]. Continuous observations are made of magnetic instabilities and the interconnection topology of magnetic field lines on the solar surface. The

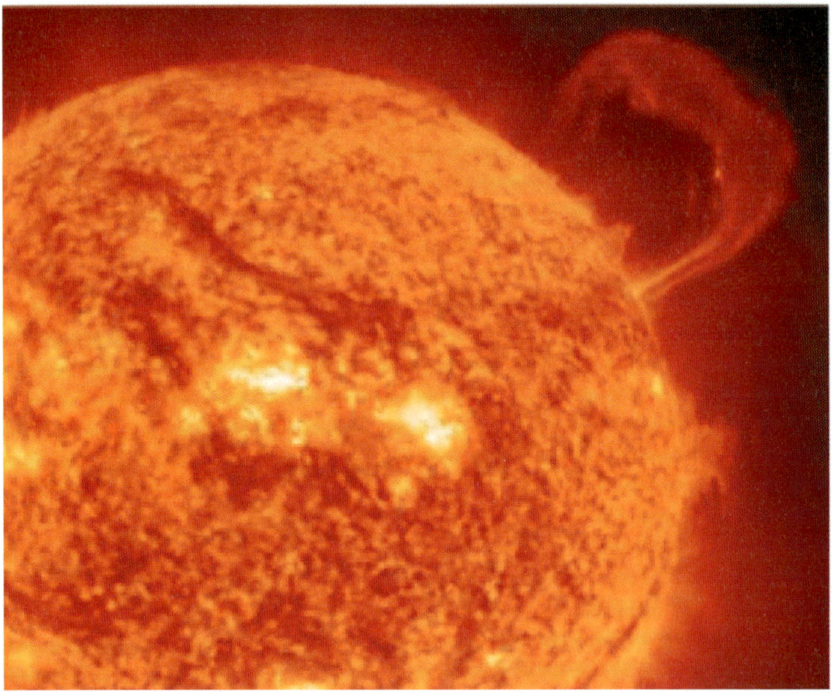

**Fig. 7.90** Plasma ejection on the Sun [9]. The plasma follows the magnetic flux lines.

dynamics of motion and compression of magnetic flux tubes is quite often transient, explosive in character, and the magnetic field strength amounts to 3000–4000 Gs in sunspots for an average magnetic induction of $\approx 1$ Gs. The temperatures in solar flares amount to $10^8$–$10^9$ K, while the plasma parameters in the solar interior are $T \approx 1.5 \times 10^7$ K, $\rho \approx 150$ g/cm$^3$, and $P \approx 0.2$ Tbar; the time of photon diffusion from the core to the solar surface is $\approx 10^5$ years. Laboratory experiments supplemented with three-dimensional mathematical simulations [149, 234] provide a basis for the explanation of complex field self-organization effects in the magnetoactive plasma.

It is significant that such effects of local energy cumulation give rise to global solar oscillations, which are recorded with a very high degree of accuracy. The characteristics of these oscillations depend on the composition and thermodynamic properties of the solar plasma, which yields a unique high-precision experimental material for verifying multicomponent plasma models and defining more precisely the elemental composition of the solar plasma [13]. The results of such a comparison are presented in Fig. 7.91, where the adiabatic exponent of the solar plasma obtained by helioseismology techniques is compared with the adiabatic exponents of several theoretical models. As in the case of neutron stars, the Sun fulfils the function of a "cosmic laboratory" for experiments in the area of high energy densities.

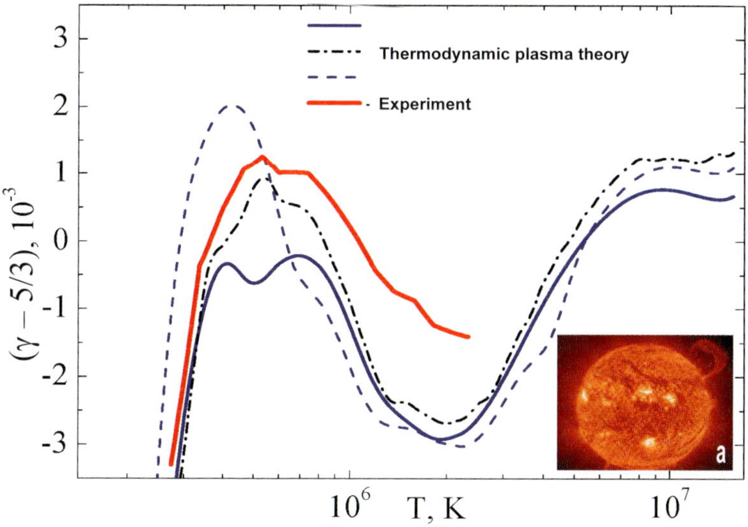


**Fig. 7.91** Solar "seismology". Comparison of experimental data on the adiabatic compressibility coefficient of the solar plasma with several theoretical plasma models [62, 71, 63, 13]. The inset (a) is a photograph of the Sun with solar prominences and bright regions of local energy release, which generate global oscillations and acoustic waves in the solar plasma.

More often than not, magnetic Reynolds numbers for astrophysical plasmas prove to be quite high, making the magnetic energy dissipation insignificant in the majority in cases [145]. But, as in aerodynamics, several important cases make an exception. These are the solar corona heating and the injection of the solar wind [149, 234], the circulation of planetary magnetospheres, and the emission from accretion disks. The rearrangement of magnetic lines of force is a special problem [149, 234]. These effects are most likely to manifest themselves in magnetospheres, double layers, and plasma jets, which are intimately related to occurrence of instabilities. In this case, kinetic processes and scale factors are as yet imperfectly understood.

The structure of magnetohydrodynamic turbulence of interplanetary and interstellar plasmas is studied from the origin of radio waves [145, 227, 190]. Measurements testify to its anisotropy. Theoretical models suggest that in the incompressible fluid limit, when passing (via a cascade) from large vortices to small ones, there occurs strengthening of the gradient of the field directed perpendicular to the background magnetic field. In this case, it is not clear if an inertial range of vortex scales forms and if the inverse passage of cascades towards large scales occurs, or if Alfv'en waves are generated, which escape from the turbulence region [145]. Laboratory and computer-assisted numerical experiments in this area would be highly interesting and informative.

## 7.4 Cosmic Rays

As far back as the beginning of the last century, it was reliably determined [142] that a stream of charged particle is incident on the Earth from space, the particle energies spanning a wide range from kiloelectronvolts to $10^{20}$ eV (Figs. 7.92 and 7.93).

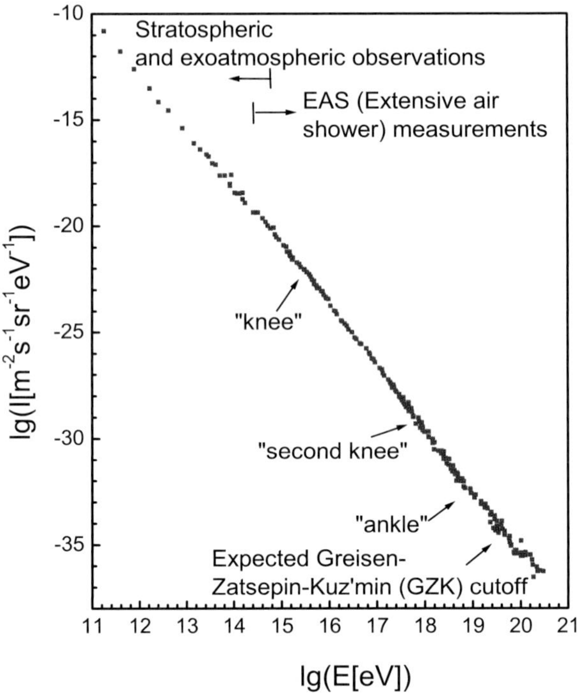


**Fig. 7.92** Observed energy spectrum of cosmic rays [227, 142].

Cosmic rays with a huge energy of $\approx 3 \times 10^{20}$ eV (which corresponds to the energy of a tennis ball, 50 J) were first recorded in 1991. Since then, tens of events with an energy of $\approx 10^{20}$ eV were detected at different facilities [142]. The events were detected by recording the fluorescence of atmospheric cascades in the atmosphere or in specifically constructed detector arrays at the Earth's surface. The spatial distribution of the incident particles was observed to be surprisingly isotropic. When protons are involved, the effect of a magnetic field on them is small, and most likely they are of extragalactic origin.

The energy of space electrons, recorded with the Fermi Gamma-Ray Space Telescope [1] and ATIC (Advanced Thin Ionization Calorimeter) [34] (Fig. 7.94), exceeds 1000 GeV. As possible sources of these particles an unknown electron source

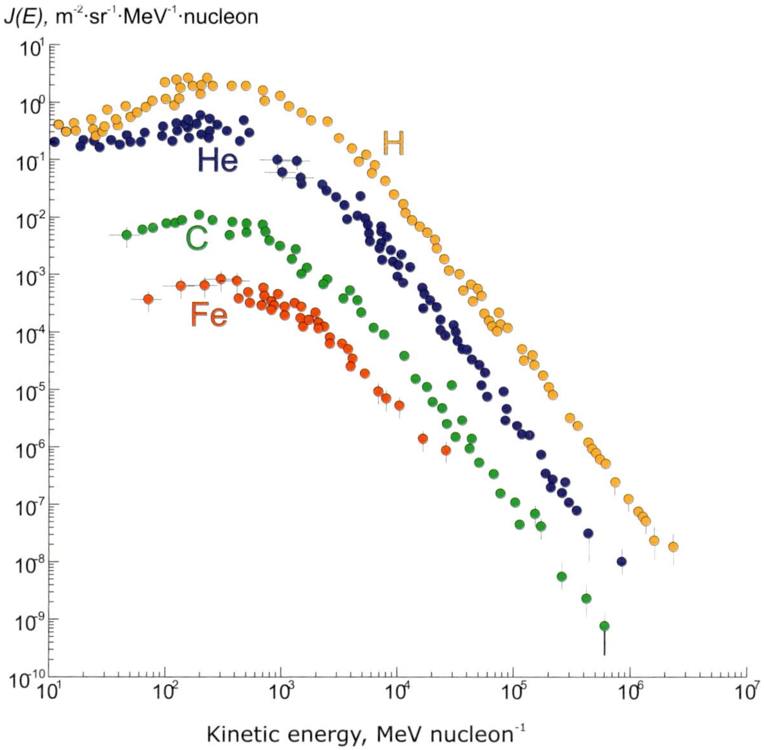


**Fig. 7.93** Differential spectra of galactic cosmic rays (CRs): protons H, helium He, carbon C, and iron Fe nuclei [202].

(perhaps a pulsar) and also a source of heavy "dark"-matter particles are being discussed.

Cosmic rays are inherently a strongly rarefied relativistic gas, whose molecules only slightly interact with each other, with a power-law energy spectrum (Fig. 7.92) rather than the Maxwellian one. At the same time, cosmic rays experience collisions with the particles of the interstellar medium and interact with the interstellar magnetic field. Although the cosmic ray flux near the Earth is low, $\approx 1$ particle/(cm$^2$·s), its energy density, $\approx 1$ eV/cm$^3$, is comparable to the density of electromagnetic radiation from all galactic stars or to the energy density of the thermal motion of the interstellar gas and the kinetic energy of the turbulent motion, as well as to the energy density of the galactic magnetic field [79]. The flux of superhigh-energy particles is extremely low, about 1 particle/km$^2$ in 100 years, but it is the origin and propagation of precisely these particles that arouses the greatest interest.

The energy limit [79, 229] arising from the ionization proton energy losses due to the scattering from relic radiation photons at an energy of $\approx 5 \times 10^{19}$ eV was revealed for protons that had traversed a distance of more than 50 Mpc. The superhigh-energy particles beyond this limit [215, 9, 149, 234, 13, 142, 208, 79, 229, 90, 209,

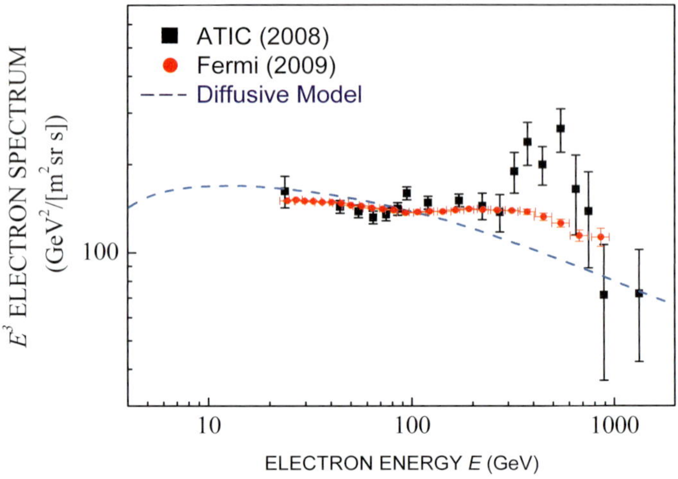


**Fig. 7.94** Observed spectrum of cosmic electrons [1, 34].

2] are being most actively studied in the search for their sources and mechanisms of acceleration [152].

According to the "proton–neutrino–proton" model [222], it is neutrinos that propagate through large distances in space and not protons, which should have ultrahigh energies at the place of production so as to eliminate attenuation arising from the scattering by relic radiation photons. This model admits the existence of superhigh-energy particle sources located at cosmological distances outside of our galaxy, which is most likely.

It is obvious that cosmic rays should be accelerated by some nonthermal mechanism, because the temperature at the center of the most massive stars does not exceed several tens of keV. And so to study these particles both the classical Fermi mechanism [59] (Fig. 7.95 [157]) and stochastic mechanisms [4, 3] are invoked, whereby the particles are accelerated in their stochastic interaction with magnetic domains or collisionless shock waves generated by supernova explosions or matter ejections from active galactic nuclei, whose turbulent zone is the site of development of diverse instabilities. However, at high energies there occur significant radiation energy losses here, even at energies below the Zatsepin–Kuz'min limit [229, 79].

Chen et al. [35] came up with the mechanism of acceleration in the field of a wake wave (see Sect. 6.3) driven by magnetohydrodynamic shock waves in the atmosphere of a gamma-ray burst. Their consideration yields high ($\approx 10^{19}$ eV), close to the Schwinger limit, wakefields in the relativistic moving plasma, which are additionally invoked to describe the gamma-ray particle spectra of these sources.

The wake acceleration mechanism may be realized in the acceleration of electrons in the jets of massive black holes [193, 77], which emit high-power X-ray radiation. It is produced by the bremsstrahlung and synchrotron mechanisms in relativistic jets from the streams of well-collimated electrons (or positrons) with an

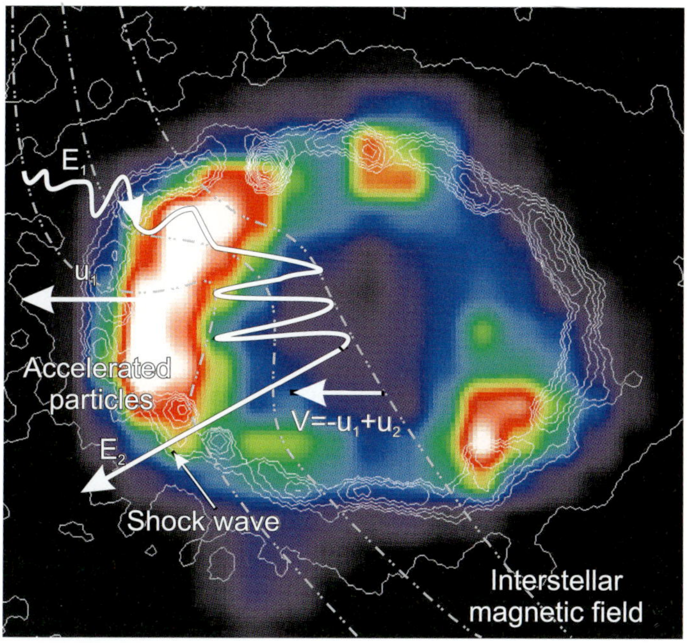


**Fig. 7.95** SN 1987 supernova explosion remnants. The picture shows the qualitative illustration of the Fermi particle acceleration mechanism of the first kind at the shock wave. The photo is from the X-ray telescope of the Chandra satellite [157].

energy of $\approx 1$ GeV. The role of Alfv'en waves in wake wave generation and their capabilities for particle acceleration are discussed in [35, 139].

The electromagnetic acceleration mechanism is related to the generation of transient electric fields, for instance, in the magnetospheres of pulsars, where the magnetic fields at the surface amount to $10^{12}$ Gs. For a rotation frequency of only $10^{-3}$ s$^{-1}$ this is sufficient to accelerate particles up to $10^{19}$ eV. However, the observed superhigh-energy particle radiation is isotropic and is evidently at variance with the location of a radiating object.

High-energy-density plasmas play a significant role in the understanding of such phenomena, and cosmic showers may be studied in plasma and accelerator laboratories. The main part is played by shock waves arising from numerous galactic sources, such as active galactic nuclei and jets, gamma-ray bursts, collisions, and galactic clusters. The generation mechanism by way of magnetic reconnection – an alternative to the shock-wave mechanism – has been elaborated in less detail.

The highest-energy particles may stem from the megnetohydrodynamic "wind", which emerges near a rapidly rotating magnetized neutron star [145, 226, 193]. These powerful magnets accelerate heavy nuclei in galaxies or their clusters with dimensions of 3 million light years. Unipolar inductors of this kind may exist in the accretion disks of massive black holes. Numerical investigations of these acceleration effects are proceeding vigorously, along with their partial physical modeling.

The problem of explaining the accelerated particle propagation through huge galactic-sized distances in the universe still remains a problem. In this case, clearly the magnitude and structure of the intergalactic fields themselves has not been adequately studied.

## 7.5 Gamma-Ray Bursts

In modern astrophysics, gamma-ray bursts are the most mysterious objects [145, 75, 227] with record energy release – excluding the Big Bang. They were accidentally discovered by the American satellites Vela 5A and B intended for monitoring nuclear explosions in the atmosphere. The bursts are detected with a frequency of more than one per day in randomly regions of the sky (Figs. 7.96, 7.97, 7.98, 7.99), and their duration ranges from milliseconds to seconds in the 0.1–100 MeV energy range. Some of them, being compact objects, are billions of light years distant and radiate an energy of $10^{51}$–$10^{53}$ erg per burst. The recorded spectra suggest that these radiators are optically thin with respect to the outgoing radiation.

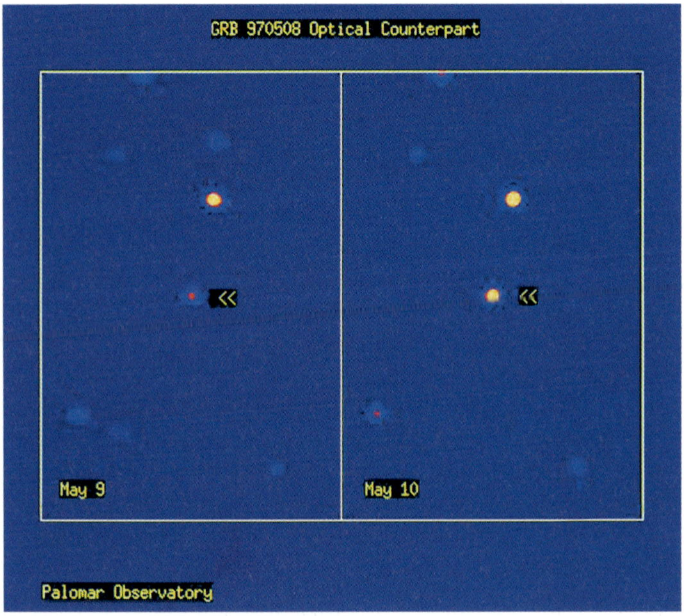


**Fig. 7.96** Gamma-ray burst GRB970508 [214]. The photograph shows the variation of the optical radiation of the GRB970508 burst when its brightness increased.

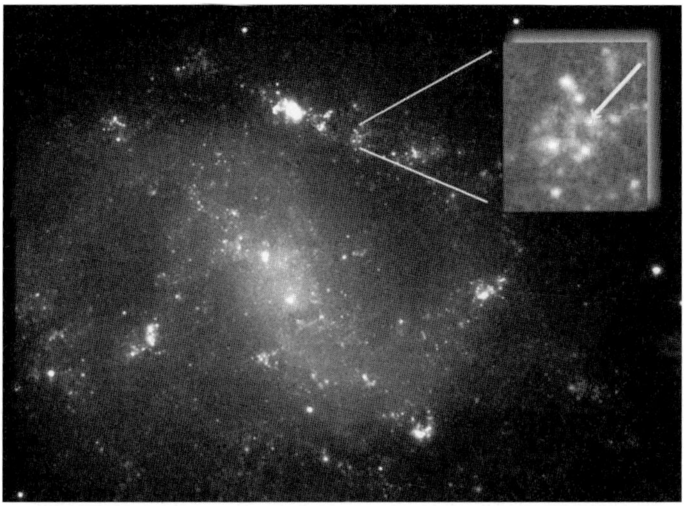

**Fig. 7.97** Explosions of stars in the galaxy ESO 184-G82. The inset shows the star formation region of size $\approx 300$ light years with a gamma-ray burst [214].

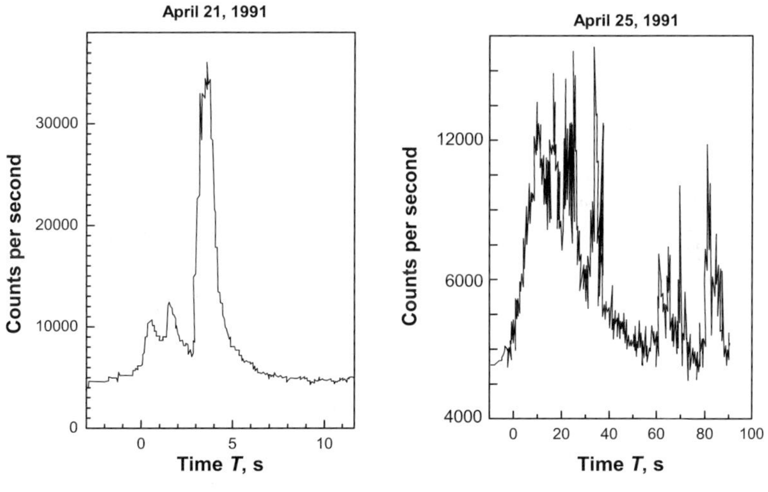


**Fig. 7.98** Temporal variation of gamma-ray burst intensity (taken from [214]).

The relativistic "fireball" model proposed for gamma-ray bursts in [132, 166] reduces to the following: as a result of the collision (coalescence) of two neutron stars [27] or the gravitational collapse of a supermassive star (Fig. 7.100) there occurs a $\approx 10^{52}$ erg energy release in a small domain filled with relativistic photons and leptons with a small admixture of baryons. This "fireball" with an initial tem-

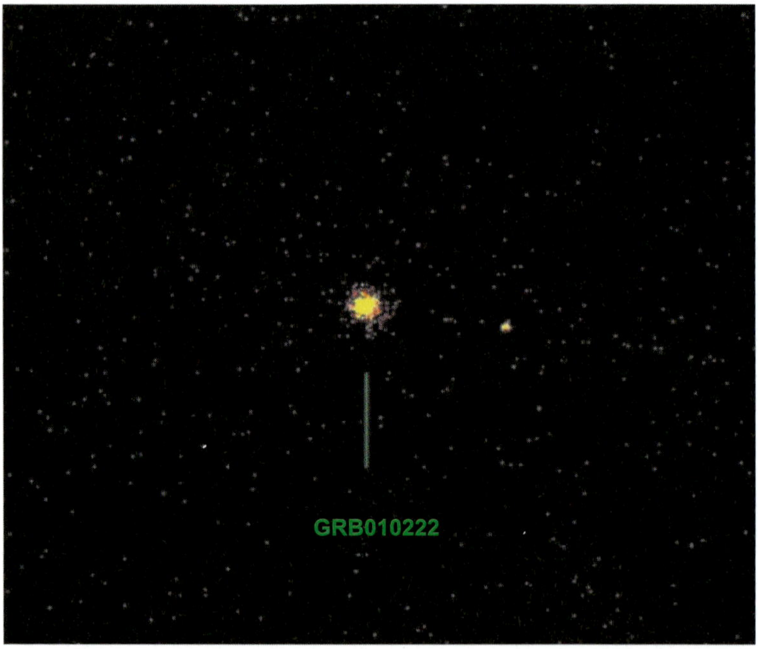


**Fig. 7.99** Gamma-ray burst GRB010222. It was observed in the gamma-ray range for only several seconds, but the explosion aftereffects in the X-ray, visible, infrared, and radio ranges were observed for several days as the expansion of a "fireball" [214].

perature of 1–10 MeV expands with relativistic velocities and generates, by way of the synchrotron or inverse Compton mechanism, gamma-ray radiation by electrons accelerated by the Fermi mechanism in optically thin shock waves. A small number of baryons are also accelerated to relativistic energies and transport part of the energy away from the "fireball", generating outside of it outward and backward shock waves, whose X-ray radiation is recorded as a gamma-ray burst (Figs. 7.99 and 7.101). A substantially longer X-ray afterglow results from diverging shock waves. The ordinary duration of this radiation is $\geq 2$ s.

It is generally agreed [227, 76, 132, 166, 8] that the predecessors of such objects are massive collapsing stars [166]. The shock waves generating gamma-ray radiation [50] emerge after the detachment of a "fireball" from the surface of a star, and the rotation of the star may transform the "fireball" into a powerful jet.

The second class of short-duration bursts with a duration $t \approx 2$ s is related to the mutual absorption of binary neutron stars (Figs. 7.102 and 7.103) or a neutron star and a black hole. Also probable in this case is the emergence of one or two jets along the axis of rotation [50]. This mutual absorption may be a source of gravitational waves; research into the detection of gravitational waves has been pursued in recent decades [82, 81].

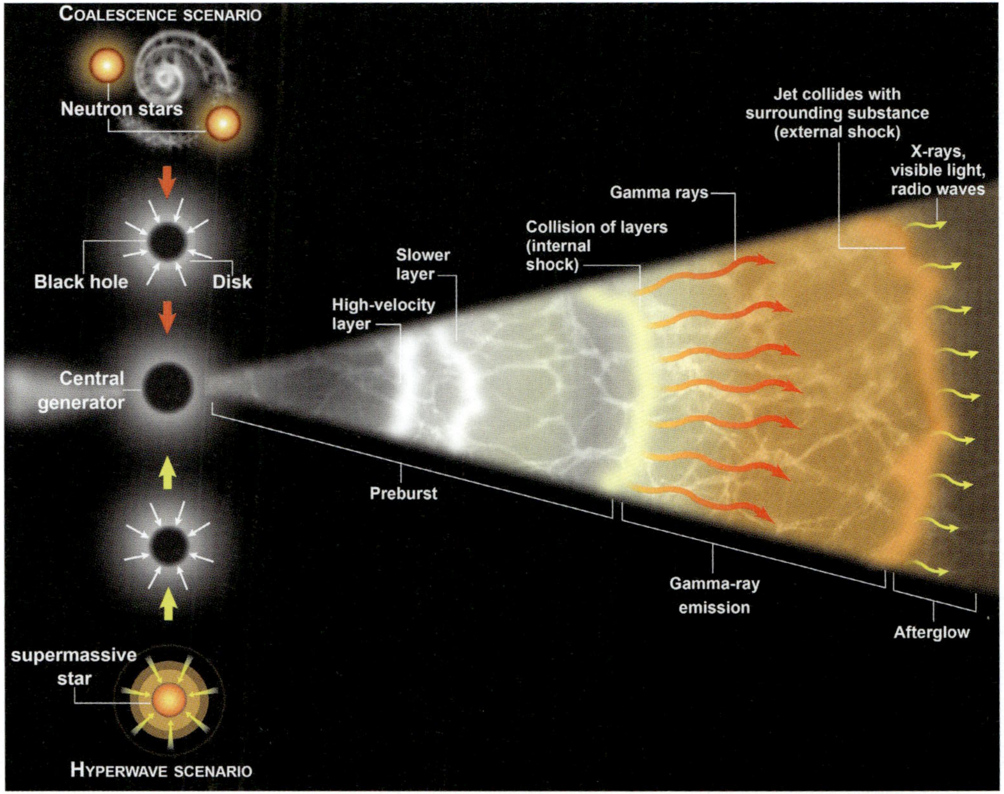


**Fig. 7.100** Gamma-ray bursts may arise from either the coalescence of two neutron stars or the collapse of a supermassive star, in either case giving rise to a black hole and a disk of matter surrounding it. This system next ejects a matter jet with shocks inside of it [73].

**Fig. 7.101** Gamma-ray burst GRB030329 in the spiral galaxy NGC 3184 is related to a supernova explosion [214].

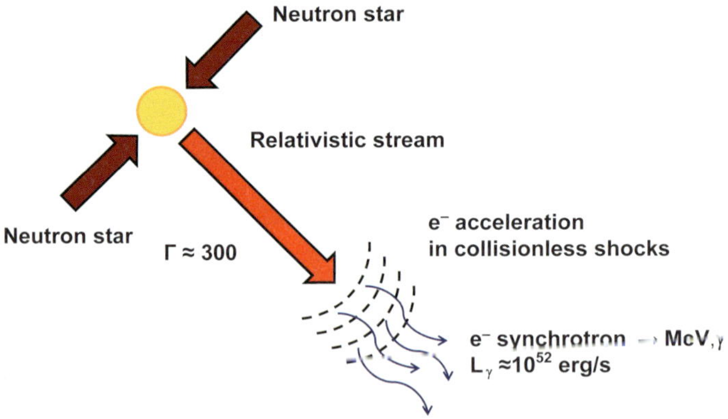


**Fig. 7.102** Gamma-ray burst scenario with "fireball" formation [221].

The typical density of matter in accretion disks and jets is $\rho \geq 10^{14}$g/cm$^3$, which corresponds to the density of nuclear matter or neutron stars. As the "fireball" expands, inelastic nuclear collisions increase in importance and the relative velocity of neutrons and protons becomes comparable to the velocity of light. Inelastic $n$- and $p$-collisions produce charged pions, gigaelectronvolt muons, and electronic neutrinos, which should be detected by spacecraft [145]. Powerful relativistic jets from gamma-ray bursts correspond to Lorentz factors $\gamma \geq 100$ and extremely high en-

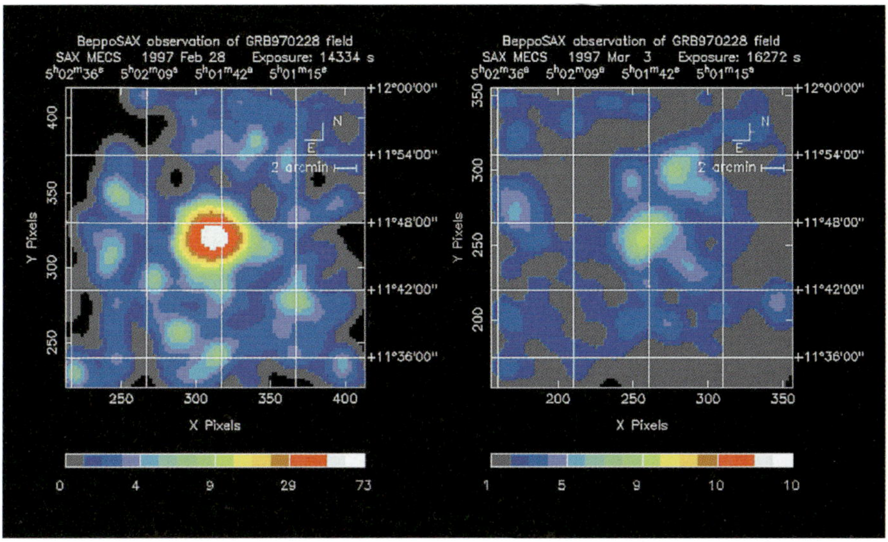


**Fig. 7.103** Development of the gamma-ray burst GRB 970228. Left: Photo made on 28.02.1997. Right: Photo made 4 days later [214].

ergy fluxes ($10^{50}$–$10^{52}$ erg/s or $10^{43}$–$10^{45}$ W). The energy density in the jet core, of the order of $10^{30}$ erg/cm$^3$, is sufficient for the spontaneous production of electron–positron pairs or nuclear reactions [105, 145, 75].

Further, at distances of $\geq 10^{11}$ cm, the electron mean (collision-)free path exceeds the characteristic jet dimension, and collisionless internal and external shocks accelerate electrons by the Fermi mechanism [59] with a relativistic distribution spectrum. Their interaction with the turbulent magnetic field generates nonthermal gamma-ray radiation, X-ray and optical radiation, which is recorded by recording instruments.

Furthermore, superimposed on the power-law continuum is the characteristic X-ray radiation arising from the interaction of the heated plasma regions with the disintegrating jets emanating from stellar surfaces.

The nonthermal spectrum in the shock-wave model of a "fireball" described above stems from the assumption that the Fermi acceleration model obeys the power law $N(j_e) \sim \gamma_e^{-p}$, with $p \approx 2$–$2.5$. This is quite consistent with observations [145]. To attain reasonable efficiency, the ratio between the accelerated electron energy and the accelerating energy should not be too small, $E_e \approx 1$, while the ratio between the magnetic energy and the total energy $E_e \leq 1$ depends on whether the synchrotron or inverse Compton spectral peak corresponds to the observed knee in the spectrum in the megaelectronvolt energy range.

In addition to that, the plasma wake acceleration mechanism [139] may also be of significance in gamma-ray bursts, because in the collision of neutron stars or the collapse of a star there emerge strong Alfvén shock or soliton waves as well as relativistic plasma ejections. In this case, to explain the gamma-ray signal shape requires taking into account the conditions of its transmission through the plasma atmosphere. Interestingly, the Alfvén (MHD) shock waves may carry an appreciable ($\approx 10^{-2}$) fraction of the "fireball" energy, i.e., up to $\approx 10^{50}$ erg. When propagating from the center, they may have a narrow front with a giant magnetic field of $\approx 10^{10}$ Gs behind it, which may compete with the Fermi particle acceleration mechanism [22].

The same mechanisms that are responsible for electron acceleration may also be applied to the acceleration of protons up to $10^{20}$ MeV.

Terawatt to petawatt short-pulse lasers (see Chap. 4, Sect. 3.5), which produce intensities of $\approx 10^{20}$ W/cm$^2$, are best suited to the laboratory modeling (at least, partial) of these processes. Impressive results have already been obtained in this area: well-collimated MeV-proton beams [128, 126, 53, 43, 127, 125], electron and positron beams with energies of 100 MeV [30, 18, 29, 168], thermally relativistic plasma with $T_e > m_e c^2$, which corresponds to "fireball" conditions, as well as superhigh (>100 MT) magnetic fields [46, 138, 139, 167, 211, 129]. These ultraextreme plasma conditions are already approaching the parameters of gamma-ray bursts.

As noted in Chap. 6, the advent of petawatt lasers opens up new avenues for the production of relativistic and electron–positron plasma in the laboratory. At intensities above $\approx 2 \times 10^{18}$ W/cm$^2$ these lasers generate electrons with an energy $kT > m_e c^2$ in laser plasmas [232, 168, 110, 115, 10, 121, 139]. Electron–positron pairs [44, 119, 30, 18] emerge in the interaction of these electron streams with heavy targets by the Bethe–Heitler mechanism.

Experiments performed with a high-power laser have shown [130, 171, 172, 86] that the yield of the pairs even exceeded the theoretical estimates. In this case, two-sided irradiation simulations suggest that the expansion of the pairs proceeds faster than the expansion of gold ions. Here, the collision of two "fireballs" may be realized for the purpose of simulating the shock-wave gamma-ray burst model and studying the "fireball" energy conversion to the internal energy and gamma-ray radiation.

The laser-produced electron–positron plasma may also be employed for modeling episodical annihilation effects in the neighborhood of black holes. In the two-sided irradiation of a gold target, two megajoule 0.1 PW lasers with a pulse duration of $\approx 10$ ns are capable [145] of producing a pair density that is several hundred times the electron background density. This would allow the theoretical limitations, $kT \approx 20mc^2$, given by Sunyaev, Zel'dovich, and Levich [203, 230] to be verified.

Observations of the SN1987 supernova explosion in the Large Magellanic Cloud yielded [93] a wealth of new and unexpected information and gave strong impetus to laboratory astrophysics [47, 184, 130, 171, 172]. In particular, calculations and experiments exposed the important role [214, 130, 171, 48, 7] of hydrodynamic instabilities and mixing in the hydrodynamics of stellar explosions, much as this is critical for laser thermonuclear fusion. At the present time, the corresponding experiments aimed at studying the Rayleigh–Taylor and Richtmyer–Meshkov instabilities in laser targets are being actively conducted in many plasma laborato-

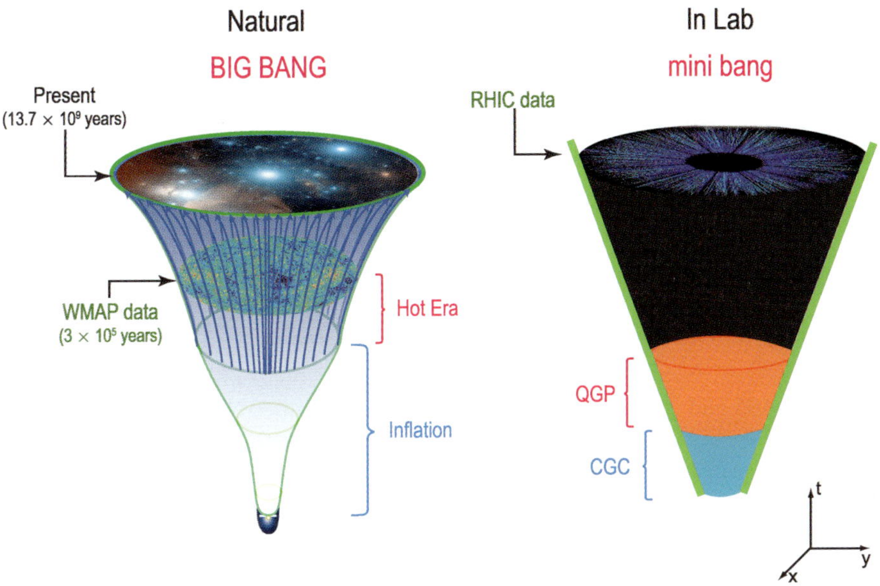


**Fig. 7.104** Analogy between the Big Bang and a "mini explosion" in relativistic nuclear collisions.

ries [47, 184, 139, 130, 171] and are yielding a wealth of data helpful for understanding transient astrophysical phenomena.

A significant part in the description of supernova explosions is assigned to one-, two-, and three-dimensional numerical simulations [47, 48, 7]. These calculations are based on the following assumption: as a result of gravitational stratification, the initial supernova structure (prior to the explosion) consists of an outer layer of hydrogen, next a helium layer, and so on up to the iron inner core. The gravitational collapse of this inner core gives rise to a radially diverging intense shock wave, which has high parameters at the helium–hydrogen interface, $\rho \approx 2.3$ g/cm$^3$, $T \approx 6$ keV, $P \approx 75$ Gbar, and leads to the development of the Richtmyer–Meshkov instability [130, 171, 48, 7, 56].

Remington et al. [172] employed the Nova laser to model a supernova explosion in the laser-to-soft-X-ray radiation conversion scheme. X-ray radiographic measurements enabled them to trace the nonlinear development stages of instabilities and to compare them with the data of numerical simulations.

It is likely that ultrarelativistic ultrahigh-energy-density plasma conditions are most amply reproduced in relativistic heavy-ion collisions. Figure 7.104 serves to illustrate the analogy between the expansion of the universe after the Big Bang and a "mini explosion" occurring in a laboratory in relativistic ion collisions in the RHIC accelerator in the domain of space $\approx$0.01–1 fm and time $\approx 10^{-23}$ s.

However, the experimental implications of ultrarelativistic effects may supposedly be recorded under presently attainable laboratory conditions, for instance, in studies of spontaneous proton decay. The point is that one of the three constituent quarks may, owing to quantum fluctuations, gain sufficiently high energy to result in proton decay. Despite the negligible probability of this process ($\approx 10^{33}$ years), for a large number of neutrons this process might be detectable in massive specimens. In a salt mine ofin Ohio State, eight thousand tons of water were used for this purpose. The absence of experimental detection of spontaneous proton decay yielded a lower bound for the proton lifetime, which is of consequence in constructing Grand Unification models [89].

## 7.6 Matter Transformation after the Big Bang

Figures 7.105 and 2.2 give a comparison of the capabilities of modern accelerators that reproduce in terrestrial laboratories the most "extreme" conditions occurring in nature – those of the "Big Bang". We see that elementary particle cosmology makes it possible to advance a long way in the measurement of the interaction of high-energy particles, which were of primary importance at the first instants of the life of the universe and defined its evolution for billions of years to come. Studying the evolution of the early universe furnishes a unique opportunity to investigate high-energy phenomena that it is impossible to reproduce in a laboratory.

According to modern notions [188, 180], the temporal evolution of the universe proceeded highly nonuniformly – whereas it is relatively slow nowadays, at the early stages it was strikingly fast, so that major qualitative changes of the state of the universe occurred in a split second. The universe we now observe emerged about 13.7–15 billion years ago from some initial singular state with infinitely high temperature and density, and since then it has uninterruptedly expanded and cooled down. According to the Big Bang theory (see Fig. 7.106, which gives the orders of magnitude of the quantities corresponding to the events occurring), the subsequent evolution will depend on the present-day expansion velocity of the universe and the average matter density in the modern universe. If $\rho$ is lower than some critical value $\rho_{cr}$, which is known from the theory and is epends on the expansion velocity, the universe will eternally expand; if $\rho > \rho_{cr}$, some day the expansion will be terminated by gravitational forces and the reverse compression phase will commence, the universe reverting to its initial singular state.

From the measurement data it follows that our universe expands by 5–10% every billion years. However, the uncertainty in universe density estimates is substantially higher. The total mass of all visible stars in galaxies is less than one hundredth of $\rho_{cr}$. The inclusion of the mass of dark matter yields an average density of $\sim 0.1\rho_{cr}$. It is therefore most likely that our universe will eternally expand, unless some new mass sources are discovered. However, if the compression stage does set in, in any case this will happen no earlier than in $\approx 10$ billion years, when the Sun has gone out and terrestrial life has terminated.

But even if sapient life survives epidemics, mutations, and wars, manages to make provision for energy, and conserves or makes acceptable a human environment, before the eyes of our distant descendants there will appear, according to

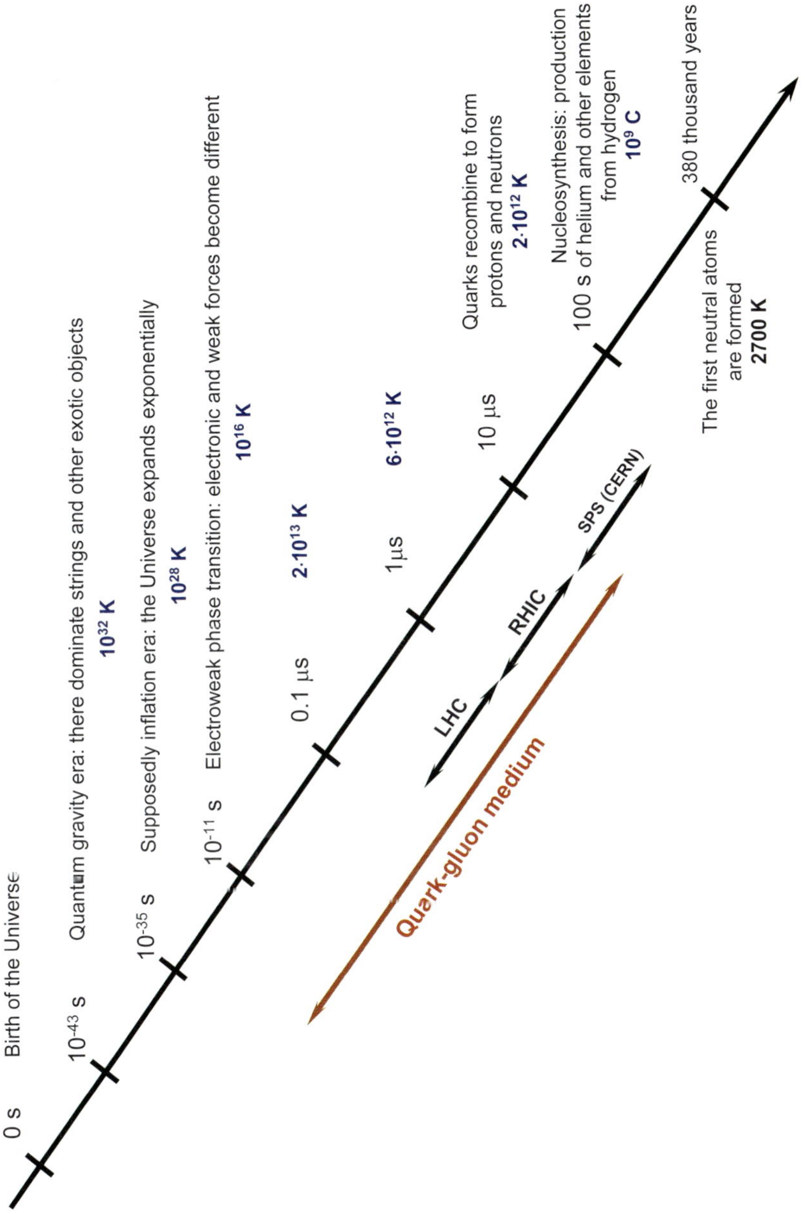


**Fig. 7.105** Cosmic time scale. Universe expansion after the Big Bang [174].

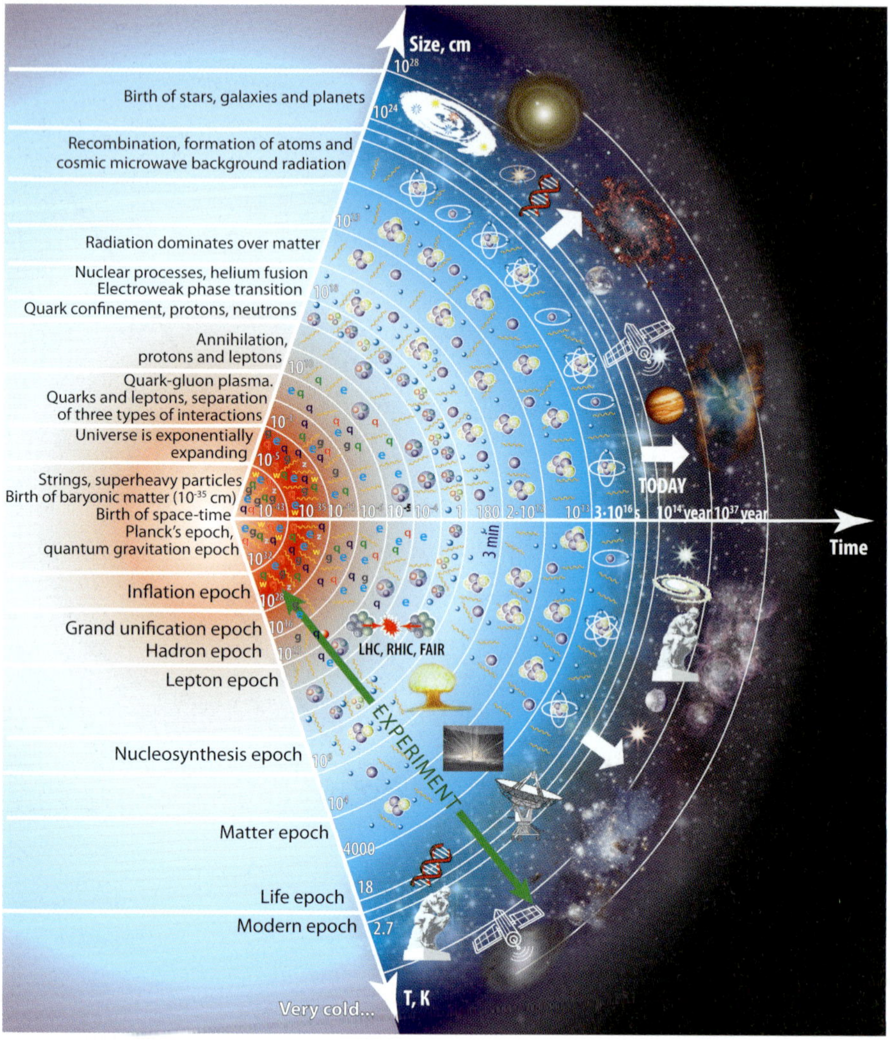


**Fig. 7.106** Time scale after the Big Bang.

the model in [114], the unattractive view of the surrounding universe – the view of a dark abyss with scarce inclusions of rare stars amidst invariable emptiness. According to Krauss and Scherrer [114], this will result from the increasingly fast expansion, which with the lapse of time will make our galaxies recede from us with a seemingly supraluminal velocity and make them completely vanish from an observer's view. This process rules out the existence of reference objects for measuring the degree of expansion and also "washes out" all characteristic features remaining after the Big Bang. In other words, all observable manifestations of the Big Bang

**Fig. 7.107** Evolution of the observable part of the universe. Reprinted, with permission, from [114].

that have ever existed, and which are so plentiful in the universe surrounding us today, will be smoothed out.

In the model in [114], the accelerated expansion of the universe may lead to the formation of a fixed "event horizon" – an imaginary sphere from outside of which neither matter nor radiation will ever reach the observer inside. In such a model the universe becomes something like an "external black hole", when matter and radiation turn out to locked outside of the event horizon and not inside it. In this case, our "local group" of galaxies (including our galaxy, the Andromeda Nebula, and several dwarf companion galaxies) will merge into a single stellar supercluster. All other galaxies will vanish from an observer's view. This process will take about 100 billion years.

The relic radiation will virtually vanish as well; its maximum will shift to a wavelength in the meter range, while its intensity will decrease by many orders of magnitude. Furthermore, the electron component of the interstellar matter will not let this radiation into our galaxy.

The amounts of hydrogen and helium, which were produced during the first three minutes after the Big Bang, will considerably decrease due to thermonuclear reactions in stars, which, according to [114], will wipe out one more important consequence of the Big Bang. It is anticipated that the observable universe will collapse into a black hole in the distant future, which will happen before that to our galaxy as well.

Figure 7.107 shows the prospects for the course of events in the future as distant as 100 trillion years. The night sky of the Earth will radically change when our galaxy, the Milky Way, merges with its closest neighbors and distant galaxies vanish from our sight. In 5 billion years the Andromeda Nebula will fill the sky almost completely. The Sun will transform to a red giant and subsequently burn out. In 100 billion years, the Milky Way, the Andromeda Nebula, and other small galaxies will make up a giant ball-shaped galaxy. In 100 trillion years, the last star goes out and black holes shine dimly, along with artificial sources (if any) contrived by supercivilizations. The supergalaxy contracts to a black hole.

The main stages of the temporal evolution of the early universe are presented in Table 7.7 [188]. Following [188, 180] in the subsequent discussion, let us consider these evolution stages.

**Table 7.7** Main epochs of the evolution of the early universe [188].

| **Name of epoch corresponding to its physical processes** | **Time elapsed after the Big Bang [s]** | **Temperature [K]** |
|---|---|---|
| Emergence of classical space-time | $10^{-43}$ | $10^{32}$ |
| Inflation stage | $\approx 10^{-42}$–$10^{-36}$ | Varies over very wide limits |
| Production of matter | $10^{-36}$ | $\approx 10^{29}$ |
| Production of a baryon excess | $10^{-35}$ | $\approx 10^{29}$ |
| Electroweak phase transition | $10^{-10}$ | $\approx 10^{16}$–$10^{17}$ |
| Quark confinement | $10^{-4}$ | $\approx 10^{12}$–$10^{13}$ |
| Primary nucleosynthesis | 1–200 | $\approx 10^{9}$–$10^{10}$ |

The very early time ranging from $10^{-43}$ s to 2–3 min is the most interesting and hard-to-describe period of the history of the universe, which saw the formation of its most essential properties. These properties now manifest themselves as the Hubble expansion, the large-scale structure of the universe, and the physical laws operating in the part of the universe observable to us.

The fundamental difficulties are aggravated in attempts to approach the "origin of the universe", more closely, i.e., to peer into the first hundredth of a second. To do this requires a profound knowledge of elementary particle physics in the superhigh-energy domain, which we do not have now, for the reason that these energies are unattainable with terrestrial accelerators. In the further advancement towards the origin of the universe there emerges the even more difficult problem of "quantization of gravity", which has no solution so far, not even a basically satisfactory one. That is why all attempts to investigate the earliest moments of the existence of our universe are in essence purely theoretical constructs.

The instant of universe's origin is the epoch of the birth of classical space-time. Generally accepted at the present time is the theory of the Big Bang, i.e., the origination of the universe from a singularity (from spatio-temporal foam, as is sometimes said) [188]. At the instant of the birth of the universe, the matter density $\rho$ and temperature $T$ amounted to Planck's values: $\rho_{\mathrm{Pl}} \approx 10^{93}$ g/cm$^3$, $T_{\mathrm{Pl}} \approx 1.3 \times 10^{32}$ K; under these conditions, the gravitational interaction is comparable with the others (strong, weak, electromagnetic) in strength and should be considered with the inclusion of quantization.

The enormous energies afforded by present-day accelerators are clearly insufficient for reproducing in laboratory experiments the conditions corresponding to the "Grand Unification" parameter domain $\approx 10^{15}$ GeV. To do this would require a Solar-System-sized accelerator.

Since the instant of the Big Bang, the universe has been continuously expanding, the matter temperature has been lowering, and the volume increasing. In the description of the birth of the universe, advantage is taken of the most general ideas of the quantum evolution of the universe as a whole. One of them asserts that the total mass of the closed universe is equal to zero. This signifies that the entire universe may come into being without energy expenses and with the initial geometrical dimension of the order of the Planck one.

As the universe expands, it borrows energy from the gravitational field to produce more matter. The positive energy of the matter is exactly balanced by the negative gravitational energy, so that the total energy balance is equal to zero.

$10^{-43}$–$10^{-42}$ s after the origin of the classical space-time, an inflation stage in the universe sets in [188, 180] (Fig. 7.108 [91]), which lasts until the stage of Grand Unification at $\approx 10^{-35}$ s, when the relic radiation emerges. Analyzing the inflation stage calls for the extrapolation of the known physical laws by 30 orders of magnitude on the spatial scale. Beyond that stage is thought to lie the limit of human philosophizing [80] and, as noted even by Tertullian – the great Christian philosopher of antiquity – this limit is precisely the boundary with the realm of theology.

The inflation stage is characterized by an ultimately high negative pressure $p = -\rho c^2$, whereby the very laws of ordinary gravitational physics are different. The matter in this state is the source of repulsion rather than attraction. It is not unlikely that the very early universe was filled not only with particles, but also with a temporal form of dark energy (with the inflaton field) with a very high density. This energy brought about the expansion of the universe with a very high acceleration, following which it decayed to form high-temperature plasma, which later divided into the usual matter and radiation. There only remained a weak trace of dark matter, which became significant only in the modern epoch. During the inflation stage, the volume of the universe increased by many orders of magnitude (by a factor of $10^{1000}$ in some models), so that the entire universe found itself in one causally connected domain, the kinetic expansion energy of the universe and its kinetic en-

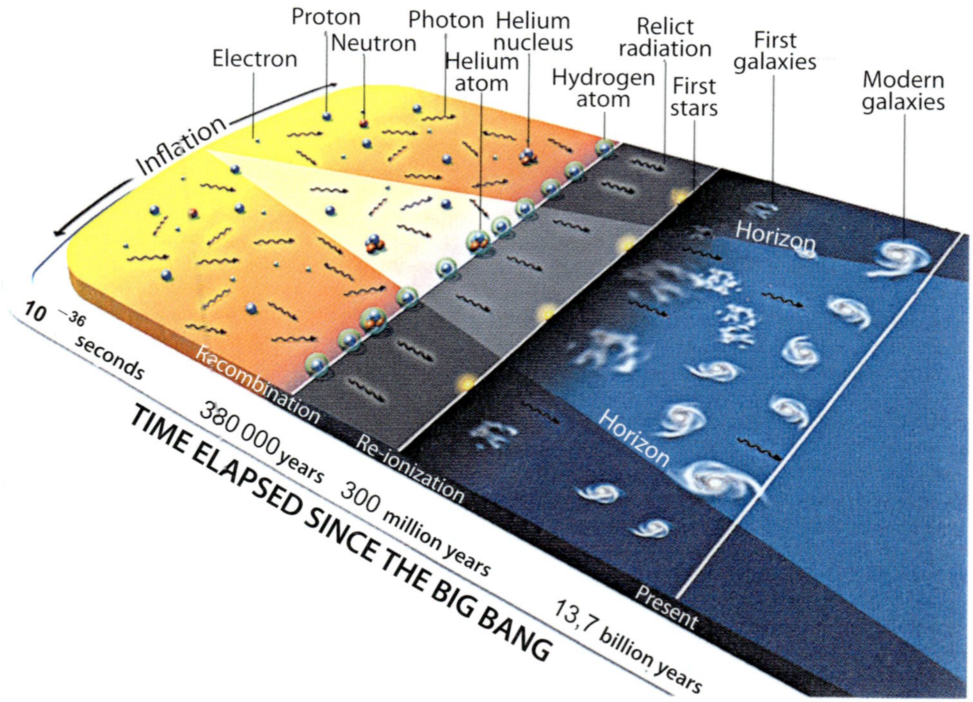


**Fig. 7.108** Temporal scale of the universe. During inflation the universe expanded rapidly and the plasma, which consisted of photons and charged particles, propagated far beyond the horizon limiting the domain within the sight of a hypothetical observer. The onset of recombination occurred 380 thousand years later: the first atoms and the relic radiation emerged. 300 million years later, the light of the first stars ionized the major part of the hydrogen and helium. Reprinted, with permission, from [91].

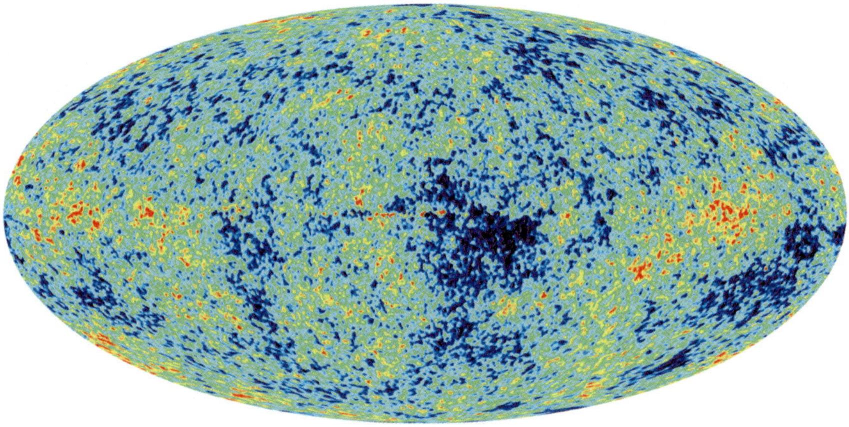

**Fig. 7.109** Structure and angular distribution of the relic radiation reflect the state of the early universe. The primary nonuniformities of matter amount to $10^{-6}$ [233].

ergy were equalized. During that stage there emerged the physical conditions which later led to the universe's expansion by the Hubble law.

Since the rate of interparticle distance variation at this stage is proportional to the distance itself, this stage is referred to as inflationary: the growth of the money supply during inflation obeys the same exponential growth law.

Unlike the explosion of chemical explosives, whenre the expansion of the detonation products is related to the pressure gradient, the expansion in the case of the Big Bang springs from the negative pressure in a homogeneous medium – antigravity – which existed early in the existence of the universe.

During the inflation, gravitational repulsive forces accelerate particles, so that subsequently they move under their own inertia and form the Hubble expansion law.

At the inflation stage there occurred yet another important process [188, 180]: the production of small density perturbations from vacuum quantum fluctuations of a scalar field and of gravitational waves from quantum metric fluctuations. The vacuum quantum fluctuations, which normally manifest themselves only on a microscopic scale, rapidly increased in the exponentially expanding universe to become cosmologically significant. Therefore, the clustering of galaxies, which emerged at a later time, and the galaxies themselves are macroscopic manifestations of quantum fluctuations in the infancy of the universe.

From the observations of relic radiation anisotropy $10^{-5}$ (Fig. 7.109) it is therefore possible to reconstruct the spectrum of primary perturbations. From the primary perturbation spectrum and the gravitational wave spectrum it is possible to reconstruct the physical laws of the inflation stage, i.e., in the $10^{16}$ GeV energy domain. This is the aim of the experiments RELICT, COBE (Cosmic Background

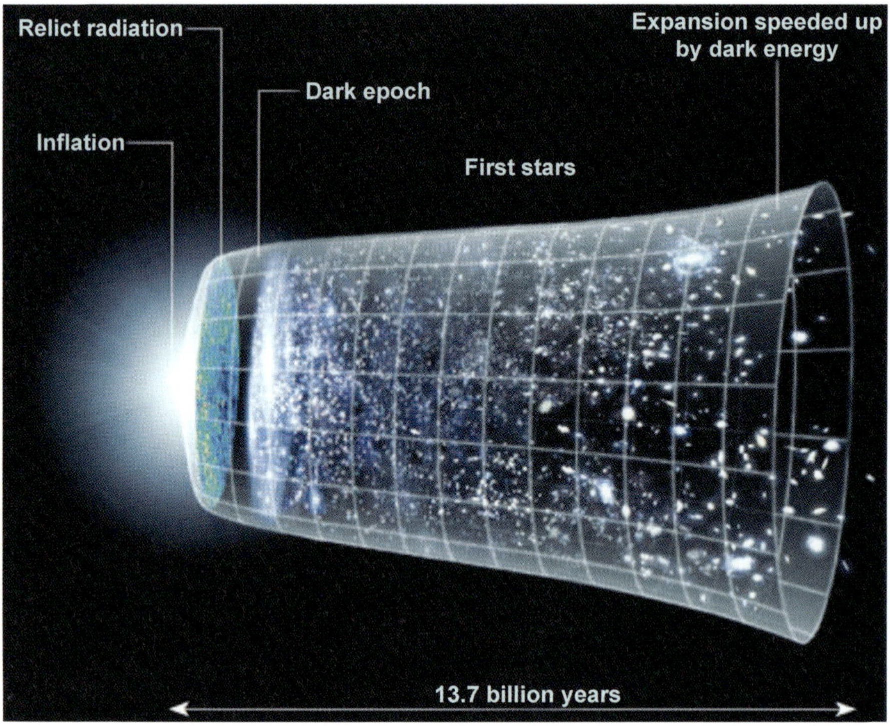


**Fig. 7.110** Expansion of the universe with the distinguished initial stage [233].

Explorer), and of the ground-based experiments Tenerife, Saskatoon, and CAT (Cosmic Anisotropy Telescope).

The difference in temperature between the domains of the early universe were sufficiently large that the extra gravity in higher-density domains terminated their infinite expansion and caused their contraction under self-gravity, with the resultant production of stars and galaxies, so that the map of Fig. 7.109 basically contains the prehistory of all structures of the universe.

The equation of state for matter with a negative pressure is unstable: it should be replaced by ordinary pressure. That is why the inflation phase of the universe's development terminated rather quickly and the stage of ordinary matter production set in (Fig. 7.110, Table 7.8 [180]).

According to Kirzhnits [106], from this point on, $\approx 10^{-37}$ s after the Big Bang, the universe underwent, after Friedman, a relatively slow expansion, and its temperature and density lowered with time. The starting temperature was about $10^{29}$ K, i.e., $\approx 10^{16}$ GeV, which exceeds the $X$- and $Y$-boson masses. The condensate of Higgs particles, which is responsible for the emergence of this mass, is absent under these conditions; the $X$- and $Y$-bosons are massless, and the state of matter is realized that corresponds to complete symmetry between the strong, weak, and electromagnetic

**Table 7.8** History of the universe: main stages [180].

| Time | Radiation temperature [K] | Comments |
|---|---|---|
| $10^{-43}$–$10^{-37}$ s | Above $10^{26}$ | Inflation. |
| $10^{-6}$ s | Above $10^{12}$ | Plasma of quarks, gluons, electrons, and their antiparticles. |
| $3\times10^{-5}$ s | $10^{12}$–$10^{9}$ | Quarks and gluons form protons and neutrons. |
| $10^{-4}$ s–3 min | $10^{12}$–$10^{9}$ | Production of deuterium, helium, and lithium nuclei. |
| 15 thousand years | $10^{4}$ | Equality of matter and radiation energy densities. |
| 300 thousand years | 4000 | Formation of atoms from light nuclei and electrons. The universe becomes transparent for photons. |
| 15 million years | 300 | Nothing particularly notable, except that the temperature of space was equal to room temperature and comfortable to human beings. |
| 1–3 billion years | 20 | Formation of the first stars and galaxies. |
| 3 billion years | 10 | Production of heavy nuclei in the explosion of the first stars; emergence of next-generation stars. |
| 3–15 billion years | 3 | Emergence of planets and sapient life. |
| $10^{14}$ years | Subsequently very cold | New stars have ceased to emerge. |
| $10^{37}$ years | | All low-luminosity objects have exhausted their energy. Black holes and elementary particles have persisted. |
| $10^{100}$ years | | All black holes have evaporated. Tenuous gas of stable elementary particles – electrons, three kinds of neutrinos, and possibly protons. |

interactions. The matter itself is a hot dense plasma made up of massless particles (quarks, leptons, photons, gluons, $W$-, $Z$-, $X$-, and $Y$-particles) and massive Higgs particles. In this case, quarks and gluons transform freely into each other, and their relative densities are defined only by the temperature.

At the point in time, $\approx 10^{-33}$ s, when the temperature had lowered to $100^{14}$ GeV = $10^{27}$ K (the $X$- and $Y$-boson mass), there occurred a phase transition with the emergence of a condensate of the corresponding Higgs particles. Production of pairs of X-mesons with a mass of the order of $10^{15}$ GeV occurred, which decayed into quarks and leptons. According to A.D. Sakharov's idea, the baryon charge was not conserved at that time and particle–antiparticle CP symmetry violation occurred.

This is how today's matter was produced with the number of protons $\approx 10^{80}$, which is close to the current value (the Eddington–Dirac number). That epoch is termed the epoch of Grand Unification of electromagnetic, weak, and strong interactions.

From that point on, the symmetry between the strong and electroweak interactions broke down, the mutual transformations of quarks and leptons strongly moderated, and a start was made on the formation of baryon matter asymmetry in the universe with the extraordinary predominance of matter over antimatter. Modern theories are able to explain (see, for instance, [180]) how this asymmetry emerges naturally in the course of evolution of the primary state in which the particle and antiparticle densities were equal.

At the point in time $\approx 10^{-12}$ s, when the temperature had lowered to $10^2$ GeV–$10^{15}$ K (the $W$- and $Z$-boson mass), there occurred the second (electroweak) phase transition, with the emergence of the second condensate of those Higgs particles that are responsible for $W$- and $Z$-boson masses. From that point on their masses became nonzero, the symmetry between the electromagnetic and weak interactions broke down, and the matter itself was a plasma made up of massive leptons, quarks, gluons, and massless photons.

The next phase transition is thought to have taken place at $\approx 10^{-5}$ s, when the temperature had lowered to a value of $\approx 100$ MeV, $10^{11}$–$10^{12}$ K (the characteristic scale of quark–gluon interaction when quark confinement occurs). As a result of this transition, the quark–gluon plasma transformed into a system of hadrons by way of quarks and antiquarks merging with the production of baryons and mesons of the corresponding resonances. At a temperature of $10^{11}$–$10^{12}$ K there occurs quark confinement ("no escape"). Quarks may exist in the free state only in a very hot plasma with a temperature $T > 10^{11}$ K. In the early universe, when the temperature far exceeded this value, there were no protons and neutrons and there existed a "quark soup". Due to the expansion of the universe the temperature lowered, quarks begin to combine to form protons and neutrons, and are no longer encountered in nature as separate particles (there is "no escape").

The technology of modern acceleration experiments enables the states in the temporal neighborhood only tenths–hundredths of a second after the Big Bang to be reproduced (Fig. 7.105); it is expected that commissioning of the LHC, RHIC, FAIR, etc. accelerators will permit advances into the milli- to microsecond range. The subsequent epochs are characterized by the absence of quarks and gluons in the free state, and leptons, hadrons, and photons are the structural constituents of the matter. In this case, with decreasing temperature there occurs decay of unstable particles and the annihilation of pairs of baryons and heavy leptons.

After the epoch of proton and neutron production, the subsequent evolution of the universe led to the epoch of nucleosynthesis $\approx 1$–$100$ s (a temperature of $\approx$ 0.1–1 MeV$\approx 10^9$–$10^{11}$ K – the characteristic nuclear scale), during which primary nucleosynthesis occurred and the lightest matter constituents with atomic weight $A < 5$ were formed. The time interval between $t \approx 1$ and $t \approx 200$ s is of considerable importance in the life of the universe. The primary light nuclei were formed during this period: $^4$He (25%), deuterium $^2$H ($3 \times 10^{-5}$ %), $^3$He ($2 \times 10^{-5}$ %), $^7$Li ($10^{-9}$ %), i.e., a start wa made on the origination of the matter we are accustomed to. The rest of the isotopes, although they were produced during this epoch, were finally formed as matter components much later.

Approximately 1 s after the Big Bang, the primary plasma temperature lowered to $10^{10}$ K, which corresponds to an energy of $\approx 1$ MeV.

With a decrease in temperature to $T = 0.7$ MeV, these weak-interaction reactions practically terminated and the neutron-to-proton number ratio became constant and equal to the number ratio at the close of the process. At that stage of the universe, protons and neutrons existed in the free form, not combined into nuclei. Later, when the temperature lowered to 100 keV, the majority of neutrons (with the exception of those which had managed to decay) were bound in the production of deuterium in the course of the reaction $p + n \rightarrow {}^2\mathrm{H} + \gamma$.

Deuterium, in its turn, efficiently captured primary plasma baryons to produce $^3$He and tritium ($^3$H). On capturing one more proton or neutron there emerged $^4$He, in which practically all undecayed neutrons ended up. The absence of suitable nuclei with a mass number $A = 5$ impeded subsequent reactions, making the production of heavier elements ($^3\mathrm{He} + {}^4\mathrm{He} \rightarrow {}^7\mathrm{Be}$, $3{}^4\mathrm{He} \rightarrow {}^{12}\mathrm{C}$, etc.) a highly improbable event.

With a further decrease in temperature, the matter consisted of H and He nuclei, electrons, photons (their density is $10^9$ times higher than that of the nuclei), and neutrinos, which can hardly interact with the rest of the matter. This epoch is referred to as "radiation-dominated" because of the predominance of the contribution of electromagnetic energy to the energy of matter. During that era (the "fireball" era), the relic radiation was the main form of matter. The closer to the origin of the universe, the higher the frequency of this radiation. For a high frequency there occurred particle–antiparticle production by photons. That is why the relic radiation in the very early universe [80] is more often used in reference to the plasma of particles and antiparticles rather than a photon gas. Such a plasma is termed an ambiplasma. For $kT = m_e c^2$, we are dealing with the leptonic era; for a higher temperature $kT \sim m_p c^2$, we are dealing with the hadronic era.

In the fireball epoch, protons, neutrons, and electrons, as well as the light elements deuterium, tritium, helium, and lithium, which were produced during the first three minutes of the universe, played only an insignificant role. The light pressure was so high that massive particles were uniformly scattered about the universe and could not form any structures such as galaxies, stars, and planets by gravitational attraction. However, dark matter, which did not interact with relic electromagnetic radiation, was something else again. Due to gravitational interaction, that dark energy was able to curl long before similar processes involved visible matter.

During the epochs under discussion, the primary plasma obeyed the radiation-dominated equation of state $p = \rho c^2/3$. In this epoch there occurred annihilation of electron–positron pairs with their transformation to photons. Experiments involving underground nuclear explosions [12, 219, 11, 212] yielded a plasma and photon temperature of $\approx 10^7$ K and an energy density of $10^9$ J/cm$^3$, corresponding to the states several minutes distant from the Big Bang, which are close to the lower bound of the radiation-dominated regime. This predominance is no longer manifested during the epoch $10^8$–$10^{10}$ s (other estimates suggest a figure of $\approx 10^{12}$ s) (the temperature $\approx 10^4$–$10^5$ K corresponds to $\approx 1$–10 eV, the characteristic atomic energy scale), when the energies of electrons and nuclei are insufficient to overcome their electromagnetic attraction and electrons recombine with nuclei to give rise to neutral atomic matter transparent to photons.

Until that time, in the "fireball" era the universe expanded by the law $R \sim \sqrt{t}$; later, when matter came to play a leading part, the expansion law approximately assumed the form $R \sim t^{2/3}$.

In the recombination epoch the light pressure lowered to the extent that particles were attracted to one another. Since all that took place gradually and the relic radiation did not cease at once to interact with visible matter by passing to the radio frequency band, there existed some period when the universe heard the "first sound" [80]! This sound is also referred to as the "Sakharov acoustic oscillations" after A.D. Sakharov, who hypothesized that they existed in the early universe. The situation is that, as noted above, between the particles of visible matter there were counteracting forces of gravitational attraction and repulsion due to electromagnetic pressure. That is why the particles alternately approached each other and moved away, and as the pressure lowered there emerged decaying oscillations of particle density similar to sound oscillations. That sound was a "whisper" – an infrasound with a wavelength of the order of the horizon in the recombination era. However, where it was heard, there emerged density perturbations, which subsequently turned into galaxies.

The curling of particles and hence higher-density domains emerged where there already existed nonuniformities caused by dark matter. Without dark matter galaxies would not have been formed by the present time. The visible matter nonuniformities caused nonuniformities in the temperature distribution of the relic radiation (Fig. 7.109), which are the reflection of the first sound and the nuclei of future galaxies.

The recombination epoch coincided with the universe "transparentization" epoch: the plasma disappeared and the matter became transparent. The temperature of this epoch is quite well known from laboratory physics: $T \approx 4500$–$3000$ K.

After recombination, photons reach the observer virtually without interacting with matter and make up the relic radiation. Nowadays its energy spectrum corresponds to the spectrum of a blackbody heated to a temperature of 2.75 K. The difference between the temperatures $\approx 3000$ and $\approx 3$ K arises from the fact that the universe has expanded by about a factor of 1000 since the transparentization epoch.

The density of relic radiation today is $\rho_{\text{rad}} \approx 10^{-34}$ g/cm$^3$, which is lower than the density of ordinary matter $\rho_{\text{sub}}$. The condition $\rho_{\text{rad}} = \rho_{\text{sub}}$ is termed the recombination era and $\rho_{\text{sub}} < \rho_{\text{rad}}$ is referred to as the "fireball" era or the "radiation-dominated era". In that era the relic radiation was seen in the optical range, while the sky was not dark as it is now is but glittering, grading into a red color with expansion to subsequently become invisible.

It is significant that the prerequisites for the production of future galaxies are found even in the "fireball" era, in which weak nonuniformities (density fluctuations) were produced, which disturb the general uniformity and isotropy of the uniform and isotropic plasma. These are the "nuclei" of future galaxies. After the recombination era, when the temperature of the universe became lower and the radiation ceased to strongly interact with matter particles and to "scatter" them in space, these "nuclei" attracted particles due to gravitation. In 1993, the first experimental data were obtained from the COBE satellite. The temperature of some regions ("spots") of the sky is slightly ($10^{-5}$–$10^{-6}$, Fig. 7.109) different from that of other areas.

After the recombination era, the "dark ages" [80] set in: the relic radiation had decoupled from matter, and the stars in galaxies had not started shining. In that era there emerged protogalaxies as gas clouds. They collided with each other to form black holes. These black holes admitted visible matter, which shone, and there emerged quasars. The relic radiation manifested itself only weakly, the dominant role being played by dark and visible matter.

Approximately 5 billion years ago there occurred yet another important event in the history of the expansion of the universe. The cosmological constant came into play [80]. Earlier it had counted for very little: being a constant small quantity, it was below the matter density of the relic radiation. The universe started expanding exponentially. Due to relative acceleration, galaxies experienced stronger repulsion from each other, which hindered their mutual collisions; this circumstance turned out to be of consequence for the possibility of the emergence of the Sun, the Solar System, the Earth, and life on it.

One more major milestone is located between the recombination epoch and our time – the formation of the large-scale structure of the universe, or the formation of galaxies. This epoch falls in the time when the temperature of relic plasma photons was $T \leq 30$ K. In the intervening period there lies the epoch of nonlinear evolution of extragalactic objects, i.e., the epoch of ordinary galaxies, quasars, and clusters and superclusters of galaxies.

At that time, the universe as a whole continued to expand and cool down. In domains with a somewhat higher density than the average, however, the matter expansion moderated due to the additional gravitational attraction and entered the phase of gravitational compression with a slow rotation. The increase of the angular velocity of this rotation in the compression of the matter has the effect that the centrifugal force comes to be of the order of the gravitational one. Here is how rotating galaxies came into being [89].

Hydrogen and helium in these galaxies were compressed by gravitational forces and heated up, fostering thermonuclear fusion reactions. The corresponding temperature growth stopped the gravitational compression, resulting in solar-type stars. Because of stronger gravity, more massive stars heated up to higher temperatures and burned up their hydrogen in a shorter time (in several hundred million years); thermonuclear transformation of helium to heavier elements such as carbon and oxygen began. After that, the central regions compressed to a superdense state like a black hole or a neutron star, while the outer regions, which contained heavy elements, were carried into interstellar space as a result of a supernova explosion to provide the material for the next generation of planets and stars. In particular, our Sun is a second- or third-generation star. The evolution of stellar objects is considered in greater detail in Sect. 7.2.

We emphasize once again that the atoms of our body were produced inside the stars, and atomic nuclei were produced even during the first three minutes of the life of the universe. Figuratively speaking, when we peer into huge distances, it is as if we are looking into the distant past because of the finiteness of the speed of light: very far away is the "place" where the atomic nuclei of our body were born and, still earlier, the elementary particles themselves [80]. In this sense the universe appears like an enormous "house", in which, according to the anthropic principle (see below), everything was prepared for the advent of human beings and which retains "photographs" from their first steps to our days.

We give the formula [227] that describes the temperature variation during the early expansion stage, down to $T \approx 10^{12}$ K:

$$T = \frac{1.3}{\xi^{1/4}\sqrt{t}}\ \text{MeV}.$$

By inverting this formula we obtain the point in time after the onset of expansion at which the temperature of relativistic particles was equal to $T$:

$$t = \frac{1.7}{(T\ [\text{MeV}])^2\,\xi^{1/2}}\ \text{s}.$$

Beginning from the atomic epoch, the photon gas has cooled independently of the rest of matter, nowadays manifesting itself as the relic radiation.

The nucleosynthesis stage is the final stage relating to the early universe. It terminated three minutes after the Big Bang.

The nucleosynthesis epoch is followed by the hidden-mass domination stage, which set in at about a temperature of $T \approx 10^6$ K, depending on the type of hidden matter carrier. Beginning with this epoch, small perturbations of matter density developed, which built up to the extent that there emerged galaxies (5–10 billion years after Big Bang), stars, and planets (10–11.5 billion years).

At this point, the cosmological stage, i.e., that relating to the universe as a whole, of matter formation of our surrounding universe may be thought of as being complete. The subsequent stages are inherently local and involve the formation of the structure of the universe – the large-scale structure (clusters and superclusters), galaxies, stars, and planets. The prevalent form of visible matter at these stages is the plasma state (cosmic plasmas: circumplanetary, interplanetary, intraplanetary plasmas, the plasma of stars and stellar atmospheres, the plasma of quasars and galactic nuclei, interstellar and intergalactic plasmas, quark–gluon plasma).

Of special significance from the standpoint of matter composition formation is the hot stellar plasma, in which there occurs the synthesis of the nuclei of light elements up to carbon (the stars of the main sequence and red giants). Heavier elements are produced in the explosion of supernovae – the catastrophic late stages of stellar evolution – by way of neutron capture with subsequent $\beta$-decay. This picture of element production allowed Carl Sagan to consider that we are all made of stellar matter.

Also produced in supernova explosions are cosmic rays – a strongly nonequilibrium component of matter. At the last stages of their evolution, stars move to the state of either a white dwarf, whose matter is an ideal, "simple" metal (a crystal lattice of nuclei surrounded by a nearly free electron gas), or a black hole, or a neutron star. The latter, as discussed above, contains a liquid (superfluid) core, made up of neutrons with a small admixture of protons and electrons (neutron matter); the inner crust, consisting of a crystal lattice of neutron-excessive nuclei, an electron gas, and a neutron liquid; and the outer crust, which is structurally similar to the matter of a white dwarf.

Therefore, we see that our universe has gone through an extremely stormy youth full of extraordinarily bright events. The "Planck temperature" $T = 10^{32}$ K was reached in $10^{-43}$ s after the beginning of the universe, $T = 10^{13}$ K was reached in $10^{-6}$ s, $T = 10^{11}$ K in 0.01 s, $T = 10^{10}$ K in 1 s, $T = 10^{9}$ K in 1 min, $T = 10^{4}$ K (change of epochs) in 100 000 years, and $T = 10^{3}$ K in 1 million years. Therefore, primary nucleosynthesis was completed within several minutes of the "birth of the universe" and the formation of atoms within 1 million years. After this rather stormy period, the initial evolution stage was completed and the routine senile expansion phase is reached, which we now observe approximately 15 billion years after the "birth of the universe".

Figure 7.111 shows the scenario of the universe's evolution discussed in [32]. According to Carroll [32], the universe sprang into existence as a highly uniform plasma and will go out of existence on turning into almost empty space, i.e., the universe evolves from a low-entropy state to a high-entropy state – the final state being termed "heat death".

The vigorous epochs of biological progress, of the origination of life and civilization that we witness nowadays fall precisely in this routine development process of the universe. In particular, the planet Earth was formed about 5 billion years ago, elementary organic life about 4 billion years ago, and DNA and the genetic code about 3.5 billion years ago (Fig. 7.112). Plants and animals emerged about 1.2 billion years ago. Early human ancestors came into existence about one million years ago and the *Homo sapiens* species about 200 000 years ago. This species began to speak only $\approx$50 000 years B.C. and write 6000–8000 years ago. Modern civilization is believed to appeared at the end of the last glacial period – about 10 thousand years B.C. The first towns emerged 6000–7000 years ago, the Christian civilization 2000 years ago, and the technical one only 200 years ago.

Since then, humans have so developed that they have become capable of posing questions about the physics of extreme states of matter. In order of magnitude this time corresponds to an appraisal made by St. Augustine of Hippo (nearly 1600 years ago) in his work *The City of God* [89]. Answering the question as to what God was occupied with before creating the universe, he said that God was preparing hell for those who put such questions [89]. The possibility of the emergence of living and sapient subjects who ask such questions and dare cognize the universe is one of the most challenging and interesting questions of the natural sciences.

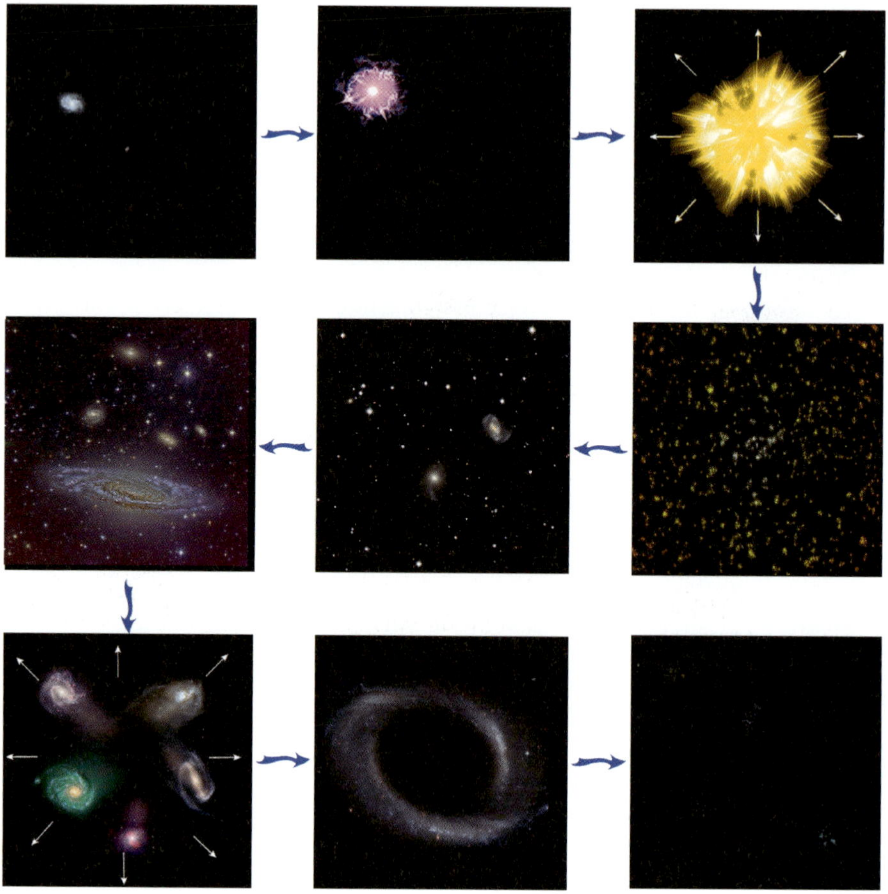

**Fig. 7.111** Evolution of the universe. Initially space is almost empty; quantum fields fluctuate in some domains. Next the space expands exponentially (inflation) and fills everywhere with the nearly uniform primary plasma. Aggregates of matter give rise to stars and galaxies, which fly apart due to accelerated expansion. Giant black holes contained at the centers of some galaxies take up everything around them and in turn slowly evaporate. The space becomes empty once again... Reprinted, with permission, from [32].

We see that in the course of evolution our universe undergoes several changes as it transforms to increasingly complex and diverse structures [80]. After the origin of classical space-time there emerge the first elementary particles, the first nuclei (3 min), atoms (recombination era) of light elements, galaxies and stars; carbon and heavy elements are synthesized inside stars and ejected into space in explosions; planets are formed and, finally, there emerges organic life and the most complex object in the universe: the human brain, consisting of $10^{11}$ neurons (of the order of the number of stars in the galaxy), each of which is much more com-

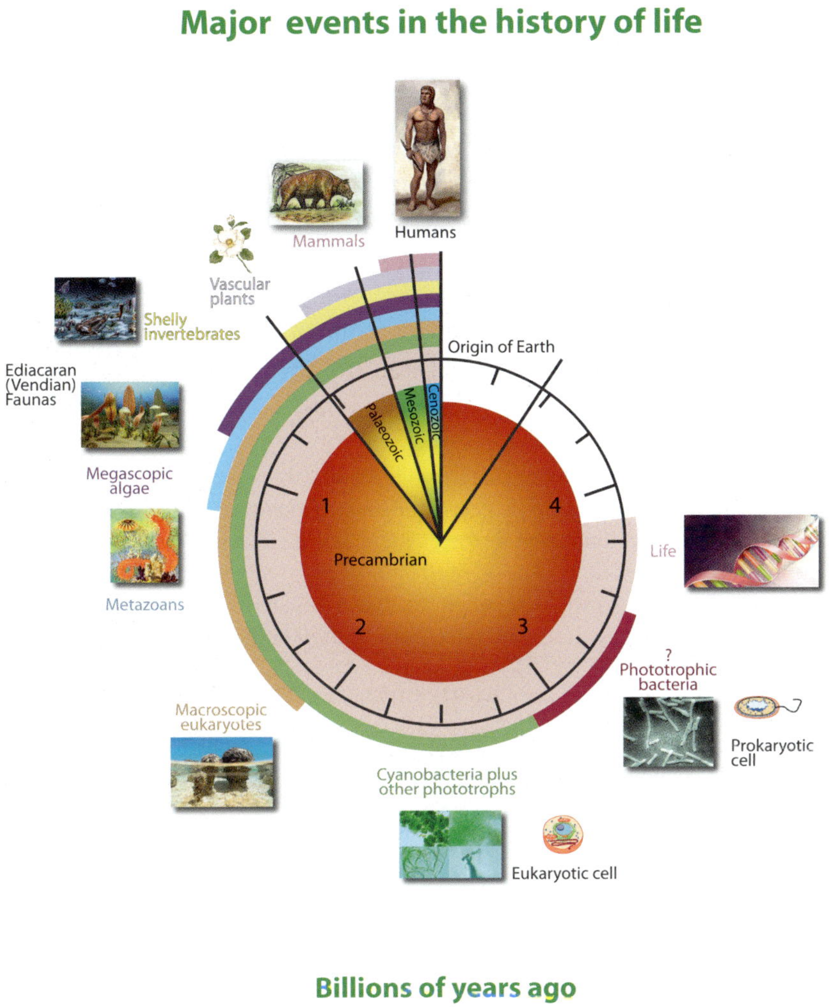


**Fig. 7.112** Main milestones in the history of life (in billions of years).

plex than a star. This took an extremely long time (about 15 billion years), during which the universe came a long way, increasing by a factor of $10^{61}$. Were the atoms of our body capable of telling their story from their birth from vacuum onwards, they would have spoken about all the eras that we analyzed in the foregoing. . . .

How is the existence of human beings related to the basic properties of the universe? Over what range may these properties vary for our life to take place? It turns out that this range is extremely narrow. We shall adduce several examples. As noted by Stephen Hawking [89], had the expansion velocity of the first-second universe differed by 1 millionth from the critical value, human beings would not have come into being in it. In one case the universe would have turned out to be too hot and would have rapidly contracted, and in the other case it would have been too cold for us, and the galaxies would not have been formed. Only some "middle" path leads to humans.

A hydrogen atom is an absolutely stable element. The reaction

$$p + e^- \to n + \nu,$$

where $p$, $e^-$, $n$, $\nu$ are the proton, electron, neutron, and neutrino, respectively, is strictly forbidden at low energies (temperatures), because

$$m_e \approx 0.511\ \text{MeV} < \Delta m_N = m_n - m_p \approx 1.3\ \text{MeV}.$$

However, should the mass $m_e$ be (ideally) increased, this reaction would be possible, which would lead to catastrophic consequences: the impossibility of galaxies and, hence, stars being formed, because the existence of neutral hydrogen is required for the formation of galaxies.

Another example is the "resonance" in the nuclear reaction involving the production of carbon from helium and beryllium accompanied by the emission of $\gamma$-ray photons:

$$2\,^4\text{He} \to {}^8\text{Be},$$

$$^4\text{He} + {}^8\text{Be} \to {}^{12}C + 2\gamma.$$

The probability of this nuclear reaction rises when the energy of $^4$He and $^8$Be is equal to 7.65 MeV, which may occur in stars at high temperatures. Were it not for this resonance, carbon would not have been produced in quantities sufficient for the origin of life [80]. Furthermore, the resultant carbon might immediately transform to oxygen by the reaction $^4\text{He} + {}^{12}\text{C} \to {}^{16}\text{O}$, and carbon-based organic life would not have been able to emerge. In this case, however, by sheer luck the figures diverged slightly and the resonance in the reaction involving the production of oxygen and the disappearance of carbon turned out to be forbidden. Indeed, the energy of $^4\text{He} + {}^{12}\text{C}$ is 7.166 MeV, while the energy of $^{16}$O is 7.118 MeV, and this reaction is impossible in the rest frame of $^{16}$O. So, an energy difference of 0.05 MeV inside a star resolved the problem of the future advent of human beings.

The biproton $^2$He is an unstable particle. This instability arises from the Coulomb repulsion of protons (its potential energy is $\approx 2$ MeV). However, estimates show that were the potential energy of nuclear attraction only 50 keV higher, the nucleus $^2$He would be stable, which would lead to the reaction

$$p + p \to {}^2\text{He} + \gamma$$

with the consequence that all protons would have burned up in the course of primary (cosmological) nucleosynthesis.

The list of such examples can be easily extended. Peculiar coincidences are encountered when comparing the magnitudes of the fundamental constants [80]:

– electromagnetic interaction constant (fine structure constant) $\alpha \approx 10^{-2}$;

– gravitational constant (dimensionless) $\alpha_G \approx 10^{-38} - 10^{-40}$;

– Fermi constant of weak interaction $\alpha_W \approx 10^{-5}$;

– strong interaction constant $\alpha_S \approx 10$;

– number of protons in the universe $N_p \approx 10^{80} \approx \alpha_G^{-2}$;

– mass of the universe $M \approx \alpha_G^{-2} m_p$, where $m_p$ is the proton mass;

– galactic mass $M \approx \alpha^4 \alpha_G^{-2} m_p$;

– stellar mass $M_{star} \approx \alpha_G^{-3/2} m_p$;

– planetary mass $M_{pl} \approx \alpha_G^{3/2} M_{star}$ (Jupiter's mass);

– mass of a human $M_{homo} = \sqrt{M_{pl} m_p}$.

Are these coincidences accidental? What is most amazing, small variations of the fundamental constants lead to the conclusion that man (like every living thing on the basis of carbon), for one reason or another, could not come into being. And the galaxies themselves would be entirely different, e.g., without hydrogen. That is why one of the formulations of the anthropic principle reduces to the statement that the metagalaxy is unstable relative to variations of the value of the fundamental constants.

The so-called convective stars, in which heat is transferred by way of convection (as in the Sun) and which have planets, exist because [80] $\alpha_G = \alpha^{20}$. Supernovae explode because $\alpha_G = (\alpha_W)^8$. Atoms exist as systems of nuclei, which are confined by the strong interaction, and electrons, which are confined in orbits due to electromagnetic interactions, because $\alpha = (\alpha_S)^{-3}$.

Lastly, the universe's expansion in the contemporary epoch is determined by the cosmological constants, which has the significance that our era commenced at about the instant the Sun came into existence. Owing to the accelerated repulsion of other galaxies from us, our galaxy is protected against collisions with them, which underlies the stability of the emerging Solar planetary system and biological evolution on the Earth.

However, an insignificant change of these constants radically changes the entire universe! The universe is like a finely tuned musical instrument!

All these coincidences and tunings are reflected in the so-called anthropic principle in cosmology, which is also referred to as the reasonability principle. Here, we are dealing with an investigation of the metagalaxy structure stability relative to variations of the numerical values of the fundamental constants $\alpha$ of the four interactions, the masses $m_p$ and $m_e$, and the dimensionality of space.

In its weak form it holds that the observable properties of the universe depend on man as the observer [80]. The point is that we as observers may exist at the time when we observe all visible objects and when all the relations we have established are fulfilled; at a different time there is "no one" to observe all this. Some coincidences are due to the properties of our observation in space and time.

The strong anthropic principle holds that the universe is so constructed that man was bound to come into being in it, which is close to the biblical anthropocentric idea regarding man as the goal of creation. The universe evolves in such a way that there emerges a creature, man, capable of unraveling the mathematical laws of evolution and seeing their hidden significance.

Of course, the coincidences of the figures given above simply concern a creature made of carbon, rather than specifically a human being – who is little different from animals in this case. However, there exists a formulation of the strong anthropic principle that implies the creation of an observer. If the universe is quantal, according to one of the versions of interpretation of quantum mechanics, the human observer defines its wave function, so that the Big Bang itself and the evolution of the universe proceed in such a way that the observer is bound to emerge. It is assumed that there exist, in addition to our universe, infinitely many – $10^{10}$ – other universes in which the fundamental constants do not obey the relations inherent in our world. The fundamental constants assume other values there. However, the point is that there are no humans in those worlds! We therefore see these relations not because we are the pinnacle of creation but because we cannot exist in other worlds!

In the further discussion of these difficult world-view problems we descend into the realms of philosophy or even theology, which is clearly beyond the scope of this book.

In the general form, this principle holds that the physical laws in the metagalaxy are not only sufficient for the main stable states (atomic nuclei, atoms, stars, galaxies), but are necessary as well.

## References


[1] Abdo, A.A., Ackermann, M., Ajello, M., et al.: Measurement of the cosmic ray $e^+ + e^-$ spectrum from 20 GeV to 1 TeV with the Fermi Large Area Telescope. Phys. Rev. Lett. **102**(18), 181101 (2009). DOI 10.1103/PhysRevLett.102.181101. URL `http://link.aps.org/abstract/PRL/v102/e181101`

[2] Abu-Zayyad, T., Belov, K., Bizel, J., et al.: Measurement of the cosmic ray composition between $10^{17}$ eV and $10^{18}$ eV (1999). Presented at the 26th International Cosmic Ray Conference, Salt Lake City, UT, USA

[3] Achterberg, A.: Particle acceleration by an ensemble of shocks. Astron. Astrophys. **231**(1), 251–258 (1990)

[4] Achterberg, A., Gallant, Y.A., Kirk, J.G., Guthmann, A.W.: Particle acceleration by ultrarelativistic shocks: theory and simulations. Month. Notices R. Astron. Soc. **328**(2), 393–408 (2001). DOI 10.1046/j.1365-8711.2001.04851.x

[5] Allard, F., Hauschildt, P.H., Alexander, D.R., Starrfield, S.: Model atmospheres of very low mass stars and brown dwarfs. Annu. Rev. Astron. Astrophys. **35**, 137–177 (1997). DOI 10.1146/annurev.astro.35.1.137. URL http://arjournals.annualreviews.org/doi/abs/10.1146/annurev.astro.35.1.137

[6] Arkani-Hamed, N., Dimopoulos, S., Dvali, G.: Phenomenology, astrophysics, and cosmology of theories with submillimeter dimensions and TeV scale quantum gravity. Phys. Rev. D **59**(8), 086004 (1999). DOI 10.1103/PhysRevD.59.086004

[7] Arnett, D.: Oxygen-burning hydrodynamics. 1: Steady shell burning. Astrophys. J. **427**(2), 932–946 (1994)

[8] Arnett, D.: Supernovae and Nucleosynthesis. Princeton University Press, Princeton, NJ (1996)

[9] Aschwanden, M.J.: Physics of the Solar Corona: An Introduction with Problems and Solutions, 2nd edn. Springer, Berlin (2006)

[10] Atzeni, S., Meyer-ter-Vehn, J.: The Physics of Inertial Fusion. Oxford University Press, Oxford (2004)

[11] Avrorin, E.N., Simonenko, V.A., Shibarshov, L.I.: Physics research during nuclear explosions. Phys. Usp. **49**(4), 432 (2006). DOI 10.1070/PU2006v049n04ABEH005958. URL http://ufn.ru/en/articles/2006/4/j/

[12] Avrorin, E.N., Vodolaga, B.K., Simonenko, V.A., Fortov, V.E.: Intense shock waves and extreme states of matter. Phys. Usp. **36**(5), 337–364 (1993). DOI 10.1070/PU1993v036n05ABEH002158. URL http://stacks.iop.org/1063-7869/36/337

[13] Ayukov, S.V., Baturin, A., Gryaznov, V.K., et al.: Analysis of the presence of small admixtures of heavy elements in the solar plasma by using the SAHA-S equation of state. JETP Lett. **80**(3), 141–144 (2004)

[14] Baade, W., Zwicky, F.: Cosmic Rays from Super-Novae. Proc. Natl. Acad. Sci. USA **20**(5), 259–263 (1934). URL http://www.pnas.org/content/20/5/259.short

[15] Balbus, S.A., Hawley, J.F.: A powerful local shear instability in weakly magnetized disks. I - Linear analysis. II - Nonlinear evolution. Astrophys. J. **376**(1), 214–233 (1991)

[16] Balega, Y.Y.: Brown dwarfs: substars without nuclear reactions. Phys. Usp. **45**(8), 883 (2002). DOI 10.1070/PU2002v045n08ABEH001191. URL http://ufn.ru/en/articles/2002/8/e/

[17] Balick, B., Frank, A.: The extraordinary deaths of ordinary stars. Sci. Am. **291**(1), 50 (2004)

[18] Bamber, C., Boege, S.J., Koffas, T., et al.: Studies of nonlinear QED in collisions of 46.6 GeV electrons with intense laser pulses. Phys. Rev. D **60**(9), 092004 (1999). DOI 10.1103/PhysRevD.60.092004. URL http://link.aps.org/abstract/PRD/v60/e092004

[19] Baturin, V.A., Mironova, I.V., Surdin, V.G.: Fizika i jevoljucija zvezd (Physics and evolution of stars). In: V.G. Surdin (ed.) Astronomiya: Vek XXI (Astronomy: XXIst Century), p. 120. Vek 2, Fryazino (2007)

[20] Baym, G., Pethick, C., Sutherland, P.: The ground state of matter at high densities: Equation of state and stellar models. Astrophys. J. **170**, 299 (1971)

[21] Begelman, M.C., Blandford, R.D., Rees, M.J.: Theory of extragalactic radio sources. Rev. Mod. Phys. **56**(2), 255–351 (1984). DOI 10.1103/RevModPhys.56.255

[22] Bell, A.R.: The acceleration of cosmic rays in shock fronts. I. Month. Notices R. Astron. Soc. **147-156**, 022105 (1978)

[23] Beskin, V.S., Gurevich, A.V., Istomin, Y.N.: Physics of the Pulsar Magnetosphere. Cambridge University Press, Cambridge (1993)

[24] Bezkrovniy, V., Filinov, V.S., Kremp, D., et al.: Monte Carlo results for the hydrogen Hugoniot. Phys. Rev. E **70**(5), 057401 (2004). DOI 10.1103/PhysRevE.70.057401. URL `http://link.aps.org/abstract/PRE/v70/e057401`

[25] Bisnovatyi-Kogan, G.S.: Stellar Physics: 1: Fundamental Concepts and Stellar Equilibrium. Springer, Berlin, Heidelberg (2001)

[26] Bleicher, M.: How to create black holes on Earth. Eur. J. Phys. **28**(3), 509–516 (2007). URL `http://stacks.iop.org/0143-0807/28/509`

[27] Blinnikov, S.I., Novikov, I.D., Perevodchikova, T.V., Polnarev, A.G.: Exploding neutron stars in close binaries. Sov. Astron. Lett. **10**(3), 177 (1984)

[28] Bondi, H., Hoyle, F.: On the mechanism of accretion by stars. Month. Notices R. Astron. Soc. **104**, 273 (1944)

[29] Bula, C., McDonald, K.T., Prebys, E.J., et al.: Observation of nonlinear effects in Compton scattering. Phys. Rev. Lett. **76**(17), 3116–3119 (1996). DOI 10.1103/PhysRevLett.76.3116. URL `http://link.aps.org/abstract/PRL/v76/p3116`

[30] Burke, D.L., Field, R.C., Horton-Smith, G., et al.: Positron production in multiphoton light-by-light scattering. Phys. Rev. Lett. **79**(9), 1626–1629 (1997). DOI 10.1103/PhysRevLett.79.1626. URL `http://link.aps.org/abstract/PRL/v79/p1626`

[31] Burrows, A., Hubbard, W.B., Lunine, J.I., Liebert, J.: The theory of brown dwarfs and extrasolar giant planets. Rev. Mod. Phys. **73**(3), 719–765 (2001). DOI 10.1103/RevModPhys.73.719. URL `http://link.aps.org/abstract/RMP/v73/p719`

[32] Carroll, S.M.: The cosmic origins of time's arrow. Sci. Am. **298**(6), 48 (2008)

[33] Chabrier, G., Baraffe, I.: Theory of low mass stars and substellar objects. Annu. Rev. Astron. Astrophys. **38**, 337–377 (2000). DOI 110.1146/annurev.astro.38.1.337. URL `http://arjournals.annualreviews.org/doi/abs/10.1146/annurev.astro.38.1.337`

[34] Chang, J., Adams Jr., J.H., Ahn, H.S., et al.: An excess of cosmic ray electrons at energies of 300–800 GeV. Nature **456**(7220), 362–365 (2008). DOI {10.1038/nature07477}

[35] Chen, P., Tajima, T., Takahashi, Y.: Plasma wakefield acceleration for ultrahigh-energy cosmic rays. Phys. Rev. Lett. **89**(16), 161101 (2002). DOI 10.1103/PhysRevLett.89.161101. URL `http://link.aps.org/abstract/PRL/v89/e161101`

[36] Cherepashchuk, A.M.: Masses of black holes in binary stellar systems. Phys. Usp. **39**(8), 759 (1996). DOI 10.1070/PU1996v039n08ABEH000160. URL `http://ufn.ru/en/articles/1996/8/a/`

[37] Cherepashchuk, A.M.: Chernye dyry v dvojnyh zvezdnyh sistemah (Black holes in double star systems). In: V.N. Soifer (ed.) Sovremennoe estestvoznanie. Entsiklopediya (Modern Natural Science. Encyclopedia), vol. 4, p. 228. Magistr-Press, Moscow (2000)

[38] Cherepashchuk, A.M.: Search for black holes. Phys. Usp. **46**(4), 335 (2003). DOI 10.1070/PU2003v046n04ABEH001282. URL `http://ufn.ru/en/articles/2003/4/a/`

[39] Cherepashchuk, A.M.: Chernye dyry vo Vselennoj (Black holes in the Universe). In: V.G. Surdin (ed.) Astronomiya: Vek XXI (Astronomy: XXIst Century), p. 219. Vek 2, Fryazino (2007)

[40] Cherepashchuk, A.M., Chernin, A.D.: Vselennaja, zhizn', chernye dyry (The Universe, Life, Black Holes). Vek 2, Fryazino (2004)

[41] Chernin, A.D.: Cosmic vacuum. Phys. Usp. **44**(11), 1099 (2001). DOI 10.1070/PU2001v044n11ABEH000962. URL `http://ufn.ru/en/articles/2001/11/a/`

[42] Chernin, A.D.: Dark energy and universal antigravitation. Phys. Usp. **51**(3), 253 (2008). DOI 10.1070/PU2008v051n03ABEH006320. URL `http://ufn.ru/en/articles/2008/3/c/`

[43] Clark, E.L., Krushelnick, K., Davies, J.R., et al.: Measurements of energetic proton transport through magnetized plasma from intense laser interactions with solids. Phys. Rev. Lett. **84**(4), 670–673 (2000). DOI 10.1103/PhysRevLett.84.670. URL `http://link.aps.org/abstract/PRL/v84/p670`

[44] Cowan, T.E., Hunt, A.W., Phillips, T.W., et al.: Photonuclear fission from high energy electrons from ultraintense laser-solid interactions. Phys. Rev. Lett. **84**(5), 903–906 (2000). DOI 10.1103/PhysRevLett.84.903. URL `http://link.aps.org/abstract/PRL/v84/p903`

[45] Cuneo, M.E., Vesey, R.A., Bennett, G.R., et al.: Progress in symmetric ICF capsule implosions and wire-array Z-pinch source physics for double-pinch-driven hohlraums. Plasma Phys. Control. Fusion **48**(2), R1–R35 (2006). DOI 10.1088/0741-3335/48/2/R01

[46] Disdier, L., Garconnet, J.P., Malka, G., Miquel, J.L.: Fast neutron emission from a high-energy ion beam produced by a high-intensity subpicosecond laser pulse. Phys. Rev. Lett. **82**(7), 1454–1457 (1999). DOI 10.1103/PhysRevLett.82.1454. URL `http://link.aps.org/abstract/PRL/v82/p1454`

[47] Drake, R.P.: High-Energy-Density Physics. Springer, Berlin, Heidelberg (2006)

[48] Drake, R.P., Leibrandt, D.R., Harding, E.C., et al.: Nonlinear mixing behavior of the three-dimensional Rayleigh–Taylor instability at a decelerating interface. Phys. Plasmas **11**(5), 2829–2837 (2004). DOI 10.1063/1.1651492

[49] Dubin, D.H.E., O'Nail, T.M.: Trapped nonneutral plasmas, liquids and crystals (the thermal equilibrium states). Rev. Mod. Phys. **71**, 87 (1999)

[50] Edens, A.D., Ditmire, T., Hansen, J.F., et al.: Studies of laser-driven radiative blast waves. Astrophys. Space Sci. **298**(1-2), 39–47 (2005). DOI 10.1007/s10509-005-3910-8

[51] Efremov, Y.N.: Zvezdnye ostrova (Star Islands). Vek 2, Fryazino (2005)

[52] Efremov, Y.N.: Spiral'naja struktura nashej galaktiki (Spiral structure of our galaxy). In: V.G. Surdin (ed.) Astronomiya: Vek XXI (Astronomy: XXIst Century), p. 313. Vek 2, Fryazino (2007)

[53] Esirkepov, T., Yamagiwa, M., Tajima, T.: Laser ion-acceleration scaling laws seen in multiparametric particle-in-cell simulations. Phys. Rev. Lett. **96**(10), 105001 (2006). DOI 10.1103/PhysRevLett.96.105001. URL `http://link.aps.org/abstract/PRL/v96/e105001`

[54] European Southern Observatory: (1999). URL `http://www.eso.org/public/outreach/press-rel/pr-1999/phot-29-99.html`

[55] Faber, S.M., Gallagher, J.S.: Exploding neutron stars in close binaries. Annu. Rev. Astron. Astrophys. **17**, 135–187 (1979)

[56] Faber, T.E.: Fluid Dynamics for Physicists. Cambridge University Press, Cambridge (1977)

[57] Fabrika, S.: The jets and and supercritical accretion disk in SS433. Astrophys. Space Phys. Rev. **12**, 1 (2004)

[58] Falize, E., Bouquet, S., Michaut, C.: Radiation hydrodynamics scaling laws in high energy density physics and laboratory astrophysics. J. Phys.: Conf. Ser. **112**(4), 042016 (4pp) (2008). URL `http://stacks.iop.org/1742-6596/112/042016`

[59] Fermi, E.: On the origin of the cosmic radiation. Phys. Rev. **75**(8), 1169–1174 (1949). DOI 10.1103/PhysRev.75.1169. URL `http://link.aps.org/abstract/PR/v75/p1169`

[60] Filinov, V.S., Bonitz, M., Levashov, P., et al.: Plasma phase transition in dense hydrogen and electron–hole plasmas. J. Phys. A **36**(22), 6069–6076 (2003). DOI 10.1088/0305-4470/36/22/332

[61] Filinov, V.S., Levashov, P.R., Bonitz, M., Fortov, V.E.: Calculation of the shock Hugoniot of deuterium atpressures above 1 Mbar by the path-integral Monte Carlo method. Plasma Phys. Rep. **31**(8), 700–704 (2005). DOI 10.1134/1.2031631

[62] Fortov, V., Iakubov, I., Khrapak, A.: Physics of Strongly Coupled Plasma. Oxford University Press, Oxford (2006)

[63] Fortov, V.E. (ed.): Entsiklopediya nizkotemperaturnoi plazmy (Encyclopedia of Low-Temperature Plasma). Nauka, Moscow (2000)

[64] Fortov, V.E.: Intense Shock Waves and Extreme States of Matter. Bukos, Moscow (2005)

[65] Fortov, V.E. (ed.): Explosive-Driven Generators of Powerful Electrical Current Pulses. Cambridge International Science, Cambridge (2007)

[66] Fortov, V.E.: Intense shock waves and extreme states of matter. Phys. Usp. **50**(4), 333 (2007). DOI 10.1070/PU2007v050n04ABEH006234. URL `http://ufn.ru/en/articles/2007/4/c/`

[67] Fortov, V.E., Gnedin, Y.N., Ivanov, M.F., et al.: Collision of comet Shoemaker–Levy 9 with Jupiter: what did we see. Phys. Usp. **39**(4), 363 (1996). DOI 10.1070/PU1996v039n04ABEH000142. URL `http://ufn.ru/en/articles/1996/4/c/`

[68] Fortov, V.E., Ilkaev, R.I., Arinin, V.A., et al.: Phase transition in a strongly nonideal deuterium plasma generated by quasi-isentropical compression at megabar pressures. Phys. Rev. Lett. **99**(18), 185001 (2007). DOI 10.1103/PhysRevLett.99.185001. URL `http://link.aps.org/abstract/PRL/v99/e185001`

[69] Fortov, V.E., Ivlev, A.V., Khrapak, S.A., et al.: Complex (dusty) plasma: current status, open issues, perspectives. Phys. Rep. **421**(1), 1–103 (2005). DOI 10.1016/j.physrep.2005.08.007

[70] Fortov, V.E., Khrapak, A.G., Khrapak, S.A., et al.: Dusty plasmas. Phys. Usp. **47**(5), 447 (2004). DOI 10.1070/PU2004v047n05ABEH001689. URL `http://ufn.ru/en/articles/2004/5/b/`

[71] Fortov, V.E., Khrapak, A.G., Yakubov, I.T.: Fizika neideal'noi plazmy (Physics of Nonideal Plasma). Fizmatlit, Moscow (2004)

[72] Fortov, V.E., Ternovoi, V.Y., Zhernokletov, M.V., et al.: Pressure-produced ionization of nonideal plasma in a megabar range of dynamic pressures. JETP **97**(2), 259–278 (2003). DOI 10.1134/1.1608993

[73] Gehrels, N., Piro, L., Leonard, P.J.T.: The brightest in the universe – gamma-ray bursts herald the birth of a black hole. Sci. Am. **287**(6), 84–91 (2002)

[74] Gelliot, T.: Understanding the evolution of giant planets: importance of equation of state (2007). Presented at the International Workshop on Warm Dense Matter, University of Rostock, Germany

[75] Ginzburg, V.L.: The Physics of a Lifetime: Reflections on the Problems and Personalities of 20th Century Physics. Springer, Berlin, Heidelberg (2001)

[76] Ginzburg, V.L.: On superconductivity and superfluidity (what I have and have not managed to do), as well as on the "physical minimum" at the beginning of the XXI century (December 8, 2003). Phys. Usp. **47**(11), 1155 (2004). DOI 10.1070/PU2004v047n11ABEH001825. URL `http://ufn.ru/en/articles/2004/11/d/`

[77] Glendenning, N.K.: Compact Stars: Nuclear Physics, Particle Physics, and General Relativity, 2nd edn. Springer, New York (2000)

[78] Grabovskii, E.V., Vorob'ev, O.Y., Dyabilin, K.S., et al.: Excitation of intense shock waves by soft x radiation from a Z-pinch plasma. JETP Lett. **60**(1), 1 (1994)

[79] Greisen, K.: End to the cosmic-ray spectrum? Phys. Rev. Lett. **16**(17), 748–750 (1966). URL `http://link.aps.org/abstract/PRL/v16/p748`

[80] Grib, A.A.: Osnovnye predstavleniya sovremennoi kosmologii (The Basic Representations of Modern Cosmology). FizMatLit, Moscow (2008)

[81] Grishchuk, L.P.: Relic gravitational waves and their detection. In: C. Lämmerzahl, C.W.F. Everitt, F.W. Hehl (eds.) Gyros, Clocks, Interferometers…: Testing Relativistic Gravity in Space. Lect. Notes Phys., Vol. 562, p. 167. Springer, Berlin, Heidelberg (2001). DOI 10.1007/3-540-40988-2

[82] Grishchuk, L.P.: Relic gravitational waves and cosmology. Phys. Usp. **48**(12), 1235 (2005). DOI 10.1070/PU2005v048n12ABEH005795. URL http://ufn.ru/en/articles/2005/12/i/

[83] Gyulassy, M.: Quark gluon plasmas: Femto cosmology with A+A @ LHC. Presented at the ExtreMe Matter Institute EMMI Kick-Off Meeting & Symposium, July 16–17, 2008, GSI, Darmstadt, Germany

[84] Haensel, P., Potekhin, A., Yakovlev, D.: Neutron Stars 1: Equation of State and Structure. Springer, New York (2007)

[85] Hands, S.: The phase diagram of QCD. Contemp. Phys. **42**(4), 209–225 (2001). DOI 10.1080/00107510110063843. URL http://pdfserve.informaworld.com/104165_762903897_713806686.pdf

[86] Hansen, J.F., Edwards, M.J., Froula, D.H., et al.: Laboratory observation of secondary shock formation ahead of a strongly radiative blast wave. Phys. Plasmas **13**(2), 022105 (2006). DOI 10.1063/1.2168157

[87] Hawke, P.S., Burgess, T.J., Duerre, D.E., et al.: Observation of electrical conductivity of isentropically compressed hydrogen at megabar pressures. Phys. Rev. Lett. **41**(14), 994–997 (1978). URL http://link.aps.org/abstract/PRL/v41/p994

[88] Hawking, S.W.: Particle creation by black holes. Commun. Math. Phys. **43**(3), 199–220 (1975). DOI 10.1007/BF02345020

[89] Hawking, S.W.: A Brief History of Time: From the Big Bang to Black Holes. Bantam Books, Toronto (1988)

[90] Hayashida, N., Honda, K., Honda, M., et al.: Observation of a very energetic cosmic ray well beyond the predicted 2.7 K cutoff in the primary energy spectrum. Phys. Rev. Lett. **73**(26), 3491–3494 (1994). URL http://link.aps.org/abstract/PRL/v73/p3491

[91] Hu, W., White, M.: The cosmic symphony. Sci. Am. **290**(2), 44–53 (2004)

[92] Ichimaru, S.: Nuclear fusion in dense plasmas. Rev. Mod. Phys. **65**(2), 255–299 (1993). DOI 10.1103/RevModPhys.65.255. URL http://link.aps.org/abstract/RMP/v65/p255

[93] Imshennik, V., Nadyozhin, D.: Supernova-1987A, and the emergence of the blast wave at the surface of the compact presupernova. Sov. Astron. Lett. **14**(6), 449–452 (1988)

[94] Imshennik, V.S.: Experimental possibilities for studying the neutronization of matter in stars. Phys. At. Nucl. **58**(5), 823–831 (1995)

[95] Imshennik, V.S.: Explosion mechanism in supernovae collapse. Space Sci. Rev. **74**(3-4), 325–334 (1995). DOI 10.1007/BF00751418

[96] Istomin, Y.N.: Electron–positron plasma generation in the magnetospheres of neutron stars. Phys. Usp. **51**(8), 844 (2008). DOI

10.1070/PU2008v051n08ABEH006596. URL http://ufn.ru/en/articles/2008/8/g/
[97] Ivanova, L.N., Imshennik, V.S., Chechotkin, V.M.: Pulsation regime of the thermonuclear explosion of a star's dense carbon core. Astrophys. Space Sci. **31**(2), 497–514 (1974). DOI 10.1007/BF00644102
[98] Jeffries, C.D., Keldysh, L.V. (eds.): Electron–Hole Droplets in Semiconductors. North-Holland, Amsterdam (1983)
[99] Kadomtsev, B.B.: Selected Works [in Russian], vol. 1. Nauka, Moscow (2003)
[100] Kando, M., Nakajima, K., Arinaga, M., et al.: Interaction of terawatt laser with plasma. J. Nucl. Mater. **248**(1), 405–407 (1997). DOI 10.1016/S0022-3115(97)00177-3
[101] Kaplan, S.A.: The Physics of Stars. Wiley, Chichester, UK (1982). [Original in Russian: Fizika Zvezd, Nauka, Moscow, 1970, 2nd edn.]
[102] Kardashev, N.S., Novikov, I.D., Shatskii, A.A.: Magnetic tunnels (wormholes) in astrophysics. Astron. Rep. **50**(8), 601–611 (2006). DOI 10.1134/S1063772906080014
[103] Karnakov, B.M., Mur, V.D., Popov, V.S.: Contribution to the theory of Lorentzian ionization. JETP Lett. **65**(5), 405–411 (1997)
[104] Kifonidis, K., Plewa, T., Janka, H.T., Müller, E.: Nucleosynthesis and clump formation in a core-collapse supernova. Astrophys. J. Lett. **531**, L123–126 (2000). DOI 10.1086/312541
[105] Kirzhnits, D.A.: Extremal states of matter (ultrahigh pressures and temperatures). Sov. Phys. – Usp. **14**(4), 512–523 (1972). DOI 10.1070/PU1972v014n04ABEH004734. URL http://stacks.iop.org/0038-5670/14/512
[106] Kirzhnits, D.A.: Lektsii po fizike (Lectures on Physics). Nauka, Moscow (2006)
[107] Klumov, B.A., Kondaurov, V.I., Konyukhov, A.V., et al.: Collision of comet Shoemaker–Levi 9 with Jupiter: what shall we see? Phys. Usp. **37**(6), 577 (1994). DOI 10.1070/PU1994v037n06ABEH000027. URL http://ufn.ru/en/articles/1994/6/c/
[108] Knudson, M.D., Hanson, D.L., Bailey, J.E., et al.: Equation of state measurements in liquid deuterium to 70 GPa. Phys. Rev. Lett. **87**(22), 225501 (2001). DOI 10.1103/PhysRevLett.87.225501. URL http://link.aps.org/abstract/PRL/v87/e225501
[109] Kocharov, G.E.: Termojadernyj kotel v nedrah solnca i problema solnechnyh nejtrino (Thermonuclear copper in the center of the sun and the problem of solar neutrinos). In: V.N. Soifer (ed.) Sovremennoe Estestvoznanie. Entsiklopediya (Modern Natural Science. Encyclopedia), vol. 4, p. 28. Magistr-Press, Moscow (2000)
[110] Kodama, R., Norreys, P.A., Mima, K., et al.: Fast heating of ultrahigh-density plasma as a step towards laser fusion ignition. Nature **412**(6849), 798–802 (2001). DOI 10.1038/35090525

[111] Kodama, R., Tanaka, K.A., Sentoku, Y., et al.: Observation of ultrahigh gradient electron acceleration by a self-modulated intense short laser pulse. Phys. Rev. Lett. **84**(4), 674–677 (2000). DOI 10.1103/PhysRevLett.84.674. URL `http://link.aps.org/abstract/PRL/v84/p674`
[112] Koester, D.: White dwarfs: Recent developments. Astron. Astrophys. Rev. **11**(1), 33–66 (2007). DOI 10.1007/s001590100015
[113] Kouveliotou, C., Duncan, R.C., Thompson, C.: Intensely magnetic neutron stars alter the quantum physics of their surroundings. Sci. Am. **288**(2), 35 (2003)
[114] Krauss, L.M., Scherrer, R.J.: The end of cosmology? Sci. Am. **298**(3), 46–53 (2008)
[115] Kruer, W.L.: The Physics of Laser Plasma Interactions. Addison-Wesley, Reading, MA (1988)
[116] Lada, C.J.: Cold outflows, energetic winds, and enigmatic jets around young stellar objects. Annu. Rev. Astron. Astrophys. **23**, 267–317 (1985). DOI 10.1146/annurev.aa.23.090185.001411
[117] Leemans, W.P., van Tilborg, J., Faure, J., et al.: Terahertz radiation from laser accelerated electron bunches. Phys. Plasmas **11**(5), 2899–2906 (2004). DOI 10.1063/1.1652834
[118] Levin, A.: Kosmicheskie bomby (Space bombs). Populyarnaya mehanika (Popular Mechanics) **8**(58), 38 (2007)
[119] Liang, E.P., Wilks, S.C., Tabak, M.: Pair production by ultraintense lasers. Phys. Rev. Lett. **81**(22), 4887–4890 (1998). DOI 10.1103/PhysRevLett.81.4887. URL `http://link.aps.org/abstract/PRL/v81/p4887`
[120] Lifshits, E.M.: Gravitational stabilities of the expanding world. JETP **16**, 587 (1946)
[121] Lindl, J.D.: Inertial Confinement Fusion. Springer, New York (1998)
[122] Lobo, F.S.N.: Phantom energy traversable wormholes. Phys. Rev. D **71**(8), 084011 (2005). DOI 10.1103/PhysRevD.71.084011. URL `http://link.aps.org/abstract/PRD/v71/e084011`
[123] Lukash, V.N., Rubakov, V.A.: Dark energy: Myths and reality. Phys. Usp. **51**(3), 283 (2008). DOI 10.1070/PU2008v051n03ABEH006567. URL `http://ufn.ru/en/articles/2008/3/d/`
[124] Lyutikov, M.: Magnetar giant flares and afterglows as relativistic magnetized explosions. Month. Notices R. Astron. Soc. **367**(4), 1594–1602 (2006). DOI 10.1111/j.1365-2966.2006.10069.x
[125] Mackinnon, A.J., Borghesi, M., Hatchett, S., et al.: Effect of plasma scale length on multi-MeV proton production by intense laser pulses. Phys. Rev. Lett. **86**(9), 1769–1772 (2001). DOI 10.1103/PhysRevLett.86.1769. URL `http://link.aps.org/abstract/PRL/v86/p1769`
[126] Maksimchuk, A., Flippo, K., Krause, H., et al.: Plasma phase transition in dense hydrogen and electron—hole plasmas. Plasma Phys. Rep. **30**(6), 473–495 (2004). DOI 10.1134/1.1768582
[127] Maksimchuk, A., Gu, S., Flippo, K., et al.: Forward ion acceleration in thin films driven by a high-intensity laser. Phys. Rev. Lett. **84**(18), 4108–4111

(2000). DOI 10.1103/PhysRevLctt.84.4108. URL http://link.aps.org/abstract/PRL/v84/p4108
[128] Mangles, S.P.D., Murphy, C.D., Najmudin, Z., et al.: Monoenergetic beams of relativistic electrons from intense laser—plasma interactions. Nature **431**(7008), 535–538 (2004). DOI 10.1038/nature02939
[129] Mason, R.J., Tabak, M.: Magnetic field generation in high-intensity-laser—matter interactions. Phys. Rev. Lett. **80**(3), 524–527 (1998). DOI 10.1103/PhysRevLett.80.524. URL http://link.aps.org/abstract/PRL/v80/p524
[130] McCray, R.: Supernova 1987A revisited. Annu. Rev. Astron. Astrophys. **31**, 175–216 (1993). DOI 10.1146/annurev.aa.31.090193.001135
[131] Mesyats, G.A.: Impul'snaya energetika i elektronika (Pulse Power and Electronics). Nauka, Moscow (2004)
[132] Meszaros, P., Rees, M.J.: Relativistic fireballs and their impact on external matter – Models for cosmological gamma-ray bursts. Astrophys. J. **405**, 278–284 (1991)
[133] Mezzacappa, A.: Ascertaining the core collapse supernova mechanism: The state of the art and the road ahead. Annu. Rev. Nucl. Part. Sci. **55**(1), 467–515 (2005). DOI 10.1146/annurev.nucl.55.090704.151608. URL http://arjournals.annualreviews.org/doi/abs/10.1146/annurev.nucl.55.090704.151608
[134] Michaut, C., Falize, E., Cavet, C., et al.: Link between laboratory and astrophysical radiative shocks. J. Phys.: Conf. Ser. **112**(4), 042013 (4pp) (2008). URL http://stacks.iop.org/1742-6596/112/042013
[135] Michel, F.C.: Theory of Neutron Star Magnetospheres. University of Chicago Press, Chicago (1991)
[136] Mihalas, D.: Stellar Atmospheres, 2nd edn. W.H. Freeman and Co., San Francisco (1978)
[137] Miller, S., Tennyson, J., Jones, H.R.A., Longmore, A.J.: Computation of frequencies and linestrengths for triatomic molecules of astronomical interest. In: U.G. Jorgensen (ed.) Molecules in the Stellar Environment. Lecture Notes in Physics, Vol. 428, pp. 296–309. Springer, Berlin, Heidelberg (1994). DOI 10.1007/3-540-57747-5_52
[138] Mima, K., Ohsuga, T., Takabe, H., et al.: Wakeless triple-soliton accelerator. Phys. Rev. Lett. **57**(12), 1421–1424 (1986). URL http://link.aps.org/abstract/PRL/v57/p1421
[139] Mourou, G.A., Tajima, T., Bulanov, S.V.: Optics in the relativistic regime. Rev. Mod. Phys. **78**(2), 1804–1816 (2006). DOI 10.1103/RevModPhys.78.309. URL http://link.aps.org/abstract/RMP/v78/p309
[140] Murray, C.A., Wenk, R.A.: Observation of order–disorder transitions and particle trajectories in a model one-component plasma: time resolved microscopy of colloidal spheres. In: H.M. Van Horn, S. Ichimaru (eds.) Strongly Coupled Plasma Physics, p. 367. University of Rochester Press, Rochester, NY (1993)

[141] Nadyozhin, D.K., Yudin, A.V.: The influence of Coulomb interaction on the equation of state under nuclear statistical equilibrium conditions. Astron. Lett. **31**(4), 271–279 (2005). DOI 10.1134/1.1896071

[142] Nagano, M., Watson, A.A.: Observations and implications of the ultrahigh-energy cosmic rays. Rev. Mod. Phys. **72**(3), 689–732 (2000). DOI 10.1103/RevModPhys.72.689. URL http://link.aps.org/abstract/RMP/v72/p689

[143] NASA: HubbleSite. URL http://hubblesite.org

[144] NASA, Jet Propulsion Laboratory: (2005). URL http://photojournal.jpl.nasa.gov/catalog/PIA03239

[145] National Research Council: Frontiers in High Energy Density Physics. National Academies Press, Washington, DC (2003)

[146] Nellis, W.J.: Shock compression of hydrogen and other small molecules. In: G.L. Chiarotti, R.J. Hemley, M. Bernasconi, L. Ulivi (eds.) High Pressure Phenomena, Proceedings of the International School of Physics "Enrico Fermi" Course CXLVII, p. 607. IOS Press, Amsterdam (2002)

[147] Nellis, W.J.: Dynamic compression of materials: metallization of fluid hydrogen at high pressures. Rep. Prog. Phys. **69**(5), 1479–1580 (2006). DOI 10.1088/0034-4885/69/5/R05

[148] Nicolaï, P., Stenz, C., Kasperczuk, A., et al.: Studies of supersonic, radiative plasma jet interaction with gases at the Prague Asterix Laser System facility. Phys. Plasmas **15**(8), 082701 (2008). DOI 10.1063/1.2963083. URL http://link.aip.org/link/?PHP/15/082701/1

[149] Nishida, A.: The Earth's dynamic magnetotail. Space Sci. Rev. **91**(3-4), 507–577 (2000). DOI 10.1023/A:1005223124330

[150] Novikov, I.D.: "Big Bang" echo (cosmic microwave background observations). Phys. Usp. **44**(8), 817 (2001). DOI 10.1070/PU2001v044n08ABEH000983. URL http://ufn.ru/en/articles/2001/8/h/

[151] Okun', L.B.: Leptony i kvarki, 2nd edn. Nauka, Moscow (1990). [English Transl.: Leptons and Quarks. North-Holland, Amsterdam (1982)]

[152] Olinto, A.V.: Ultra high energy cosmic rays: the theoretical challenge. Phys. Rep. **333–334**, 329–348 (2000). DOI 10.1016/S0370-1573(00)00028-4

[153] Olling, R.P., Merrifield, M.R.: Two measures of the shape of the dark halo of the Milky Way. Month. Notices Royal Astron. Soc. **311**(2), 361–369 (2000). DOI {10.1046/j.1365-8711.2000.03053.x}

[154] Öpik, E.J.: Stellar associations and supernovae. Irish Astron. J. **2**(8), 219–233 (1953)

[155] Page, D., Applegate, J.H.: The cooling of neutron stars by the direct Urca process. Astrophys. J. **394**, L17–L20 (1992)

[156] Palmer, D.M., Barthelmy, S., Gehrels, N., et al.: A giant gamma-ray flare from the magnetar SGR 1806-20. Nature **434**(7037), 1107–1109 (2005). DOI {10.1038/nature03525}

[157] Panasyuk, M.I.: Stranniki vselennoj ili jeho Bol'shogo Vzryva (Wanderers of the Universe or a Big Bang Echo). Vek 2, Fryazino (2005)

[158] Parker, E.N.: Dynamics of the interplanetary gas and magnetic fields. Astrophys. J. **128**, 664 (1958)

[159] Partridge, H., Schwenke, D.W.: The determination of an accurate isotope dependent potential energy surface for water from extensive ab initio calculations and experimental data. J. Chem. Phys. **106**(11), 4618–4639 (1997). DOI 10.1063/1.473987. URL `http://link.aip.org/link/?JCP/106/4618/1`

[160] Pavlovski, A.I., Boriskov, G.V., et al.: Isentropic solid hydrogen compression by ultrahigh magnetic field pressure in megabar range. In: C.M. Fowler, R.S. Caird, D.T. Erickson (eds.) Megagauss Technology and Pulsed Power Applications, p. 255. Plenum, New York (1987)

[161] Perlmutter, S., Aldering, G., Della Valle, M., et al.: Discovery of a supernova explosion at half the age of the Universe. Nature **391**(6662), 51–54 (1998). DOI {10.1038/34124}

[162] Phinney, E.S.: Black hole-driven hydromagnefic flows flyweels vs. fuel. In: A. Ferrari, A.G. Pacholczyk (eds.) Astrophysical Jets. Reidel, Dordrecht (1983)

[163] Pieranski, P.: Colloidal crystals. Contemp. Phys. **24**(1), 25–70 (1983). DOI 10.1080/00107518308227471

[164] Popov, S.B., Prohomov, M.E.: Zvezdy: zhizn' posle smerti (Stars: a life after death). In: V.G. Surdin (ed.) Astronomiya: Vek XXI (Astronomy: XXIst Century), p. 183. Vek 2, Fryazino (2007)

[165] Popov, V.S., Karnakov, B.M., Mur, V.D.: Quasiclassical theory of atomic ionization in electric and magnetic fields. Phys. Lett. A **229**(5), 306–312 (1997). DOI 10.1016/S0375-9601(97)00140-0

[166] Price, P.A., Berger, E., Reichart, D.E., et al.: GRB 011121: A massive star progenitor (2002). ArXiv:astro-ph/0203467v1

[167] Pukhov, A.: Strong field interaction of laser radiation. Rep. Prog. Phys. **66**(1), 47–101 (2003). DOI 10.1088/0034-4885/66/1/202

[168] Pukhov, A., Meyer-ter-Vehn, J.: Laser wake field acceleration: the highly non-linear broken-wave regime. Appl. Phys. B **74**(4-5), 355–361 (2002). DOI 10.1007/s003400200795

[169] Quintenz, J., Sandia's Pulsed Power Team: Pulsed power team. In: Proc. 13th Int. Conf. on High Power Particle Beams. Nagaoka, Japan (2000)

[170] Rebolo, R., Martin, E.L., Zapatero Osorio, M.R. (eds.): Brown Dwarfs and Extrasolar Planets, *Astronomical Society of the Pacific Conference Series*, vol. 134. ASP, San Francisco (1998)

[171] Remington, B.A., Arnett, D., R. Paul, D., Takabe, H.: Modeling astrophysical phenomena in the laboratory with intense lasers. Science **284**(5419), 1488–1493 (1999). DOI 10.1126/science.284.5419.1488. URL `http://www.sciencemag.org/cgi/content/abstract/284/5419/1488`

[172] Remington, B.A., Kane, J., Drake, R.P., et al.: Supernova hydrodynamics experiments on the Nova laser. Phys. Plasmas **4**(5), 1994–2003 (1997). DOI 10.1063/1.872341

[173] Richer, J., Michaud, G., Rogers, F., et al.: Radiative accelerations for evolutionary model calculations. Astrophys. J. **492**(2, Part 1), 833–842 (1998). DOI {10.1086/305054}

[174] Riordan, M., Zajc, W.A.: The first few microseconds. Sci. Am. **294**(5), 34A–41 (2006)

[175] Rodionova, Z.F., Surdin, V.G.: Planety solnechnoj sistemy (Planets of the Solar System). In: V.G. Surdin (ed.) Astronomiya: Vek XXI (Astronomy: XXIst Century), p. 34. Vek 2, Fryazino (2007)

[176] Rubakov, V.A.: Large and infinite extra dimensions. Phys. Usp. **44**(9), 871 (2001). DOI 10.1070/PU2001v044n09ABEH001000. URL http://ufn.ru/en/articles/2001/9/a/

[177] Rubakov, V.A.: Introduction to cosmology. Proc. Sci. **RTN2005** (2005). URL http://pos.sissa.it/archive/conferences/019/003/RTN2005_003.pdf

[178] Rubakov, V.A.: Hierarchies of fundamental constants (to items Nos 16, 17, and 27 from Ginzburg's list). Phys. Usp. **50**(4), 390 (2007). DOI 10.1070/PU2007v050n04ABEH006240. URL http://ufn.ru/en/articles/2007/4/i/

[179] Rubakov, V.A., Tinyakov, P.G.: Infrared-modified gravities and massive gravitons. Phys. Usp. **51**(8), 759 (2008). DOI 10.1070/PU2008v051n08ABEH006600. URL http://ufn.ru/en/articles/2008/8/a/

[180] Rubin, S.G.: Ustroistvo nashei vselennoi (The Constitution of Our Universe). Vek 2, Fryazino (2006)

[181] Rubin, V.: Seeing dark matter in the Andromeda Galaxy. Phys. Today **59**(12), 8–9 (2006). DOI 10.1063/1.2435662

[182] Rubin, V.C., Ford Jr, W.K., Krishna Kumar, C.: Stellar motions near the nucleus of M31. Astrophys. J. **181**, 61–78 (1973)

[183] Russel, W.B., Saville, D.A., Schowalter, W.R.: Colloidal Dispersions. Cambridge University Press, Cambridge (1989)

[184] Ryutov, D.D., Remington, B.A., Robey, H.F., Drake, R.P.: Magnetodynamic scaling: from astrophysics to the laboratory. Phys. Plasmas **8**(5), 1804–1816 (2001). DOI 10.1063/1.1344562

[185] Salpeter, E.E.: Nuclear reactions in the stars. I. Proton–proton chain. Phys. Rev. **88**(3), 547–553 (1952). DOI 10.1103/PhysRev.88.547. URL http://link.aps.org/abstract/PR/v88/p547

[186] Samus', N.N.: Peremennye zvezdy (Variable stars). In: V.G. Surdin (ed.) Astronomiya: Vek XXI (Astronomy: XXIst Century), p. 162. Vek 2, Fryazino (2007)

[187] Sandage, A., Tammann, G.A., Saha, A., et al.: The Hubble constant: A summary of the Hubble Space Telescope program for the luminosity calibration of type Ia supernovae by means of Cepheids. Astrophys. J. **653**(2), 843–860 (2006). DOI 10.1086/508853. URL http://iopscience.iop.org/0004-637X/653/2/843

[188] Sazhin, M.V.: Kosmologija rannej vselennoj (Cosmology of the Early Universe). In: V.N. Soifer (ed.) Sovremennoe estestvoznanie. Entsiklopediya (Modern Natural Science. Encyclopedia), vol. 4, p. 253. Magistr-Press, Moscow (2000)
[189] Schatz, T., Schramm, U., Habs, D.: Crystalline ion beams. Nature **412**(6848), 717–720 (2001). DOI 10.1038/35089045
[190] Schekochihin, A.A., Cowley, S.C., Dorland, W.: Interplanetary and interstellar plasma turbulence. Plasma Phys. Control. Fusion **49**(5A), A195–A209 (2007). DOI 10.1088/0741-3335/49/5A/S16
[191] Schertlera, K., Greinera, C., Schaffner-Bielichc, J., Thoma, M.H.: Quark phases in neutron stars and a third family of compact stars as signature for phase transitions. Nucl. Phys. A **677**(1-4), 463–490 (2001). DOI 10.1016/S0375-9474(00)00305-5
[192] Schramm, U., Schatz, T., Bussmann, M., Habs, D.: Cooling and heating of crystalline ion beams. J. Phys. B **36**(3), 561–571 (2003). DOI 10.1088/0953-4075/36/3/314
[193] Shapiro, S.L., Teukolsky, S.A.: Black Holes, White Dwarfs, and Neutron Stars. Wiley, New York (1983)
[194] Shara, M.: When stars collide. Sci. Am. **287**(5), 44–51 (2002)
[195] Shashkin, A.A.: Metal—insulator transitions and the effects of electron—electron interactions in two-dimensional electron systems. Phys. Usp. **48**(2), 129 (2005). DOI 10.1070/PU2005v048n02ABEH001944. URL `http://ufn.ru/en/articles/2005/2/b/`
[196] Shatskii, A.A., Novikov, I.D., Kardashev, N.S.: A dynamic model of the wormhole and the Multiverse model. Phys. Usp. **51**(5), 457 (2008). DOI 10.1070/PU2008v051n05ABEH006581. URL `http://ufn.ru/en/articles/2008/5/c/`
[197] Shevchenko, V.V.: Solnechnaja sistema (The Solar System). In: V.N. Soifer (ed.) Sovremennoe estestvoznanie. Entsiklopediya (Modern Natural Science. Encyclopedia), vol. 4, p. 125. Magistr Press, Moscow (2000)
[198] Shevchenko, V.V.: Priroda planet (The nature of planets). In: V.G. Surdin (ed.) Astronomiya: Vek XXI (Astronomy: XXIst Century), p. 93. Vek 2, Fryazino (2007)
[199] Shinkai, H., Hayward, S.A.: Fate of the first traversible wormhole: Black-hole collapse or inflationary expansion. Phys. Rev. D **66**(4), 044005 (2002). DOI 10.1103/PhysRevD.66.044005. URL `http://link.aps.org/abstract/PRD/v66/e044005`
[200] Spielman, R.B., Deeney, C., Chandler, G.A., et al.: Tungsten wire-array Z-pinch experiments at 200 TW and 2 MJ. Phys. Plasmas **5**(5), 2105–2111 (1998). DOI 10.1063/1.872881
[201] Stefani, F., Gundrum, T., Gerbeth, G., et al.: Experimental evidence for magnetorotational instability in a Taylor–Couette flow under the influence of a helical magnetic field. Phys. Rev. Lett. **97**(18), 184502 (2006). DOI 10.1103/PhysRevLett.97.184502. URL `http://link.aps.org/abstract/PRL/v97/e184502`

[202] Stozhkov, Y.I.: Kosmicheskie luchi (Cosmic rays). In: V.N. Soifer (ed.) Sovremennoe estestvoznanie. Entsiklopediya (Modern Natural Science. Encyclopedia), vol. 4, p. 191. Magistr-Press, Moscow (2000)

[203] Sunyaev, R.A., Zeldovich, Y.B.: Distortions of the background radiation spectrum. Nature **223**(5207), 721–722 (1969). DOI 10.1038/223721a0

[204] Surdin, V.G.: Rozhdenie zvezd (Star Production). Editorial URSS, Moscow (1999)

[205] Surdin, V.G.: Fundamental'nye vzaimodejstvija (Fundamental Interactions). In: V.G. Surdin (ed.) Astronomiya: Vek XXI (Astronomy: XXIst Century), p. 8. Vek 2, Fryazino (2007)

[206] Surdin, V.G.: Mlechnyj put' (The Milky Way). In: V.G. Surdin (ed.) Astronomiya: Vek XXI (Astronomy: XXIst Century), p. 267. Vek 2, Fryazino (2007)

[207] Takabe, H.: Hydrodynamic instability, integrated code, laboratory astrophysics and astrophysics. In: H. Hora, G.H. Miley (eds.) Edward Teller Lectures: Lasers and Inertial Fusion Energy, p. 313. Imperial College Press, London (2005)

[208] Takahashi, Y., Hillman, L.W., Tajima, T.: Relativistic lasers and high-energy astrophysics: Gamma ray bursts and highest energy acceleration. In: T. Tajima, K. Mima, H. Baldis (eds.) High-Field Science, p. 171. Kluwer/Plenum, New York (2000)

[209] Takeda, M., Hayashida, N., Honda, K., et al.: Extension of the cosmic-ray energy spectrum beyond the predicted Greisen–Zatsepin–Kuz'min cutoff. Phys. Rev. Lett. **81**(6), 1163–1166 (1998). DOI 10.1103/PhysRevLett.81.1163. URL http://link.aps.org/abstract/PRL/v81/p1163

[210] Tassoul, J.L.: Theory of Rotating Stars. Princeton University Press, Princeton, NJ (1978)

[211] Tatarakis, M., Watts, I., Beg, F.N., et al.: Laser technology: Measuring huge magnetic fields. Nature **415**(6869), 280 (2002). DOI 10.1038/415280a

[212] Trunin, R.F.: Shock compressibility of condensed materials in strong shock waves generated by underground nuclear explosions. Phys. Usp. **37**(11), 1123 (1994). DOI 10.1070/PU1994v037n11ABEH000055. URL http://ufn.ru/en/articles/1994/11/d/

[213] Tzortzakis, S., Mechain, G., Patalano, G., et al.: Coherent subterahertz radiation from femtosecond infrared filaments in air. Opt. Lett. **27**(21), 1944–1946 (2002). DOI 10.1364/OL.27.001944. URL http://www.opticsinfobase.org/abstract.cfm?URI=ol-27-21-1944

[214] Vacca, J.R. (ed.): The World's 20 Greatest Unsolved Problems. Prentice Hall PTR, Englewood Cliffs, NJ (2004)

[215] Vainshtein, S.I., Zeldovich, Y.B., Ruzmaikin, A.A.: The Turbulent Dynamo in Astrophysics. Nauka, Moscow (1980)

[216] Velikhov, E.P.: Stability of a plane Poiseuille flow of an ideally conducting fluid in a longitudinal magnetic field. Zh. Eksp. Teor. Fiz. **36**(4), 1192–1202 (1959)

[217] Velikhov, E.P.: Stability of an ideally conducting liquid flowing between rotating cylinders in a magnetic field. Zh. Eksp. Teor. Fiz. **36**(5), 1398–1404 (1959)

[218] Vinci, T., Loupias, B., Koenig, M., et al.: Laboratory astrophysics using high energy lasers: need for 2D simulation. J. Phys.: Conf. Ser. **112**(4), 042012 (4pp) (2008). URL `http://stacks.iop.org/1742-6596/112/042012`

[219] Vladimirov, A.S., Voloshin, N.P., Nogin, V.N., et al.: Shock compressibility of aluminum at $p > 1$ Gbar. JETP Lett. **39**(2), 82 (1984)

[220] Wakker, B.P., Richter, P.: Our growing, breathing galaxy. Sci. Am. **290**(1), 38–47 (2004)

[221] Waxman, E.: Gamma-ray bursts and collisionless shocks. Plasma Phys. Control. Fusion **48**(12B), B137–B151 (2006). DOI 10.1088/0741-3335/48/12B/S14

[222] Weiler, T.J.: Cosmic-ray neutrino annihilation on relic neutrinos revisited: a mechanism for generating air showers above the Greisen–Zatsepin–Kuzmin cutoff. Astropart. Phys. **11**(3), 303–316 (1999)

[223] Willingale, L., Mangles, S.P., Nilson, P.M., et al.: Collimated multi-MeV ion beams from high-intensity laser interactions with underdense plasma. Phys. Rev. Lett. **96**(24), 245002 (2006). DOI 10.1103/PhysRevLett.96.245002. URL `http://link.aps.org/abstract/PRL/v96/e245002`

[224] Woolsey, N.C., Ash, A.D., Cortois, C., et al.: Collisionless plasma astrophysics simulation experiments using lasers. AIP Conf. Proc. **827**, 365–375 (2006)

[225] XFEL Project Group at DESY: The European X-ray laser project XFEL. URL `http://xfel.desy.de/`

[226] Yakovlev, D.G., Levenfish, K.P., Shibanov, Y.A.: Cooling of neutron stars and superfluidity in their cores. Phys. Usp. **42**(8), 737 (1999). DOI 10.1070/pu1999v042n08ABEH000556. URL `http://ufn.ru/en/articles/1999/8/a/`

[227] Zasov, A.V., Postnov, K.A.: Obshchaya astrofizika (General Astrophysics). Vek 2, Fryazino (2006)

[228] Zasov, A.V., Surdin, V.G.: Raznoobrazie galaktik (A variety of galaxies). In: V.G. Surdin (ed.) Astronomiya: Vek XXI (Astronomy: XXIst Century), p. 329. Vek 2, Fryazino (2007)

[229] Zatsepin, G.T., Kusmin, V.A.: Top boundary of cosmic ray spectrum. JETP Lett. **4**(3), 78 (1966)

[230] Zel'dovich, Y.B., Levich, E.V., Syunyaev, R.A.: Stimulated Compton interaction between Maxwellian electrons and spectrally narrow radiation (in Russian). Zh. Eksp. Teor. Fiz. **62**(4), 1392–1408 (1972)

[231] Zel'dovich, Y.B., Novikov, I.D.: Relativistic Astrophysics (in Russian). Nauka, Moscow (1967). [English Transl.: Relativistic Astrophysics. University of Chicago Press, Chicago (1971)]

[232] Zel'dovich, Y.B., Raizer, Y.P.: Fizika udarnykh voln i vysokotemperaturnykh gidrodinamicheskikh yavlenii, 2nd edn. Nauka, Moscow (1966). [English

Transl.: Physics of Shock Waves and High-Temperature Hydrodynamic Phenomena. Dover, Mineola, NY (2002)]
[233] Zelenyi, L.M.: Private communication (2007)
[234] Zelenyi, L.M., Verigin, M.I., Zakharov, A.V., et al.: The heliosphere and the interaction of the terrestrial planets with the solar wind. Phys. Usp. **48**(6), 615 (2005). URL `http://ufn.ru/en/articles/2005/6/i/`
[235] Zubko, V., Dwek, E., Arendt, R.G.: Interstellar dust models consistent with extinction, emission, and abundance constraints. Astrophys. J. Suppl. Ser. **152**(2), 211–249 (2004). DOI 10.1086/382351

# Chapter 8
# Conclusion

The science of the structure of matter and cosmic physics are closely related and interwoven [2]. On the one hand, the solution of almost any problem in cosmic physics is impossible without invoking data about the material structure of the corresponding celestial object. On the other hand, cosmic physics furnishes nuclear and subnuclear physical information, which will substantially supplement the data obtainable in terrestrial laboratories. This applies, for instance, to the characteristics

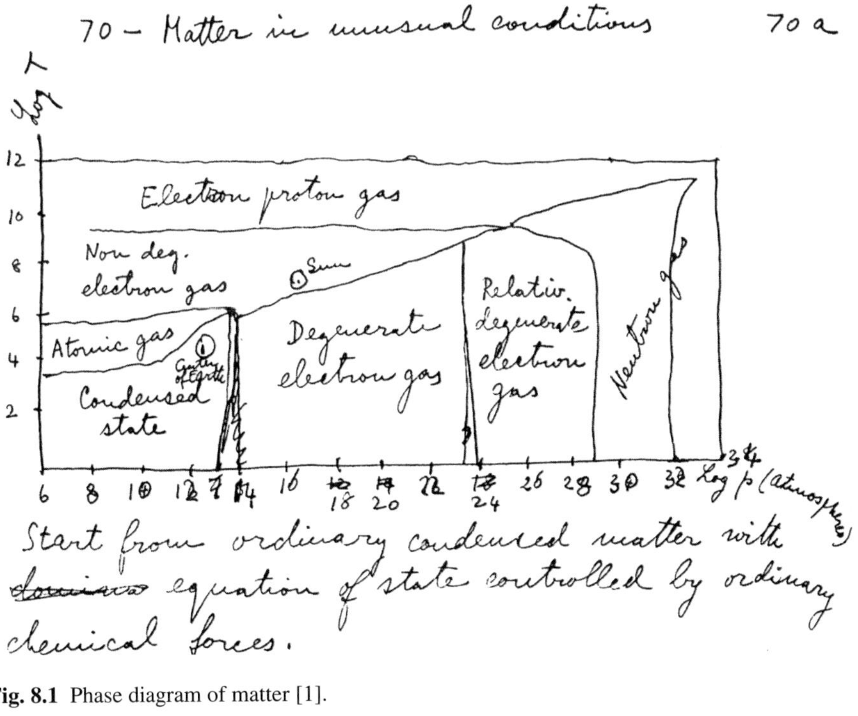


**Fig. 8.1** Phase diagram of matter [1].

V.E. Fortov, *Extreme States of Matter*, The Frontiers Collection,
DOI 10.1007/978-3-642-16464-4_8, © Springer-Verlag Berlin Heidelberg 2011

of nuclear forces (pulsar data) and to the number of neutrino families (cosmological data). The role of space (especially, of the universe as a whole) as a source of fundamental data can be expected to rise in importance in the future as well. This is so because the limits of the potential of accelerator physics are already in sight. At the same time, the constantly emerging new opportunities in experimental high-energy-density physics give hope that the ultraextreme states of matter that are so typical of the universe around us can be reproduced under laboratory conditions.

We conclude our exposition by demonstrating the phase diagram of matter in the form of a sketch drawn by the hand of the genius Enrico Fermi (Fig. 8.1 [1]) – with only one "experimental" point: the Sun. It only remains to wonder at how much it has been possible to achieve and comprehend in a mere 50 years in the area of extreme pressures and temperatures, which is so hostile to human beings and incompatible with their life, and which at the same time is the basis of their life.

## References

[1] Fermi, E.: Notes on Thermodynamics and Statistics. University of Chicago Press, Chicago (1966)
[2] Kirzhnits, D.A.: Lektsii po fizike (Lectures on Physics). Nauka, Moscow (2006)

# THE FRONTIERS COLLECTION



**Information and Its Role in Nature**
By J.G. Roederer

**Relativity and the Nature of Spacetime**
By V. Petkov

**Quo Vadis Quantum Mechanics?**
Edited by A.C. Elitzur, S. Dolev, N. Kolenda

**Life – As a Matter of Fat**
The Emerging Science of Lipidomics
By O.G. Mouritsen

**Quantum–Classical Analogies**
By D. Dragoman and M. Dragoman

**Knowledge and the World**
Edited by M. Carrier, J. Roggenhofer, G. Küppers, P. Blanchard

**Quantum–Classical Correspondence**
By A.O. Bolivar

**Mind, Matter and Quantum Mechanics**
By H. Stapp

**Quantum Mechanics and Gravity**
By M. Sachs

**Extreme Events in Nature and Society**
Edited by S. Albeverio, V. Jentsch, H. Kantz

**The Thermodynamic Machinery of Life**
By M. Kurzynski

**The Emerging Physics of Consciousness**
Edited by J.A. Tuszynski

**Weak Links**
Stabilizers of Complex Systems from Proteins to Social Networks
By P. Csermely

**Mind, Matter and the Implicate Order**
By P.T.I. Pylkkänen

**Quantum Mechanics at the Crossroads**
New Perspectives from History, Philosophy and Physics
Edited by J. Evans, A.S. Thorndike

**Particle Metaphysics**
A Critical Account of Subatomic Reality
By B. Falkenburg

**The Physical Basis of the Direction of Time**
By H.D. Zeh

**Asymmetry: The Foundation of Information**
By S.J. Muller

**Mindful Universe**
Quantum Mechanics and the Participating Observer
By H. Stapp

**Decoherence and the Quantum-To-Classical Transition**
By M. Schlosshauer

**Quantum Superposition**
Counterintuitive Consequences of Coherence, Entanglement, and Interference
By Mark P. Silverman

**The Nonlinear Universe**
Chaos, Emergence, Life
By A. Scott

**Symmetry Rules**
How Science and Nature Are Founded on Symmetry
By J. Rosen

**Entanglement, Information, and the Interpretation of Quantum Mechanics**
By G. Jaeger

Printing and Binding: Stürtz GmbH, Würzburg